T0144553

Circuits and Systems for Security and Privacy

Devices, Circuits, and Systems

Series Editor
Krzysztof Iniewski
Emerging Technologies CMOS Inc.
Vancouver, British Columbia, Canada

PUBLISHED TITLES:

PUBLISHED TITLES:

FORTHCOMING TITLES:

FORTHCOMING TITLES:

Nanoelectronics: Devices, Circuits, and Systems
Nikos Konofaos

Power Management Integrated Circuits and Technologies
Mona M. Hella and Patrick Mercier

Radio Frequency Integrated Circuit Design
Sebastian Magierowski

Semiconductor Devices in Harsh Conditions
Kirsten Weide-Zaage and Malgorzata Chrzanowska-Jeske

Smart eHealth and eCare Technologies Handbook
Sari Merilampi, Lars T. Berger, and Andrew Sirkka

Structural Health Monitoring of Composite Structures Using Fiber Optic Methods
Ginu Rajan and Gangadhara Prusty

Tunable RF Components and Circuits: Applications in Mobile Handsets
Jeffrey L. Hilbert

Circuits and Systems for Security and Privacy

Edited by
Farhana Sheikh
Portland, Oregon, USA

Leonel Sousa
INESC-ID, Instituto Superior Técnico
Universidade de Lisboa, Portugal

Krzysztof Iniewski MANAGING EDITOR
Emerging Technologies CMOS Inc.
Vancouver, British Columbia, Canada

CRC Press
Taylor & Francis Group
Boca Raton London New York

CRC Press is an imprint of the
Taylor & Francis Group, an **informa** business

CRC Press
Taylor & Francis Group
6000 Broken Sound Parkway NW, Suite 300
Boca Raton, FL 33487-2742

© 2016 by Taylor & Francis Group, LLC
CRC Press is an imprint of Taylor & Francis Group, an Informa business

No claim to original U.S. Government works

Printed on acid-free paper
Version Date: 20151027

International Standard Book Number-13: 978-1-4822-3688-0 (Hardback)

This book contains information obtained from authentic and highly regarded sources. Reasonable efforts have been made to publish reliable data and information, but the author and publisher cannot assume responsibility for the validity of all materials or the consequences of their use. The authors and publishers have attempted to trace the copyright holders of all material reproduced in this publication and apologize to copyright holders if permission to publish in this form has not been obtained. If any copyright material has not been acknowledged please write and let us know so we may rectify in any future reprint.

Except as permitted under U.S. Copyright Law, no part of this book may be reprinted, reproduced, transmitted, or utilized in any form by any electronic, mechanical, or other means, now known or hereafter invented, including photocopying, microfilming, and recording, or in any information storage or retrieval system, without written permission from the publishers.

For permission to photocopy or use material electronically from this work, please access www.copyright.com (http://www.copyright.com/) or contact the Copyright Clearance Center, Inc. (CCC), 222 Rosewood Drive, Danvers, MA 01923, 978-750-8400. CCC is a not-for-profit organization that provides licenses and registration for a variety of users. For organizations that have been granted a photocopy license by the CCC, a separate system of payment has been arranged.

Trademark Notice: Product or corporate names may be trademarks or registered trademarks, and are used only for identification and explanation without intent to infringe.

Library of Congress Cataloging-in-Publication Data

Names: Sheikh, Farhana, editor. | Sousa, Leonel A. (Leonel Augusto) editor. | Iniewski, Krzysztof, 1960- editor.
Title: Circuits and systems for security and privacy / edited by Farhana Sheikh, Leonel Sousa, Krzysztof Iniewski.
Description: Boca Raton : CRC Press, 2016. | Series: Devices, circuits and systems | Includes bibliographical references and index.
Identifiers: LCCN 2015040706 | ISBN 9781482236880 (alk. paper)
Subjects: LCSH: Computers--Circuits. | Computer security. | Integrated curcuits--Design and construction. | Data encryption (Computer science)
Classification: LCC TK7888.4 .C57 2016 | DDC 005.8--dc23
LC record available at http://lccn.loc.gov/2015040706

**Visit the Taylor & Francis Web site at
http://www.taylorandfrancis.com**

**and the CRC Press Web site at
http://www.crcpress.com**

Contents

    Ricardo Chaves, Leonel Sousa, Nicolas Sklavos, Apostolos P. Fournaris,
    Georgina Kalogeridou, Paris Kitsos, and Farhana Sheikh

    Nathaniel Pinckney, David Money Harris, Nan Jiang, Kyle Kelley, Samuel
    Antao, and Leonel Sousa

8 Side-Channel Attacks 287

Josep Balasch, Oscar Reparaz, Ingrid Verbauwhede, Sorin A. Huss, Kai Rhode, Marc Stöttinger, and Michael Zohner

Preface

Security and privacy are fundamental issues for consumers as personal information processing moves from secure servers to mobile devices such as smartphones, PDAs, and medical devices. This book on Circuits and Systems for Security and Privacy helps address the challenges of designing secure circuits and systems that can be simultaneously efficient and resistant to attacks. The indubitable interest of this book also arises from the fact that the weakest link in the design of secure cryptosystems are not the algorithms adopted to protect the information but their implementations on circuits and systems, which are currently vulnerable to malicious attacks.

The book has been organized and written having in mind graduate students, scientists, engineers, all professionals whose activity is related to information security and privacy. As much as possible, this book has been prepared in one self-contained form. It starts by providing the mathematics underlying cryptography. Building from this ground, all the aspects concerning the design of efficient and secure circuits and systems are covered in this book. Topics that are extensively covered in this book include symmetric key cryptography, block ciphering, secure hashing, public key cryptosystems which underpin various internet standards, physically unclonable functions, random number generators, and side channel attacks.

This book would not have seen the daylight without the initial challenge of Kris Iniewski, the valuable contribution of the authors that have written the various book chapters, and the support and patience of the CRC press staff. During the two years that it took preparing this book, we have experienced difficulties and faced different problems, both at personal and professional levels. However, the support of our families and colleagues made this journey possible and painless, bringing the book to a successful conclusion.

We really hope you appreciate the book as much as we have enjoyed preparing it. We would be very proud if it becomes a useful piece of work for your professional development in a company or at a university. We also expect that this book may inspire the investigation of new algorithms, architectures, and circuits that can lead to more secure systems and improved preservation of data and information privacy.

Farhana Sheikh, Portland, Oregon
Leonel Sousa, Lisboa, Portugal

October 27th, 2015

Editors

Farhana Sheikh (IEEE SM'14, IEEE M'93) received the B.Eng. degree in Systems and Computer Engineering with high distinction, also receiving the Chancellor's Medal from Carleton University, Ottawa, Canada in 1993 and the M.Sc. and Ph.D. degrees in Electrical Engineering and Computer Sciences from the University of California, Berkeley, in 1996 and 2008, respectively. In 1993, Farhana was selected as one of 100 Canadians to Watch in Canada's national magazine, Macleans. From 1993 to 1994 she worked at Nortel Networks as a software engineer in firmware and embedded system design. From 1996 to 2001, she was at Cadabra Design Automation, now part of Synopsys as software engineer and senior manager. She joined Intel Corporation after her Ph.D. in 2008 where she worked as a Senior Research Scientist and Staff Scientist in Circuit Research Lab, part of Intel Labs from 2008 to 2014. In 2014, she became Senior Staff Scientist and Technical Manager of Digital Communications Lab, in Wireless Communications Research, Intel Labs and is presently leading a team of researchers in the area of digital communications and signal processing architectures and circuits. Her research interests include low-power digital CMOS design, energy-efficient high-performance circuits for next generation wireless systems, wireless security, cryptography, and signal/image processing. Farhana has co-authored over 30 publications in VLSI circuit design and has filed over 10 patents on accelerators for wireless communication, cryptography, and 3-D graphics. Dr. Sheikh was a recipient of the Association of Professional Engineers of Ontario Gold Medal for Academic Achievement, the NSERC'67 scholarship for graduate studies, and the Intel Foundation Ph.D. Fellowship. In 2012, she received the ISSCC Distinguished Technical Paper award.

Leonel Sousa received a Ph.D. degree in Electrical and Computer Engineering from the Instituto Superior Tecnico (IST), Universidade de Lisboa (UL), Lisbon, Portugal, in 1996, where he is currently Full Professor. He is also a Senior Researcher with the R&D Instituto de Engenharia de Sistemas e Computadores (INESC-ID). His research interests include VLSI architectures, computer architectures, parallel computing, computer arithmetic, and signal processing systems. He has contributed to more than 200 papers in journals and international conferences, for which he got several awards, including: DASIP'13 Best Paper Award, SAMOS'11 'Stamatis Vassiliadis' Best Paper Award, DASIP'13 Best Poster Award, and the Honorable Mention Award UTL/Santander Totta for the quality of the publications in 2009. He has contributed to the organization of several international conferences, namely as program chair and as general and topic chair, and has given keynotes in some of them. He has edited three special issues of international journals, and he is currently Associate Editor of the IEEE Transactions on Multimedia, IEEE Transactions on Circuits and Systems for Video Technology, IEEE Access, Springer JRTIP and IET Electronics Letters, and Editor-in-Chief of the Eurasip JES. He is a Fellow of the IET, a Distinguished Scientist of ACM, and a Senior Member of IEEE.

Contributors

Leonel Sousa
University of Lisbon
Lisbon, Portugal

Farhana Sheikh
Intel Corporation
Portland, OR, USA

David O. Novick
Intel Corporation
Hillsboro, OR, USA

Deniz Toz
K. U. Leuven
Leuven, Belgium

Josep Balasch
K. U. Leuven
Leuven, Belgium

Ricardo Chaves
University of Lisbon
Lisbon, Portugal

Nicholas Sklavos
University of Patras
Patras, Greece

Apostolos P. Fournaris
Technological Educational Institute
    of Western Greece
Patras, Greece

Georgina Kalogeridou
University of Cambridge
Cambridge, UK

Paris Kitsos
Technological Educational Institute
    of Western Greece
Patras, Greece

Nathaniel Pinckney
University of Michigan, Ann Arbor
Ann Arbor, MI, USA

David M. Harris
Harvey Mudd College
Claremont, CA, USA

Nan Jiang
Harvey Mudd College
Claremont, CA, USA

Kyle Kelley
Harvey Mudd College
Claremont, CA, USA

Samuel Antao
University of Lisbon
Lisbon, Portugal

Raghavan Kumar
University of Massachusetts,
    Amherst
Amherst, MA, USA

Xiaolin Xu
University of Massachusetts,
    Amherst
Amherst, MA, USA

Wayne Burleson
University of Massachusetts,
    Amherst
Amherst, MA, USA

Sami Rosenblatt
IBM Systems and Technology Group
NY, USA

Toshiaki Kirihata
IBM Systems and Technology Group
NY, USA

Viktor Fischer
Jean Monnet University
Saint-Etienne, France

Patrick Haddad
ST Microelectronics
France

Oscar Reparaz
K. U. Leuven
Leuven, Belgium

Ingrid Verbauwhede
K. U. Leuven
Leuven, Belgium

Sorin A. Huss
Technical University of Darmstadt
Darmstadt, Germany

Kai Rhode
Technical University of Darmstadt
Darmstadt, Germany

Marc Stottinger
Nanyang Technological University
Singapore

Michael Zohner
Center for Advanced Security
    Research
Darmstadt, Germany

1

Introduction

Farhana Sheikh

Intel Corporation

Leonel Sousa

University of Lisbon

CONTENTS

Security and privacy are increasingly important concerns for the consumer as personal information processing moves from secure servers to a growing set of mobile devices such as smart phones, PDAs, medical devices, and other embedded portable devices. The storage of data is moving from hard disks on our personal computers to more centralized servers (i.e., cloud) which require secure and efficient management of sensitive data. The mathematical strength of cryptographic algorithms that are used to protect information is no longer the weakest link: it is the implementation that is more vulnerable to malicious attacks.

Recently, the challenges in implementing cryptography and security into integrated circuits and systems were outlined in [440]. Area, power, energy, throughput, and latency are all important optimization goals that must be considered when building cryptography and security into all types of data transfer, communication, and storage systems. Implementations have to be resistant to physical attacks and countermeasures in addition to being able to withstand large amounts of computing power that can potentially break an algorithm. The goal of this book is to give the circuit and system designer an overview of security and cryptography from an algorithm and implementation perspective. The book focuses on efficient implementations that optimize power, area, and throughput and are also secure such that they resist attacks such as side-channel attacks.

The book is divided into nine chapters. This first chapter gives an overview of all of the chapters. It reviews the basic building blocks of a cryptographic system: symmetric key algorithms, public key algorithms, hash functions, random number generators (RNGs), and physically uncloneable functions (PUFs). The second chapter gives the reader a brief introduction to the mathematics underlying cryptography and includes review of basic arithmetic used in algorithms presented in the subsequent chapters. The third chapter presents symmetric key algorithms and implementation challenges of the Advanced Encryption Standard (AES) which is in wide use today. The fourth chapter focuses on hashing algorithms and their implementations: SHA-1, SHA-2, and SHA-3. The fifth chapter reviews public key cryptography where modular multiplication is the work-horse of RSA and Elliptic Curve Cryptography (ECC). A detailed review of energy-efficient implementations of large scale Montgomery multiplication followed by a deep-dive into reconfigurable architectures for public key cryptography implementations is presented in this chapter. The sixth chapter focuses on physically unclonable functions that leverage randomness to implement low-energy unique ID mechanisms for large-scale and small-scale systems. The seventh chapter continues the theme of randomness and discusses in detail the implementations of true random number generators, a key ingredient necessary in all processing elements. The eighth chapter presents a detailed treatment of side channel attacks and how to make systems resistent to them. The book concludes with some conclusions and remarks about the future of cryptographic systems and their implementations.

1.1 Mathematics and Cryptography

Cryptography has existed since the time humans found it necessary to keep secrets from each other. In the past, it was simply enough to shift characters in a message by a fixed amount to hide a message but as the digital age dawned, ubiquitous computing has made it necessary to rely on complex mathematics to aid in secure transmission of data. Reliance on probability theory which studies events and their likelihood plays a central role in cryptography as it helps answer questions such as "What is the likelihood that two hash values are identical?". Following this, combinatorics which is the study of counting allows a cryptographer to understand the design, capabilities, and limits of a cryptographic system. Number Theory enables the understanding of prime numbers, integer factorization, and finding the greatest common divisor which are the key ingredients of Diffie-Hellman key exchange protocols, the RSA algorithm, digital signatures, and encryption protocols. Efficient integer arithmetic lays the foundation for Montgomery multiplication and Karatsuba multiplication which enable efficient RSA and ECC implementations. Abstract algebra is focused on defining, describing, and manipulating mathematical objects and their structure. The study of abstract algebra has led to many new algorithms and ideas relating to cryptography, including some of the most famous and widely used such as AES. Abstract algebra defines the infrastructure for both information theory and cryptography algorithms. Elliptic curves have been the study of mathematicians going back to the 3rd century AD and are now becoming the new wave in public key cryptography. We review this material also in Chapter 2. In the current age of a proliferation of computing devices, the need for robust cryptographic solutions that protect these devices from malicious attacks has become mandatory. Thankfully, mathematics allows us to lay the foundation for these solutions.

1.2   Symmetric Key Cryptography and Block Ciphers

Public key cryptography, also known as asymmetric key cryptography, allows users to communicate securely without prior access to a shared secret key. Two keys are required: one is a private key and one is a public key. These two keys are mathematically related. The public key is used to encrypt the message and the private key is used to decrypt the message. It is very difficult to ascertain the private key. The sender of a message encrypts using the recipient's public key; this message can only be decrypted using the recipient's paired private key. It is very difficult to obtain the private key from the public key if high-quality algorithms are used. Examples of quality public key cryptography algorithms include: RSA, Diffie-Hellman, and Elliptic Curve.

In contrast, private key cryptography, also known as symmetric key cryptography uses a trivially related (or the same) key for both encryption and decryption. The message or information can be processed in either streams or blocks. Private key cryptography algorithms are usually used in conjunction with public key cryptography for fast distribution of the private key. Some examples of such systems are PGP and SSL. Symmetric key algorithms are typically much less computationally intensive as compared to asymmetric key algorithms; however, symmetric key algorithms almost always operate in real-time environments. Hence, they require implementations that must be computationally efficient.

The Advanced Encryption Standard (AES) is an example of a symmetric key algorithm and is a block cipher. AES operates on blocks of 128 bits. It is based on the Rijndael cipher algorithm which was invented by Joan Daeman and Vincent Rijmen [94]. The predecessor to AES was the Data Encryption Standard (DES) [305] invented by IBM. DES is now deemed insecure for many applications due to its relatively short key length. The Advanced Encryption Standard was first published on November 26, 2001 and has been effective since May 26, 2002 [307]. As mentioned earlier, AES operates on a fixed block size of 128 bits; the key size can be 128 bits (AES-128), 192 bits (AES-192), or 256 bits (AES-256). It uses a substitution-permutation network to encrypt and decrypt the data and all operations are carried out in a specified finite field. Chapter 3 discusses implementation challenges and examples for AES.

1.3   Secure Hashing

Hashing is the transformation of a string of characters into a usually shorter fixed-length value or key that represents the original string. Hashing is used to index and retrieve items in a database because it is faster to find the item using a shorter hashed key than to find it using the original value. It is also used in many encryption algorithms. The hashing algorithm is called the hash function (and probably the term is derived from the idea that the resulting hash value can be thought of as a "mixed up" version of the represented value). In addition to faster data retrieval, hashing is also used to encrypt and decrypt digital signatures (used to authenticate message senders and receivers). The digital signature is transformed with the hash function and then both the hashed value (known as a message-digest) and the signature are sent in separate transmissions to the receiver. Using the same hash function as the sender, the receiver derives a message-digest from the signature and compares it with the message-digest it also received. They should be the same. Hashing functions can be used to transmit and receive encryption keys securely.

The Secure Hash Standard (SHA) was first published by NIST in 1993. By 1995, this algorithm was revised [309] in order to eliminate some of the SHA standard's initial weaknesses. The revised algorithm is usually referenced as SHA-1. In 2001, the SHA-2 hashing algorithm was proposed to address larger digest messages, making it more resistant to possible attacks and allowing it to be used with larger data inputs, up to 2^{128} bits in the case of SHA512.

In 2005 Xiaoyun Wang, Yiqun Lisa Yin, and Hongbo Yu [449] unexpectedly discovered vulnerabilities in the widely used SHA-1 hash algorithm [309]. This attack called into question the practical security of SHA-1 when used in digital signatures and other applications requiring collision resistance. The SHA-2 family, which was designed to replace SHA-1, enjoys a similar structure, leading to concerns that it might as well fall to elaborations or variations of Wang's attack.

To respond to these concerns the United States government agency National Institute of Standards and Technology (NIST) is sponsoring an international competition to create a replacement algorithm, which will be called SHA-3 [308]. The competition drew 64 submissions, and to constitute a first round NIST identified 51 that met its minimum submission criterion. Based on public comments and internal reviews of the 51 first-round candidates, NIST later narrowed the field to a more manageable number of 14 semi-finalists, to enable deeper analysis. From these 14 semi-finalists, 5 finalists were chosen, and a winner was selected in October 2012. The Keccak [40] algorithm was chosen to constitute the SHA-3 standard. Chapter 4 reviews SHA-1, SHA-2, and SHA-3 standards in addition to reviewing some of the SHA-3 finalists in terms of implementation tradeoffs showing why Keccak was eventually selected to become the SHA-3 standard.

1.4 Public Key Cryptography

Almost all forms of data communication in the modern day require security transmission and storage. This can be due to requirements to secure data residing on the application platform or due to the support of communication protocols. In particular, private key and public key cryptography is used to guarantee secure communication of information. As mentioned earlier, public key cryptography allows users to communicate securely without prior access to a shared secret key. Two keys are required: one is a private key and one is a public key which are mathematically related. The public key is used to encrypt the message and the private key is used to decrypt the message. Public key cryptography algorithms support digital signatures that can be used to validate data from a given signature generated by the sender. Two popular public key cryptography algorithms that are in wide use today are the Rivest-Shamir-Adleman (RSA) algorithm and the Elliptic Curve Cryptography (ECC) algorithm. Efficient implementations of public key cryptography algorithms are continuously being explored and implemented largely as high-performance hardware accelerators to support the large computational demands. These accelerators typically take advantage of efficient parallelization and programmability to enable flexibility and configuration. Montgomery multiplication introduced in Chapter 2 is considered the work-horse in the implementation of these algorithms and efficient hardware implementations of Montgomery multiplication and configurable RSA and ECC accelerator platforms are presented in Chapter 5.

1.5   Physically Unclonable Functions

The ubiquitous nature of computing has seen an exponential increase in computing devices ranging from tablets to small embedded RFID tags which all have data communication capabilities coupled with an increase in security and privacy issues. The increasing need for protection of sensitive data stored or communicated using these devices has brought about the need for unique secret identification keys for each individual computing device or element since most classical cryptography primitives are based on the concept of *secret binary keys* embedded on computing and communication devices. The fact that these unique identifiers must be stored in non-volatile memory opens up the device to physical (invasive, non-invasive, and side channel) and software attacks. Physically Unclonable Functions (PUFs) provide a mechanism to produce a unique, non-replicatable identifier based on mapping a set of external inputs to some response. For example, silicon PUFs use process variations to generate unique identifiers. Chapter 6 explains how these PUFs operate and how they can be implemented efficiently. It also outlines how intrinsic chip identifiers can be created using bitmaps of an embedded DRAM macro.

1.6   Random Number Generators

Random number generators (RNGs) are computational or physical functions that generate a sequence of bits or symbols that do not feature any pattern. The generated bits or symbols are independent and uniformly distributed. Random number generators are integral components of microprocessors and modern computing devices since they provide high-entropy keys for data encryption, secure wireless and wireline communication, digital watermarking, media content protection, and secure content storage. They can be used to generate keys or identifiers which can be used in various cryptography applications. The can also be used to intentionally generate noise in a communication system. Chapter 7 introduces random number generators and discusses how they can be efficiently implemented in digital devices.

1.7   Side Channel Attacks

Modern cryptographic algorithms are typically designed around computationally hard problems that generally require the use of one or more secret keys. When a cryptographic system is broken by a malicious attack, it is generally implied that the attacker has found the secret key(s). A side channel attack happens when an adversary or attacker decides to exploit the physical characteristics of the system implementation to extract the secret key. This method is sometimes easier than trying to solve a computationally hard problem. The side channel attack analyzes physically observable information leaked by cryptographic devices when performing some operation. Contrary to theoretical models employed in cryptanalysis in which an adversary tries to break cryptographic constructions using mathematical models and input/output message pairs, practical realizations of cryptography are susceptible to more dangerous attacks enabled by the existence of side channels. Time, power consumption, or electromagnetic emanations are the most common examples of observable phenomenons that an attacker can exploit to gain knowledge of secret cryptographic keys. For example, system power may change in a specific fashion when a computation involves a secret key. The attacker may be able to repeatedly analyze this behavior to glean information regarding the secret key by capturing system power profiles. Chapter 8 presents an overview of side channel analysis and then presents mechanisms to design side channel resistant hardware modules.

1.8 Final Remarks

We hope that the previous sections have given a flavor of what is to come in the rest of this book. We encourage the reader to consider each chapter as a brief introduction to a rich subject area within the broader topic of cryptographic systems and circuits. References have been provided to allow the reader to explore each subject area in a deeper fashion than is presented in this treatment. There is a vast amount of literature generated in this space in the past 15 years and this book is a humble attempt at highlighting some of this work in one self-contained form.

2

Mathematics and Cryptography

David O. Novick

Intel Corporation

CONTENTS

2.1   Introduction

2.1.1   Mathematics: The Toolchest of Modern Cryptography

2.1.1.1   History: From old tools to new

Cryptography has existed since the time people had secrets to keep or lacked confidence that their communication with another party would not be tampered with en route. Many accounts of the historical approaches to cryptography are captured in works like Simon Singh's "The Code Book" [397] and David Kahn's "The Codebreakers" [194]. While simply shifting characters in a message by a fixed amount, hiding a message on the previously shaven head of a messenger, or affixing a wax seal to a letter was sufficient (sometimes) at one time, those days are long gone. With the dawning of the digital age, reliance on the hope that the methods used to hide the secrets will not be discovered by those seeking to compromise those secrets is regrettably naive. Furthermore, while previously used techniques were at least partially reliable because they took a human being an inordinate time to break; today's ubiquitous computing resources (including your cell phone) demand a much firmer foundation if secrets are still to be kept or message integrity is to be assured.

2.1.1.2   A treasure trove of techniques

While the transition from old methods to new was not instantaneous, a few pivotal events serve as signposts for those who wish to understand how this new set of techniques, capable enough to meet the challenges of a new age emerged. Alan Turing's work on cryptography at Bletchley Park in the 1940s made use of sophisticated statistical methods to break codes [227]. Pioneers like Claude Shannon [386] were able to mathematically quantify various properties of an electronic communications channel such as its capacity and its error detection rate. Shannon also made great strides in quantifying information content itself through the use of mathematics and applied these various methods to cryptographic systems [385]. All of these served as early indicators of what was to become the new foundation for modern cryptography. The 1976 paper, "New Directions in Cryptography" [103], published by Whitfield Diffie and Martin Hellman, established a visible turning point in the cryptographic sciences. It provided a foreshadowing that the mathematical knowledge gathered over the last 2+ millennia was soon to become the dispensary of the tools required to meet the cryptographic needs of the digital age. Somewhat ironically, this knowledge was thought and even hoped by many to be of no practical value. In spite of that, cutting edge cryptographic research continues to draw from this boundless well of knowledge we call mathematics.

2.1.1.3 Purpose of this chapter

The sheer vastness of the subject matter relevant to modern cryptography contained in pure mathematics precludes any chance of one book, let alone this one chapter, providing even a modest overview of the mathematical underpinnings of modern and emergent cryptographic algorithms to those seeking to understand and correctly implement them. Nevertheless, it is has been the diligent effort of this chapter's author to lay a suitable foundation in those areas of mathematics encountered in the rest of this book as well as those the reader will certainly encounter as they work on security and cryptographic systems' design. Where a particular subject lacks sufficient coverage in this chapter, it is the author's hope that what is provided will not only serve as a reasonable introduction to areas of study possibly not encountered before, but that it will also whet the reader's appetite to consult the comprehensive set of references provided at the end of this chapter. The references section also includes references to important topics not covered in this chapter. One of the most important items that is elided from this chapter are the many proofs behind the theorems that are just stated as true. The author encourages the reader to consult the proofs found in the references, for it is only when one understands the proof of a theorem that one can fully appreciate that theorem and its applications. It is the opinion of this author as well as many other mathematicians that the mathematics behind cryptography is beautiful and worthy of study in its own right; the fact that it has so many applications to solving real computational problems in modern cryptography makes it a marvel deserving of never-ending intellectual pursuit. May this chapter serve as a delectable hors d'oeuvre for the mind.

2.1.2 Notation Summary

This chapter is by necessity, dense with respect to mathematical notation. Depending on the reader's background, some of this notation may be unfamiliar. A complete table of notation may be found in Appendix A.

Common numeric fields and rings such as the real numbers are represented using their normal double-struck font. $\mathbb{C}$, $\mathbb{R}$, $\mathbb{Q}$, $\mathbb{Z}$ are used for the complex, real, rational, and integer numbers, respectively. Polynomials over these fields and rings utilize brackets, which enclose the indeterminates. For example, $\mathbb{Q}[x]$ represents the ring of polynomials with rational coefficients and one indeterminate denoted x.

For a complete list, please consult Appendix A.

2.2 Probability Theory

2.2.1 Introduction

Probability theory is the study of events and their likelihood. Probability theory plays a central role in cryptography as it helps to answer questions, such as "What is the likelihood that two cryptographic hashes of different plaintexts will yield a collision (the hash values are the same)?", "What is the probability that an attacker will be able to guess the value of the private key using cryptographic primitive A?", or "What is the probability that an attacker will be able to create a forged (i.e., it is accepted as valid by the recipient) signature using algorithm B?" Many of the decisions that must be made in the design of cryptographic primitives are tradeoffs based on the likelihood of specific events that can be quantified using the tools of probability theory.

Probability theory can be split into two major branches; discrete probability theory and continuous probability theory. The primary difference between the two is the makeup of the sample space. This section will focus on the principles of discrete probability.

2.2.2 Basic Concepts

We begin with a concept we call a *random experiment*. This is a process that may be executed repeatedly, and whose outcome is (at least in part) believed to be unpredictable. A simple example is the selection of a card from a shuffled deck. Each time we execute the experiment, an *elementary outcome* is obtained (e.g., the 10 of diamonds is chosen). An elementary outcome is referred to as ω. The set of all possible elementary outcomes for a specific experiment is called the *sample space* and is denoted by Ω.

> **Definition 2.2.1** *An event is a set of elementary outcomes, i.e., a subset of the sample space.*

In our example, the selection of any spade is an event. An event that consists of a single elementary outcome is referred to as a *simple event*, while one that is comprised of two or more elementary outcomes is called a *compound event*. Events are usually notated as capital letters, e.g., $A = \{$The event where a card with a value less than 6 is chosen.$\}$

Definition 2.2.2 *The probability of an event* $A \subset \Omega$, *denoted by* $P[A]$ *is a real number in the interval* $[0,1]$ *generated by the function* $P[X]$ *that assigns a value to A from this interval.*

When each elementary outcome in an event is equally probable, $P[A]$ is calculated as the ratio of the total number of elementary outcomes in A to the total number of events in the sample set. When some elementary outcomes in an event are more probable than others, $P[A] = \sum_{r_i} P[\{r_i\}]$, where r_i are the individual events of A. This ratio serves as an indicator of the likelihood of the event occurring. For any sample space Ω, comprised of n elementary outcomes, $\{\omega_1, \omega_2, \omega_3, \ldots, \omega_n\}$, $\sum_{\omega_i} P[\{\omega_i\}] = 1.0$. An example will help to solidify these concepts and terminology.

Consider again the experiment of selecting a card from a fair, standard 52 card deck. Let us define the event A as that of selecting a black card. The probability $P[A] = 0.5$, as there are 26 black cards in each 52 card deck and $\frac{26}{52} = 0.5$.

Two events, A and B, are called *incompatible*, if their intersection $= \varnothing$, i.e., they have no elementary outcomes in common. In this case, the probability that either event A or event B occurs, notated as $P[A] \cup P[B]$ is equal to $P[A] + P[B]$. However, if A and B share one or more elementary outcomes, then $P[A] \cup P[B]$ is equal to $P[A] + P[B] - P[A \cap B]$, i.e., the sum of their probabilities minus the probability of those outcomes they have in common. For example, what is the probability of selecting a red card or a jack from our deck of cards? Let $P[A]$ = the probability of selecting a red card and $P[B]$ = the probability of selecting a jack. $P[A] = \frac{13}{52}$ and $P[B] = \frac{4}{52}$. Since these outcomes overlap, the probability of selecting a red card or a jack is $P[A] + P[B] - P[A \cap B]$ or $\frac{13}{52} + \frac{4}{52} - \frac{2}{52} = \frac{15}{52}$, since there are 13 red cards, 4 jacks, and 2 red jacks in a full deck. This same concept can be extended to larger sets of events.

It is often the case that computing the probability of an event depends on the probability of another event occurring. This is known as conditional probability. For this scenario to be meaningful, the two events will not be independent. We denote the probability of event A occurring, assuming event B has occurred as $P[A|B]$, and it is computed as:

$$P[A|B] = \frac{P[A \cap B]}{P[B]} \text{ if } P[B] > 0 \qquad (2.1)$$

We close this short section with a very important and useful formula in probability known as *Bayes' Theorem*, that can be used to compute one conditional probability, when its "reverse" is known. Bayes' Theorem is often used when evaluating the efficacy of a particular test providing accurate predictive capabilities. For example, if 80% of patients diagnosed with pancreatic cancer (event B) obtain a positive result when administered a particular test (event A) - i.e., $P[A|B] = 0.80$, what is the likelihood that a positive result on this test is a reliable indicator that the patient has pancreatic cancer - i.e., what is $P[B|A]$? Bayes' formula is computed as:

$$P[B|A] = \frac{P[A|B] \times P[B]}{P[A]} \text{ if } P[A] > 0 \qquad (2.2)$$

2.2.3 The Birthday Problem

The Birthday Problem, which appears frequently in the analysis of cryptographic algorithms, and is at the root of numerous attacks against cryptographic primitives, especially cryptographic hash functions, begins with a simple question: How many randomly chosen people need to be in a room before the probability that any two of them shares the same birthday (month and day) is greater than 50%? Intuition might suggest that you would need somewhere around 100 people. However, the answer, calculated through the use of probability theory and combinatorics we arrive at a very unexpected answer, and therefore this problem is also known as *The Birthday Paradox*. Quite surprisingly, in a room containing a mere 23 people, there is a greater than 50% chance that two people will share the same birthday. In a room with 40 people, there is an almost 90% chance that two share the same birthday! For details on calculating these values as well as a discussion on this problem's application to cryptography, please see [171].

2.3 Combinatorics

2.3.1 Introduction

From the beginning, mankind has been counting. One of the first mental exercises we teach small children is that of how to count: fingers, toes, pieces of candy, birds in a tree, etc. As a child's concepts of justice and fairness develop, we quickly see the two skills intertwined, enabling a child to compare countings, and vehemently protest, "Hey, that's not fair, she got more than me!"

Although it is an over-simplification, combinatorics is the study of counting. Certainly, one major branch of combinatorics, that of enumerative combinatorics, is focused on the various ways one can count items [2, 273, 408]. Other branches, including analytic combinatorics, graph theory, design theory, order theory, and algebraic combinatorics have elements of counting in them as well, besides their own unique areas of specialization. What seems like a trivial exercise is in fact a most interesting and active area of mathematical research. What may have at one time primarily found its outlet in popular mathematics (e.g., Martin Gardiner's famed mathematical puzzles), today's world, filled with myriad logistical and optimization problems, calls on combinatorics to solve real problems. The study of combinatorics provides us with the tools needed to solve these ever more complex questions of counting.

How many ways can we arrange five distinct pieces of fruit and four indistinct baskets of nuts in a circular pattern? How many distinct ways can we seat fifteen people, 9 male and 6 female, at four tables which seat four each, without seating more than 3 members of any single gender at one table? How many combinations of five letters can be formed with the English alphabet? What if duplicates are not allowed? What if each combination must begin with a consonant or end with a vowel? How many cryptographic keys of length ℓ can be generated by a particular function? How many ways can the set of byte arrays of length greater than zero and less than 1024 be reduced to a byte array of length 32?

A solid grounding in combinatorics is needed if one is to understand the design, capabilities, and limits of different cryptosystems and cryptographic primitives. Combinatorics is central to the study of the analysis of algorithms, including cryptographic ones. The study of combinatorics is closely related to the subject of coding theory which has close ties to information theory, a fundamental topic in the study of theoretical cryptography. Furthermore, combinatorics is such a rich and diverse field of study that connections to cryptography can be found even in unexpected places. For example, while writing this chapter, the author was introduced to a research paper written by a team of cryptographers in Morocco who used both elliptic curve cryptography and Catalan numbers (a combinatorial concept) in the development of a new cryptosystem [6].

What follows is a brief introduction to a few concepts from combinatorics that are most relevant to cryptography.

2.3.2 Permutations

Definition 2.3.1 *A permutation is a 1 : 1 function from a set onto itself.*

For example, consider the set of the three colors $C = \{red, blue, green\}$. Given an ordered sequence of the elements of C, the permutation that sends *red $\rightarrow$ blue, blue $\rightarrow$ green*, and *green$\rightarrow$ red* rearranges the sequence [*red, blue, green*] to [*blue, green, red*]. The study and use of permutations is a fundamental element in today's cryptographic primitives. For example, nearly all modern block ciphers (e.g., DES, AES, Twofish) make use of permutations in what is called their round functions. These permutations are often referred to as *P-boxes*, and with few exceptions are much greater in size than the simple example given above. Below are some of the major concepts regarding permutations and their representation.

2.3.2.1   Representations

Permutations may be represented in a number of different ways. Here is a sampling of the most common representations used in the literature. For each representation type, we will use the finite set $\mathcal{E} = \{1,2,3,4,5,6,7,8\}$ as our starting point.

Two-line representation

In many ways, a two-line representation, attributed by some to Arthur Cayley [457], is one of the most intuitive. In this representation, a permutation is displayed in a format similar to a $2 \times n$ matrix, where $n = |\mathcal{E}|$ (the cardinality, or number of elements in $\mathcal{E}$). The first row lists the elements of $\mathcal{E}$ (usually ordered, if a natural order exists); the second lists the image of each element under the given permutation.

Consider the following two-line representation:

$$\sigma = \begin{pmatrix} 1 & 2 & 3 & 4 & 5 & 6 & 7 & 8 \\ 3 & 6 & 2 & 1 & 8 & 7 & 4 & 5 \end{pmatrix} \tag{2.3}$$

This permutation, takes $1 \rightarrow 3$, $2 \rightarrow 6$, $3 \rightarrow 2$, etc. Given the sequence $a = (\ 3\ \ 5\ \ 4\ \ 1\ \ 7\ \ 2\ \ 8\ \ 6\ )$ of the elements of $\mathcal{E}$, the application of the permutation σ on a, $\sigma(a) = (\ 2\ \ 8\ \ 1\ \ 3\ \ 4\ \ 6\ \ 5\ \ 7\ )$.

Permutation lists

A very different approach that looks much like the two-line representation (and can lead to confusion) is that of permutation lists. This approach is used by some computer algebra systems, including *Mathematica*. In this representation, a permutation is represented less like a function that maps an element $a \in \mathcal{E}$, to its image $\sigma(a) \in \mathcal{E}$ under the permutation, but instead focuses on the positional influencing aspects of the permutation.

The permutation list $\ell = \{4,7,5,3,8,1,2,6\}$, acts on a sequence in the following manner. Beginning with the first element of the list, the value in position 1 indicates the new position of the element that is currently in the first position of the given sequence. In this example, whatever is in the first position of the sequence will be moved to position 4. Likewise, the element in position 2 will now be in position 7, that in position 3 will be in position 5, etc. Given the permutation list ℓ and the sequence a defined above, application of ℓ to a will result in the sequence $(\ 2\ \ 8\ \ 1\ \ 3\ \ 4\ \ 6\ \ 5\ \ 7\ )$.

Cycle notation

The two previous representation methods are completely usable when working with permutations. However, cycle notation is much more compact and also helps when studying the various properties of permutations. Much like the permutation list representation, cycle notation focuses on how the position of sequence elements are changed under the given permutation. Cycle notation takes advantage of the fact that all permutations consist of a set of "movement" items, including that which does not move an element, a transposition of two elements, and an orbit of a subset of elements, where movement from element to element results in a return to the first element of the orbit.

Consider again, the permutation list above, $\ell = \{4, 7, 5, 3, 8, 1, 2, 6\}$. From the description, we know that this means that the element in the first position moves to the fourth position, the element in the fourth position moves to the third position, etc. In cycle notation, we can succinctly capture this cyclic movement as (1 4 3 5 8 6). Notice that the placement of 6 at the end of the cycle, implies that it moves to the position found at the beginning of the cycle; in this case, 1 (i.e., the first position). Now, while we created a cycle consisting of six elements of $\mathcal{E}$, we still lack two elements. Neither 7 nor 2 appear in this cycle. When one examines the permutation list, we find that the element in the second position moves to the seventh position, and the element in the seventh position, moves to the second position. This is a simple transposition of elements, represented in cycle notation as (2 7). Therefore, the complete permutation is represented in cycle notation as (1 4 3 5 8 6) (2 7). Applying this permutation to the sequence $a = (\ 3\ \ 5\ \ 4\ \ 1\ \ 7\ \ 2\ \ 8\ \ 6\ )$ results in the sequence $(\ 2\ \ 8\ \ 1\ \ 3\ \ 4\ \ 6\ \ 5\ \ 7\ )$, as expected. When for a given permutation, an element does not move, it need not appear in the list of cycles. Also, the order in which the elements of a particular cycle appear does not change the meaning of the cycle. For example, the cycles (1 4 3 5 8 6) and (3 5 8 6 1 4) are semantically identical, demonstrating that representations of permutations using the cycle notation are not unique.

2.3.2.2 Operations on permutations

Permutations allow two major operations; composition and taking of inverses. In addition, for every set, there exists a neutral element. Below, when we consider group theory, we will find that these properties enable us to define a group over a set of permutations. The study of permutation groups, their use in the representation of groups, their applicability to other algebraic subjects such as Galois Theory, and many other topics are still an active mathematical research area.

Combining two or more permutations through the operation of composition is simply the application of one permutation, followed by that of the next. Using cycle notation, the combination of two permutations is denoted by juxtaposing the cycle list of each permutation, separated either by a · or a ∘. Application of permutations is applied from right to left.

Given the permutation above, which we shall call $p_1 = (1\ 4\ 3\ 5\ 8\ 6)\ (2\ 7)$, we also define the permutation $p_2 = (3\ 2)\ (5\ 4\ 1)\ (7\ 8)$. Combining $p_2 \circ p_1$, we obtain a new permutation $p_3 = (4\ 2\ 8\ 6\ 5\ 7\ 3)$.

The neutral element of a permutation which we shall denote e, is that permutation which moves no items.

The inverse of a permutation is the permutation denoted p^{-1} such that $p \circ p^{-1} = e = p^{-1} \circ p$. When using cycle notation, creating the inverse of a permutation is a simple matter of reversing the order of the elements in each of the cycles of the permutation. For $p_1 = (1\ 4\ 3\ 5\ 8\ 6)\ (2\ 7)$, we define the inverse $p_1^{-1} = (6\ 8\ 5\ 3\ 4\ 1)\ (2\ 7)$. Combining these through the composition operation yields the neutral permutation e. Note that the inverse of a transposition is the transposition itself.

2.3.2.3 Properties of permutations

For any finite set S of order n, the number of permutations on S is $n!$. This is easily demonstrated by considering the process of choosing elements for a permutation σ in the two-line representation. When one begins, they have n choices for which element is the image of 1, i.e., $\sigma(1)$ in this permutation. Once that element is selected, it may no longer be used, as permutations are 1:1. Therefore, for $\sigma(2)$, there are $(n-1)$ choices. This continues until there is only one item left for assignment to the image of the last element in the permutation, $\sigma(n)$. Putting this all together, we have $(n)(n-1)(n-2)(n-3)\ldots(2)(1) = n!$.

Every permutation of a finite set can be represented as a composition of transpositions (i.e., cycles of length two). This means that in spite of the fact that some of the permutations used above consisted of cycles of length greater than 2, it is possible to take each of those cycles and decompose it into a set of transpositions, so that the permutation is comprised of nothing more than a set of transpositions. Each permutation admits various representations as a set of transpositions. However, it is an invariant property of a permutation that if one representation of a permutation is comprised of an even (resp. odd) number of transpositions, all representations via transpositions will contain an even (resp. odd) number of transpositions.

A derangement (also known as a complete permutation) is a permutation that leaves none of the items in the set in their original position. There are numerous formulae for computing the number of derangements of a set of order n, $d(n)$. One will suffice:

$$d(n) = n! \sum_{i=0}^{n} \frac{(-1)^i}{i!} \tag{2.4}$$

So, for a set of order 3,

$$d(3) = 3! \sum_{i=0}^{3} \frac{(-1)^i}{i!} = 6\left(1 + (-1) + \frac{1}{2} + \left(-\frac{1}{6}\right)\right) = 2 \tag{2.5}$$

These two derangements in cycle notation are $(1,2,3)$ and $(1,3,2)$.

2.3.2.4 Subset permutations

We now know that the number of permutations of size n on a set with n elements is $n!$. However, what if we need to compute the number of permutations of r elements, where $r < n$, drawn from a set of n elements, while still maintaining the restriction that an item may only appear once in a permutation. Below is a formula for calculating this value:

$$\frac{n!}{(n-r)!} \tag{2.6}$$

For example, if presented with the question of: How many ways can we select boys for a math competition that requires a team of 3, with each member assigned a number from 1 to 3 (i.e., order matters here), from the set of boys $B = \{$Albert, Bob, Charles, David, Edward$\}$. Since $n = |B| = 5$, and $r = 3$, the answer is:

$$\frac{5!}{(5-3)!} = \frac{5!}{2!} = \frac{120}{2} = 60 \text{ ways} \tag{2.7}$$

2.3.3 Binomial Coefficients

While understanding permutations helps one answer a large number of combinatorial (counting) problems, there is another class of problems in which permutations do not apply. Permutations assume that all of the items in the set are ordered (i.e., they are treated as a sequence). Consider a small modification to the question above: How many ways can we select a team of 3 boys for a math competition from the set of boys $B = \{$Albert, Bob, Charles, David, Edward$\}$? Notice in this case, there is no order enforced on the team's composition. Therefore, in this case, a team comprised of Bob, David, and Edward, would be the same as a team comprised of David, Edward, and Bob, which is not true of the question asked in the last section. When the selection does not require ordering, using binomial coefficients is the tool of choice for computing an answer.

Binomial coefficients get their name from the fact that they are the coefficients in the polynomials generated within the context of the binomial theorem [i.e., those of the form $(x+y)^n$]. Binomial coefficients are represented as $\binom{n}{k}$ and are read as "n choose k." The binomial coefficient $\binom{n}{k}$ represents the number of combinations of k elements that can be selected from a set of order n. The binomial coefficient $\binom{n}{k}$ is computed as:

$$\frac{n!}{k!(n-k)!} \tag{2.8}$$

Returning to our question regarding the number of ways to select 3 boys from a set of 5, without respect for order; using the formula above we find:

$$\frac{5!}{3!(5-3)!} = \frac{5!}{3!2!} = \frac{120}{6(2)} = \frac{120}{12} = 10 \text{ ways} \tag{2.9}$$

Comparing this result of 10 with the result of 60 above, it is fairly straight-forward to understand this difference of a factor of 6. Consider the team comprised of {Bob, David, Edward}. In the last section, this team would be considered different than the team {Edward, Bob, David}, as each team member needed to have a number associated with them; order mattered. There are 6 permutations of the set comprised of {Bob, David, Edward}. In this section, all such permutations are considered to be equivalent. Therefore, when order does not matter and each combination of 3 boys "collapses" into a single team definition, the total number of possible teams is $\frac{1}{6}$ of that computed when order does matter.

2.3.4 Recurrence Relations

Definition 2.3.2 *A recurrence relation (sometimes referred to as a difference equation) is a function that is defined in terms of other invocations of the same function.*

The most familiar recurrence relation, known to almost all, is that of the Fibonacci sequence. The Fibonacci numbers are a sequence of numbers $\mathcal{F}$ where each element of the sequence is calculated as a function of previous elements. Specifically:

$$\mathcal{F}_n = \mathcal{F}_{n-1} + \mathcal{F}_{n-2} \tag{2.10}$$

Each Fibonacci number is the sum of the two previous Fibonacci numbers. The first 15 Fibonacci numbers are 1, 1, 2, 3, 5, 8, 13, 21, 34, 55, 89, 144, 233, 377, and 610. $\mathcal{F}_{16} = \mathcal{F}_{15} + \mathcal{F}_{14}$ which is equal to $610 + 377 = 987$.

Numerous types of recurrence relations with differing properties and levels of complexity exist. In addition, there are various methods for converting a recurrence relationship, which can be inefficient to compute, to a closed form expression or to a generating function; both of which can be used to efficiently compute elements of the sequence.

Recurrence relations have been used in various areas of cryptography, including prime testing and integer factorization algorithms [118]. Various recurrence relations have also been used in the creation of new cryptographic primitives [415]. Probably of greatest notoriety in the connection between recurrence relations and cryptography is the Montgomery multiplication algorithm (described below). Montgomery multiplication provides a very fast way to perform multiplication modulo some $n \in \mathbb{Z}^+$. This technique is very useful in numerous cryptosystems, including RSA, which performs modular exponentiation. In many implementations, Montgomery multiplications makes use of recursively defined sequences of integers (i.e., recurrence relations) [92].

2.4   Number Theory

2.4.1   Introduction

Mathematics is the queen of sciences and arithmetic (number theory) is the queen of mathematics. She often condescends to render service to astronomy and other natural sciences, but in all relations she is entitled to the first rank.
— Carl Friedreich Gauss as quoted in *Gauss zum Gedächtniss* [444]

The study of simple counting numbers has captured the minds of the inquisitive since before the time of Euclid (circa 300 B.C.). Euclid's work, *The Elements*, comprised of thirteen books believed to be not only the work of Euclid but also the collection of earlier works by other authors, contains not only the foundations of geometry as it has been studied for most of the last two millennia, but also contains abundant material dealing with number theory. Book VII defines Euclid's algorithm for finding the greatest common divisor between two natural numbers (the same algorithm we teach today) and defines both prime and composite numbers. As one progresses through Books VIII and IX, they find that these mathematicians of antiquity were able to prove the Fundamental Theorem of Arithmetic (every number > 1 has a unique factorization into primes) and the fact that there exists an infinite number of primes. The study of numbers has continued unabated since then, history being replete with the names of famous amateur and professional mathematicians such as Diophantus, Fibonacci, Fermat, The Bernoulli family, Euler, Gauss, Dedekind, Hardy, and Ramanujan, who have left a legacy of discovery in the study of this set of numbers we teach to toddlers. In addition, as number theory has evolved, mathematicians have learned how to extend, compare, and contrast the properties of simple counting numbers to those of more sophisticated number systems. Within this great corpus of knowledge, we find many of the properties of the counting numbers and their extensions serve at the core of modern cryptography. In effect, "the Queen" has once again condescended to render her services; this time to that of modern cryptography. Without the understanding of the prime numbers and integer factorization into primes that we possess today, there would be no RSA algorithm for encryption and digital signatures. Without the discrete logarithm problem (though in part an algebraic one), there would be no Diffie-Hellman key exchange protocol and no El Gamal cryptosystem. In addition, certain zero-knowledge proof and oblivious transfer techniques would be without foundation.

2.4.2 Greatest Common Divisor

Definition 2.4.1 *The greatest common divisor (GCD) between two natural numbers $a, b \in \mathbb{N}$, represented as (a, b) or $GCD(a, b)$, is the largest natural number d that divides (represented as $d|a$) both a and b.*

Euclid's Elements contains a simple (it only requires basic arithmetic) polynomial time algorithm for computing the GCD. While more efficient algorithms exist, Euclid's algorithm is still used today and is most instructive.

2.4.2.1 Euclid's algorithm

Euclid's algorithm for finding the GCD of two natural numbers $a, b \in \mathbb{N}$ is likely to be familiar to most readers. It consists of a sequence of repeated steps wherein at each step the larger of the two inputs, a is expressed as a quotient of b with some remainder $0 <= r < b$ (i.e., $a = bq + r$), which finally reduces to 0 in the last step. In each successive step, b and r from the previous step become a and b respectively for the current step. The GCD is the remainder r of the next to the last step. A simple example of this algorithm should suffice. Let us compute $(3780, 3234)$:

$$
\begin{aligned}
3780 &= 3234 \times 1 + 546 \\
3234 &= 546 \times 5 + 504 \\
546 &= 504 \times 1 + 42 \\
504 &= 42 \times 12 + 0
\end{aligned}
\tag{2.11}
$$

Therefore, $(3780, 3234) = 42$. This algorithm is known to be in the complexity class $\mathfrak{P}$; however, there are parallelizable algorithms that perform better than this [76].

Bezout's Theorem states that for integers $a, b \in \mathbb{Z}$, there exist integers $x, y \in \mathbb{Z}$ such that $ax + by = d$, where $d = (a, b)$. Through the use of the Extended Euclidean Algorithm, one can easily calculate the values of x and y for any a and b. For details of this algorithm, see any standard text on general number theory, including [351] and [392]. Although the details of the computation have been left out, through the use of the Extended Euclidean Algorithm, we find that $(3780 \times 6) + (3234 \times (-7)) = 42$.

2.4.3 Modular Arithmetic

Definition 2.4.2 *We say that two integers a and $b \in \mathbb{Z}$ are congruent modulo n if each, on division by n, have the same remainder.*

For example, $26 \equiv 61$ (mod 7), because both have remainder 5 when divided by 7. It is easy to confirm that for all n, every integer is congruent to some value $0 \leq r < n - 1$ modulo n. In fact, congruence modulo n forms an equivalence relation over the integers. It is often the case, especially in cryptography, that one wishes to perform calculations, using only the smallest member of each equivalence class modulo n. For example, modulo 7, all arithmetic is performed using only integers between 0 and 6. This is known as modular arithmetic. This approach to arithmetic is generally not taught early to children, but is in fact known early to them. Consider what we commonly call "clock arithmetic." With non-military time, this is nothing more than arithmetic modulo 12. When asked at 10:00, what time it will be in 6 hours, we do not say, "It will be 16 o'clock." Rather we "mod out" 12, and respond with "It will be 4 o'clock." This is because $10 + 6 \equiv 4$ (mod 12).

As simple as modular arithmetic is, it is quite useful in practical applications and has yielded numerous theorems regarding its characteristics. Consider the following theorem:

For modulus n, we define a^{-1} as the multiplicative inverse of a if $a^{-1} \equiv 1$ (mod n). Let $a, n \in \mathbb{Z}$ with $n > 0$. Then a has a multiplicative inverse modulo n if and only if $(a, n) = 1$ (i.e., they are relatively prime).

As an example, (mod 14), the multiplicative inverse of 5, which is relatively prime to 14, is 3 [i.e., $5 \times 3 \equiv 1$ (mod 14)]. However, 4, which has 2 as a common factor, does not have a multiplicative inverse (mod 14). From this theorem, we can deduce that modulo a prime p, each integer $a \in \mathbb{Z}$ where $0 <= a < p$, has a multiplicative inverse, as all values of a are coprime to p. As we shall see when we consider abstract algebra, the set of integers modulo n, where n is not necessarily prime, forms an algebraic structure known as a commutative ring, while the set of integers modulo p, where p is prime, form another algebraic structure, known as a field. Both rings and fields based upon modular arithmetic are very important in today's cryptography.

2.4.4 The Chinese Remainder Theorem

Definition 2.4.3 *A linear congruence is an equation of the form $ax \equiv b$ (mod n), where $a, b \in \mathbb{Z}$ and $n \in \mathbb{Z}^{+}$.*

The Linear Congruence Theorem states that a linear congruence has a solution for x if and only if $d = (a, n) | b$ (i.e., if the GCD d of a and n divides b). Under these conditions, there are exactly d solutions to the linear congruence, each of the form $\{ x_0 + k\frac{n}{d} | 0 <= k < d \}$ where x_0 is the smallest solution for x in the linear congruence. For example, take the linear congruence: $12x \equiv 8$ (mod 28).

First, does it have a solution? $(12, 28) = 4$, and $4|8$, so the answer is yes. By the statement above we know that there are a total of 4 solutions. By simple trial and error, we find that the smallest solution is 3. Using the equation above, we first calculate $\frac{n}{d} = \frac{28}{4} = 7$, and then use this value to calculate the full set of solutions $\{3, 10, 17, 24\}$.

Now consider the case where one has a system of k simultaneous linear congruences:

$$
\begin{aligned}
x &\equiv a_1 \pmod{n_1} \\
x &\equiv a_2 \pmod{n_2} \\
&\cdots \\
x &\equiv a_k \pmod{n_k}
\end{aligned} \tag{2.12}
$$

If $n_1, n_2, n_3, \ldots, n_k$ are pairwise coprime, then there exists a unique solution x for this set of linear congruences. This is known as the Chinese Remainder Theorem. In addition, any $x' \in \mathbb{Z}$ is also a solution to this system if and only if $x \equiv x' \pmod{n}$ where $n = \prod_{i=1}^{k} n_i$. Among other cryptographic uses, the Chinese Remainder Theorem is used to optimize the algorithm for performing RSA decryption. What follows is an example of the Chinese Remainder Theorem:

$$
\begin{aligned}
x &\equiv 9 \pmod{7} \\
x &\equiv 21 \pmod{13} \\
x &\equiv 44 \pmod{29} \\
x &\equiv 50 \pmod{47}
\end{aligned} \tag{2.13}
$$

Because each of the moduli $\{7, 13, 29, 47\}$ is a prime number, we know that they are pairwise coprime, and therefore a solution exists. Numerous polynomial time algorithms exist for computing the result. Here is one algorithm:

Algorithm 2.1 Polynomial time algorithm for computing Chinese Remainder Theorem result.

Let $c_i = \frac{n}{n_i}$, for each $i = 1, 2, \ldots, k$ and n is the product of the individual moduli: $n_1, n_2, n_3, \ldots, n_k$.
Let d_i equal the solution x to the linear congruence, $c_i x \equiv 1 \pmod{n_i}$, for each $i = 1, 2, \ldots, k$.
Let $x_0 = a_1 c_1 d_1 + a_2 c_2 d_2 + \ldots + a_k c_k d_k$.
The solution x is equal to $x_0 \pmod{n}$.

Eliding the details of the computation on the equations above, $x_0 = 11685942$. $11685942 \equiv 26840 \pmod{n}$. Therefore, the solution to the system of simultaneous congruences above is 26840. This is easily confirmed:

$$
\begin{array}{llllll}
26840 &\equiv 2 \pmod{7} & \text{and} & 2 &\equiv 9 \pmod{7} & \\
26840 &\equiv 8 \pmod{13} & \text{and} & 8 &\equiv 21 \pmod{13} & \\
26840 &\equiv 15 \pmod{29} & \text{and} & 15 &\equiv 44 \pmod{29} & (2.14) \\
26840 &\equiv 3 \pmod{47} & \text{and} & 3 &\equiv 50 \pmod{47} &
\end{array}
$$

2.4.5   Quadratic Residues and Quadratic Reciprocity

In the last section, we focused our attention on linear congruences. What about congruences of higher degree? Consider, for example, the congruence $b^2 \equiv a \pmod{n}$. For which values of b, a, and n does this congruence have a solution, if any? Are there multiple solutions? In what follows, we set out to answer these questions as well as others related to this topic of *quadratic residues*, which play an important role in today's cryptographic algorithms.

For positive integers a and n, with $n \neq 0$ and $(a, n) = 1$, a is called a *quadratic residue* mod n if the congruence $b^2 \equiv a \pmod{n}$ has at least one solution, $b \in \mathbb{Z}$. If no solution exists, a is called a *quadratic non-residue*. We will focus most of our attention on the case where n is an odd prime.

For the remainder of this discussion, let p be an odd prime and let $a \in \mathbb{Z}^+$ such that $p \nmid a$. We denote the *Legendre symbol* of a with respect to p as:

$$\left(\frac{a}{p} \right) = \begin{cases} 1 & \text{if } a \text{ is a quadratic residue} \quad (\text{mod } p) \\ -1 & \text{if not} \end{cases} \tag{2.15}$$

First, it is clear that all perfect squares are quadratic residues $(\text{mod } p)$. What about the rest of the positive integers $< n$? The following theorem provides a partial answer.

Theorem 2.4.4 *Let p be an odd prime. There are then $\frac{(p-1)}{2}$ quadratic residues in the set $\mathcal{S} = \{0, 1, \ldots, (n-1)\}$.*

The Legendre symbol has numerous properties that are valuable when computing with it as well as seeking to determine if a certain value a is a quadratic residue (mod p). Below is a list of properties.

1. *Euler's Criterion*: If p is an odd prime and $a \in \mathbb{Z}$ with $p \nmid a$, then $a^{\frac{(p-1)}{2}} \equiv \left(\frac{a}{p} \right) \pmod{p}$.

2. If $a \equiv b \pmod{p}$, then $\left(\frac{a}{p} \right) = \left(\frac{b}{p} \right)$.

3. $\left(\frac{a_1, a_2, \ldots a_n}{p} \right) = \left(\frac{a_1}{p} \right) \left(\frac{a_2}{p} \right) \ldots \left(\frac{a_n}{p} \right)$. In other words, the Legendre symbol is multiplicative.

4. $\left(\frac{-1}{p} \right) = (-1)^{\frac{p-1}{2}}$.

5. For any odd prime p, $\left(\frac{2}{p} \right) = (-1)^{\frac{p^2-1}{8}}$.

6. *Law of Quadratic Reciprocity*: If p and q are distinct odd primes, then $\left(\frac{q}{p} \right) \left(\frac{p}{q} \right) = (-1)^{\frac{p-1}{2} \frac{q-1}{2}}$.

This last one, attributed to Gauss, is considered by many [8] to be the primary result regarding the Legendre symbol.

Using the theorems and properties above, let us work through the following example: let $p = 13$. How many elements of the set $S = \{1, 2, \ldots, 12\}$ are quadratic residues, (mod 13), and what are they? We will let Q_r represent the set of quadratic residues. First, by the first theorem above, we know that there are $\frac{13-1}{2} = 6$ quadratic residues, (mod 13): $|Q_r| = 6$. To determine which elements of S are quadratic residues, we start with the perfect squares. We know that, being perfect squares, $\{1, 4, 9\} \in Q_r$. This accounts for 3, we need 3 more. Note that $3^6 \pmod{p} = 1$. Therefore 3 is a quadratic residue, based on Euler's criterion. The same is true of 10 and 12. Therefore, the entire set of quadratic residues mod 13 is $Q_r = \{1, 3, 4, 9, 10, 12\}$.

The cases where n is a prime power or a composite number have similar sets of properties that can be used for computing quadratic residues. Some of them make use of the properties defined for primes. For more details, please consult one of the many references on Number Theory at the end of this chapter. Chapter 12 of [392], which is available online as a free PDF download, is an especially good resource and provides coverage for all three cases.

2.4.6 The Prime Numbers

There are two facts about the distribution of prime numbers of which I hope to convince you so overwhelmingly that they will be permanently engraved in your hearts. The first is that, despite their simple definition and role as the building blocks of the natural numbers, the prime numbers belong to the most arbitrary and ornery objects studied by mathematicians: they grow like weeds among the natural numbers, seeming to obey no other law than that of chance, and nobody can predict where the next one will sprout. The second fact is even more astonishing, for it states just the opposite: that the prime numbers exhibit stunning regularity, that there are laws governing their behaviour, and that they obey these laws with almost military precision. — Don Zagier, 1977

The author assumes the reader's familiarity with the prime numbers. However, for sake of completeness, a brief definition of prime numbers under the integers $\mathbb{Z}$ is found below.

Definition 2.4.5 *A prime number p is a positive integer, greater than 1 such that p has no divisors other than 1 and itself.*

The first twenty primes are 2, 3, 5, 7, 11, 13, 17, 19, 23, 29, 31, 37, 41, 43, 47, 53, 59, 61, 67, and 71. 2 is the only even prime. As banal as the study of the prime numbers may seem to some, it turns out that as Zagier intimated (see quote above), the primes are deserving of, and have been the subject of almost unending study continuing to this very day. Some of mathematics still unlocked secrets, such as the Riemann Hypothesis, are intimately intertwined with the properties of the prime numbers. Unfortunately, we can only highlight a few of the interesting questions and properties of the primes in this section.

2.4.6.1 Distribution of the primes

As stated in the introduction to this section on number theory, Euclid, in Book IX of *The Elements* established that there exists an infinite number of primes. Since then, numerous additional proofs of the same theorem have been defined. Paulo Ribenboim [344] gives eleven different proofs in one book and as recently as 1995, new proofs continue to be developed [365]. Immediately, other questions follow. Very early in Hardy and Wright's *An Introduction to the Theory of Numbers* [161], they dedicate a section to asking the following four questions:

1. "Is there a simple general formula for the *n*-the prime p_n?"

2. "Is there a simple general formula for the prime which follows a given prime?"

3. "Is there a rule by which, given any prime p, we can find a larger prime q?"

4. "How many primes are there less than a given number x?"

These questions, are followed by extensive treatment of the subject, including an entire chapter in the latter part of Hardy and Wright's classic, dealing only with the properties of the series of primes. It is left to the inquisitiveness of the reader to seek out the answers to the first three questions. However, the fourth question has a clear answer in what is known as the *prime counting function*. This question generalizes into another, "How are the primes distributed among the integers?" Both have been explored extensively by mathematicians and both have reasonably satisfying answers.

The prime counting function, denoted $\pi(x)$ (which has nothing to do with the transcendental number $\pi \approx 3.1415926535$), returns, for each positive real number, the number of prime numbers less than or equal to x. A cursory scan of the literature will show that there is no known single function that computes $\pi(x)$ with no margin of error. Instead, there are numerous formulae and algorithms that produce very accurate results, especially when x is large. Ján Mináč gives the following compact formula (though quite far from the best performing) for $\pi(x)$ which Ribenboim provides a proof of in [343]:

$$\pi(x) = \sum_{j=2}^{x} \left\lfloor \frac{(j-1)! + 1}{j} - \left\lfloor \frac{(j-1)!}{j} \right\rfloor \right\rfloor \tag{2.16}$$

where $\lfloor x \rfloor$ is the standard floor function. Studying the growth rate of $\pi(x)$

leads to the *Prime Number Theorem* which addresses the more general question of the overall distribution of the prime numbers.

Theorem 2.4.6 *The Prime Number Theorem states that:*

$$\pi(x) \sim \frac{x}{\ln x} \tag{2.17}$$

which can also be stated as

$$\lim_{x \to \infty} \frac{\pi(x)}{x / \ln x} = 1 \tag{2.18}$$

Informally, as x grows, the difference between $\pi(x)$ and $\frac{x}{\ln x}$ diminishes to 0, though not quickly enough to use it as a formula for $\pi(x)$. However, in spite of this seeming regularity in the distribution of the primes, one will find that the distribution of the primes when examined for short intervals is anything but predictable. For more details, please consult one of the many texts written that focus only on the primes, their characteristics, and their distribution such as [91], [343], and [424].

2.4.6.2 Primality testing

Detecting and selecting large prime numbers is a common operation in many of today's cryptographic algorithms. For example, the modulus of RSA, n, is comprised of two carefully selected large primes p and q. Therefore, having efficient methods for prime testing is essential. At a high level, there are two major types of algorithms.

The first is probabilistic, in that it does not guarantee that it always returns a correct answer. If it returns false for an input p, p is guaranteed not to be prime. However, if the algorithm returns true, there is a small probability that p is not prime, even though the algorithm has indicated that it is. Many of these algorithms have parameters that allow for an "accuracy dial" to be supplied as input. The larger the value of this parameter, the greater the probability that the algorithm does not incorrectly report that a specific composite value is prime. As a rule, this increased accuracy comes at the cost of reduced performance.

In contrast, deterministic prime testing algorithms are guaranteed to always return correct results. However, much like the parameterized probabilistic algorithms when used with large accuracy parameters, this results in reduced performance.

Selection of the right prime testing algorithm is dependent on the specific requirements of the application. In some cases, a combination of prime testing algorithms may be employed. Numerous resources found in the References section of this chapter provide details on the many probabilistic and deterministic algorithms that have been used and continue to be used today [91, 171, 285, 345].

2.4.7 Euler's Totient Function

> **Definition 2.4.7** *Euler's Totient Function, also known as the Euler-ϕ (phi) Function is the function $\phi : \mathbb{N} \rightarrow \mathbb{N}$, such that on any input $n \in \mathbb{N}$, $\phi(n)$ returns the count of natural numbers less than n that have no common factors with n.*

Two natural numbers, $a, b \in \mathbb{N}$, that share no common factors besides 1 (i.e., $(a, b) = 1$) are called *coprime*. Therefore, $\phi(n)$ returns the number of natural numbers less than n that are coprime to n. Examples are listed below:

1. $\phi(6) = 2$: Each of the natural numbers $\{2, 3, 4, 6\}$ shares at least one common factor with 6, except for 1 and 5.

2. $\phi(14) = 6$: Each of the natural numbers $\{2, 4, 6, 8, 10, 12, 14\}$ share common factors with 14. The set $\{1, 3, 5, 9, 11, 13\}$ of cardinality 6, contains all natural number less than 14 that are coprime to 14.

3. $\phi(13) = 12$: Notice that in the case where $p \in$ prime numbers, that $\phi(p)$ is always equal to $p - 1$. This is easy to see, as by definition, prime numbers have no factors besides 1 and themselves, and therefore will be coprime to every natural number less than p.

The Euler-ϕ function appears in many different number theoretic and cryptographic theorems and proofs. Consider the following properties of the Euler-ϕ function.

- Two functions f and g are considered *multiplicative* if for two coprime inputs a and b, $f(ab) = f(a)f(b)$. The Euler-ϕ function is multiplicative, i.e., if $(a, b) = 1$ then $\phi(a)\phi(b) = \phi(ab)$. For example:

$$
\begin{aligned}
(20, 9) &= 1 \text{ i.e., 20 and 9 are coprime} \\
\phi(20) &= 8 \\
\phi(9) &= 6 \\
\phi(180) &= 48 = 8 \times 6
\end{aligned}
$$

- For any prime p, and any positive integer k, $\phi\left(p^k\right) = p^k - p^{k-1}$.

- For any positive integer n, $\sum_{d|n} \phi(d) = n$.

- For any positive integer n which is factored into powers of distinct primes, i.e.,

$$n = \prod_{p} p^{e_p}$$

$$\phi(n) = n \prod_{p|n} \frac{p-1}{p} \tag{2.19}$$

2.4.8 Fermat's Little Theorem

Fermat's Little Theorem is a fundamental result of basic number theory, and also serves as the backdrop to why RSA encryption and decryption work as designed. It is defined as follows.

> **Theorem 2.4.8** *Fermat's Little Theorem: For any prime number p, and any integer $a \in \mathbb{Z}$, the following equality holds: $a^p \equiv a \pmod{p}$.*

For example, $4^5 \equiv 4 \pmod 5$. This is easily confirmed with basic arithmetic: $4^5 = 1024$ and 1024 divided by 5 is equal to 204 with a remainder of 4. If a is not divisible by p, dividing both sides of the equality by a yields a useful restatement of the theorem: $a^{(p-1)} \equiv 1 \pmod{p}$.

The following generalization of Fermat's Little Theorem, leads to the proof of correctness of RSA encryption: given a prime p and $\ell, m \in \mathbb{Z}^+$ (the positive integers), if $\ell \equiv m \pmod{(p-1)}$, then for each $c \in \mathbb{Z}$, $c^\ell \equiv c^m \pmod{p}$.

2.4.9 Euler's Theorem

Euler's Theorem is a generalization of Fermat's Little Theorem and states the following.

> **Theorem 2.4.9** *Euler's Theorem: For $a, n \in \mathbb{Z}$ where $(a, n) = 1$, $a^{\phi(n)} \equiv 1 \pmod{n}$ where $\phi(n)$ is the Euler totient function defined earlier.*

The converse of this theorem is true as well, i.e., if $a^{\phi(n)} \equiv 1 \pmod{n}$, then $(a, n) = 1$.

2.5   Efficient Integer Arithmetic

2.5.1   Introduction

Arithmetic of large integers, including arithmetic modulo $n \in \mathbb{Z}$ can often be much more compute intensive than standard arithmetic. For example, modular multiplication, which is an operation often used in the implementation of cryptographic primitives, is more complicated than standard multiplication. Performing modular multiplication requires multiplication first, and then reduction modulo n. Reduction modulo n can be an expensive operation, especially if one utilizes standard division in its implementation. As a result, alternative approaches to modular arithmetic are essential to efficient cryptographic implementations. A few of these techniques are described below. For detailed information on the implementation of these algorithms, as well as comparisons of them please see not only the original papers referenced below, but also Chapter 14 of [285] and [53].

2.5.2   Montgomery Multiplication

Montgomery multiplication makes use of a technique known as Montgomery reduction [293]. Montgomery reduction provides an efficient method for performing modular multiplication and exponentiation without needing to perform the (often costly) standard modular reduction operation. Standard modular reduction often utilizes a processor's division instruction, which as a rule, is computationally expensive.

First let us consider the Montgomery reduction operation. Select three positive integers: n, $\mathcal{R}$, and $\mathcal{T}$ that satisfy the following conditions:

- **Condition 1**: $\mathcal{R} > n$ and $(n, \mathcal{R}) = 1$ (i.e., they are coprime).

- **Condition 2**: $\mathcal{R}^{-1}$ is the multiplicative inverse of $\mathcal{R}$ (mod n). In other words, $\mathcal{R}\mathcal{R}^{-1} = \mathcal{R}^{-1}\mathcal{R} \equiv 1$ (mod n).

- **Condition 3**: $0 \leq \mathcal{T} < n\mathcal{R}$. $\mathcal{T}$ is the value that will be reduced using Montgomery reduction.

The value $\mathcal{T}\mathcal{R}^{-1}$ (mod n) is called a Montgomery reduction of $\mathcal{T}$ modulo n with respect to $\mathcal{R}$. By carefully selecting a value for $\mathcal{R}$ (one method is described below), this reduction can be computed more efficiently than the standard algorithm for modulo reduction. Selecting $\mathcal{R}$, we consider the base b and length ℓ of the representation of n under the base b. In doing so, a typical value for $\mathcal{R} = b^{\ell}$. For example, let $n = 2049$; n's binary representation is 100000000001 ($2049 = 2^{11} + 1$). In this case, the value $b = 2$ and $\ell = 12$, and therefore $\mathcal{R} = 4096$ which is greater than n and coprime to n, as required by the conditions stated above. Note that this approach to the selection of an efficient $\mathcal{R}$ is only one approach. The important thing to consider when selecting $\mathcal{R}$ is that it is chosen in such a way as to make the Montgomery reduction more efficient than standard modular reduction.

Now, given n and $\mathcal{R}$ with the constraints stated above, we state the following as factual, though justification can be found in Section 14.29 of [285]:

1. Let $n' = -n^{-1}$ (mod $\mathcal{R}$), and $\mathcal{T}$ be any integer in the range defined above.

2. Let $\mathcal{U} = \mathcal{T}n'$ (mod $\mathcal{R}$).

3. Then $\frac{(\mathcal{T}+\mathcal{U}n)}{\mathcal{R}} \equiv \mathcal{T}\mathcal{R}^{-1}$ (mod n), i.e., the Montgomery reduction of $\mathcal{T}$ modulo n with respect to $\mathcal{R}$.

Given the above, once $\mathcal{R}$ has been selected, we can move on to the actual computation of the Montgomery reduction and its use in multiplication. With $n = 2049$ and $\mathcal{R} = 4096$, let $x = 1224$ and $y = 2362$. Let us compute xy (mod n) using Montgomery reduction. Let $\mathcal{T} = xy = 2891088$ and $n' = -n^{-1}$ (mod $\mathcal{R}$) $= -2049$; $\mathcal{U} = (2891088)(-2049)$ (mod 4096) $= 688$. Computing $\frac{(\mathcal{T}+\mathcal{U}n)}{\mathcal{R}} = \frac{(2891088+(688\times2049))}{4096} = 1050 = \mathcal{T}\mathcal{R}^{-1}$ (mod n). Multiplying this result by $\mathcal{R}$ in order to cancel out the $\mathcal{R}^{-1}$ and leave $\mathcal{T}$ as well as reducing (mod n), we get 1998 which is equal to 2891088 (mod 2049) as expected. Notice that in this computation, the only modular reduction that was computed modulo n was the one-time computation of $\mathcal{R}^{-1}$ (mod n). All others were computed (mod $\mathcal{R}$). With $\mathcal{R} = 4096$, a power of two, this reduction can be computed easily using only right shifts.

2.5.3 Barrett Reduction

Barrett reduction [31] was originally designed as part of an algorithm for efficiently performing RSA encryption on a digital signal processor, where division and, therefore, modular reduction is time intensive. Barrett reduction provides an efficient mechanism to compute $r = x \mod n$, given x and n. Because it requires precomputation of a value $\mu = \left\lfloor \frac{b^{2k}}{n} \right\rfloor$, where b is the base used to represent x and n, and k is equal to the number of base b digits in n, this approach is most useful when many modular reductions with a single n must be performed, such as in RSA. The steps of the algorithm are as follows in Algorithm 2.2.

2.5.4 Karatsuba Multiplication

The classical algorithm for multiplying two n digit numbers requires n^2 single digit multiplications. The Karatsuba algorithm [197] greatly reduces the number of single digit multiplications required when n is large and thus speeds up multiplication significantly. The Karatsuba algorithm can also be adapted for use in the multiplication of polynomials [453]. There are also

---

Algorithm 2.2 Barrett reduction algorithm.

Input: Positive integers x and n represented in base b and μ as defined above.

Output: $r = x \ (\mathrm{mod} \ n)$

$q_1 = \left\lfloor \frac{x}{b^{k-1}} \right\rfloor, q_2 = q_1 \mu, q_3 = \left\lfloor \frac{q_2}{b^{k+1}} \right\rfloor$

$r_1 = x \ (\mathrm{mod} \ b^{k+1}), r_2 = q_3 \left(n \ (\mathrm{mod} \ b^{k+1}) \right), r = (r_1 - r_2)$

If $r < 0$ then $r = r + b^{k+1}$

While $r \geq n$ do $r = (r - n)$

Return r.

---

generalizations to this algorithm that under certain circumstances can provide even greater optimization. The following outlines the major efficiency gain provided by Karatsuba multiplication:

Let $x, y \in \mathbb{Z}$. The Euclidean algorithm guarantees that both x and y can be represented in the form $a_1 \mathcal{B}^n + a_0$ for some base $\mathcal{B}$, say $x = (x_1 \mathcal{B}^n + x_0)$ and $y = (y_1 \mathcal{B}^n + y_0)$. The multiplication $(x_1 \mathcal{B}^n + x_0)(y_1 \mathcal{B}^n + y_0)$, using the standard multiplication algorithm requires 4 multiplications:

$$
\begin{aligned}
z_2 &= x_1 y_1 \\
z_1 &= x_1 y_0 + x_0 y_1 \text{ (Note that this requires 2 multiplications.)} \\
z_0 &= x_0 y_0
\end{aligned}
$$

The Karatsuba algorithm computes the same values z_0, z_1, z_2 using only 3 multiplications; z_2 and z_0 are computed as above, but z_1 is computed as:

$$
z_1 = (x_1 + x_0)(y_1 + y_0) - z_0 - z_2
$$

which only requires a single multiplication.

2.6   Abstract Algebra

2.6.1   Introduction

Abstract algebra is a deep and broad field of study, containing numerous intertwined branches and sub-branches of study; all focused on defining, describing, and manipulating various mathematical objects and their structure. Just as some do not appreciate certain styles of art or music, many do not see the intrinsic beauty that an algebraist sees in the patterns, symmetry, and interworkings of algebraic structures. Now, while it is true that many mathematicians spend their entire research careers studying the intricacies of a single type of algebraic structure with no expectation of practical use and while many will plumb its depths simply motivated by the wonder it invokes in them, abstract algebra has emerged as a rich source of structure, fruitful ideas, and practical challenges that are directly applicable to modern cryptography. This section can only scratch the surface of what is both a deep and broad area of ongoing research. As researchers in this field continue to explore new and existing algebraic structures and their interrelationships, it should prove no surprise when abstract algebra continues to fill the cryptographic idea pipeline with new and challenging topics for practical application.

2.6.2 Algebraic Structures: An Overview

The number and variety of different algebraic structures that exist in this field of mathematics is virtually unbounded. From the structurally simple Magma to the countless generalizing constructs of Category Theory to the unexpected relationships between groups and fields elucidated in the study of Galois Theory, the study of algebraic structures provides the means to rigorously define, compare, contrast, and reason about these structures, most of which the majority of us have unknowingly used since our first contact with mathematics. If you have performed clock arithmetic, you have used groups, if you have performed polynomial arithmetic, you have used rings, and if you have performed the most common of tasks, that of performing standard arithmetic over the positive and negative numbers and their fractions, you have used fields. The study of algebraic structures provides the framework to better understand these common, yet deeply fascinating mathematical objects.

2.6.3 Groups

2.6.3.1 Group structure and operations

Definition 2.6.1 *A group, $\mathcal{G}$, consists of a set (possibly countably or uncountably infinite), a single binary operation we will denote as $+ : \mathcal{G} \times \mathcal{G} \to \mathcal{G}$, and a special element we will denote e.*

In order to be considered a group, $\mathcal{G}$ and its elements must conform to the following rules:

1. The group $\mathcal{G}$ is closed under the binary operation $+$. This means that for all $a, b \in \mathcal{G}, a + b \in \mathcal{G}$.

2. The binary operation $+$ is associative. In other words, for any $a, b, c \in \mathcal{G}, (a + b) + c = a + (b + c)$.

3. The special element e is called the identity, or neutral, element, such that for all $a \in \mathcal{G}, a + e = e + a = a$. A simple proof demonstrates that for any group $\mathcal{G}$, this element is unique.

4. For each element $a \in \mathcal{G}$, there exists a unique element $a^{-1} \in \mathcal{G}$, known as the inverse of the element, such that $a + a^{-1} = a^{-1} + a = e$.

5. In addition, a group may possess the following optional property, known as commutativity: for all $a, b \in \mathcal{G}, a + b = b + a$.

Not all groups possess the commutative property. Those that do are known as Abelian groups.

2.6.3.2   Common properties of groups

In the study of groups and other algebraic structures, one finds that numerous properties are defined over these structures, which help us to better understand, categorize, and compare them. The following is a summary of some of the most common used in group theory:

Definition 2.6.2 *The order of a group is the number of elements contained in that group.*

While many groups possess an infinite number of elements, group theory possesses an unending storehouse of groups with finite order. The order of a group is represented by $|\mathcal{G}|$.

Definition 2.6.3 *A subgroup of a group is a subset of the elements of the group that form a group under the same binary operation as the entire group.*

In other words, if two elements of the subgroup are combined using the group's binary operator, another member of the subgroup will result. Every subgroup S of G contains the identity element $e \in G$ as its identity element, as well as the inverse of each subgroup member. A subgroup $\mathcal{S}$ of $\mathcal{G}$ is notated as $\mathcal{S} \leq \mathcal{G}$.

Theorem 2.6.4 *Lagrange's Theorem, named after the 18th century French mathematician Joseph-Louis Lagrange, states that for any finite group $\mathcal{G}$, the order of every subgroup $\mathcal{S} \leq \mathcal{G}$, divides the order of $\mathcal{G}$.*

Definition 2.6.5 *A generating set of a group or subgroup is a subset of the group elements by which every element of the group or subgroup can be represented as a finite combination (using the group's binary operator) of these elements and their inverses.*

A subset $S \subseteq G$ is called a *generator of G* if it generates the entire group, G. The group or subgroup generated by a subset $S \subseteq G$ is denoted $\langle S \rangle$. If the generating set of a group or subgroup consists of a single element $g \in G$, G is called a *cyclic group* or *subgroup*. In a cyclic group, every element $a \in G$, may be generated by a finite number of combinations of the *group generator $g \in G$* with itself or its inverse. The cyclic group generated by $g \in G$, is denoted $\langle g \rangle$.

Based on Lagrange's Theorem, one can conclude that the subgroup generated by any of the non-identity elements of a group with prime order is the entire group. Therefore, one can prove that *every group of prime order is cyclic*.

Definition 2.6.6 *The order of a group element $g \in G$ in a finite group is the minimum number of times it must be combined with itself, using the group operation, in order to generate the identity element, e.*

For all groups G with binary operation $+$, and elements $a, b \in G$, $a + b^{-1} = b^{-1} + a^{-1}$. Note that except for Abelian groups, the order of operations is significant.

2.6.3.3 Examples of groups

One of the most beautiful properties of groups and group theory is that they naturally model that ubiquitous and aesthetically pleasing property we know as symmetry. Nature abounds with symmetry, and as such, nature abounds with groups; not just those that are strictly numerical, but also those that are geometrical or in some other way demonstrate symmetry. In each case, although the set of elements may be different and the specific mechanics of the binary operation $+$ may be different, each group is unified with all others through its adherence to the group properties defined above. Here are a few examples.

Integers with addition

The set $\mathbb{Z}$ of all integers with the binary operation of standard addition, the identity element 0, and for each $n, -n \in \mathbb{Z}$, $n + (-n) = 0$ (i.e., the negative of an element is its inverse) forms a group. Because addition is commutative over the $\mathbb{Z}$, this group is also Abelian.

Non-zero rationals with multiplication

Take the set $\mathbb{Q}\backslash 0$ (the rational numbers, excluding 0) with the binary operation of standard multiplication, the identity element 1, and for each $\frac{a}{b}, \frac{b}{a} \in \mathbb{Q}\backslash 0$, $\frac{a}{b} \times \frac{b}{a} = 1$. This forms another Abelian group.

Non-negative integers with modular addition

Using the principles of modular arithmetic described above, one can form a group by choosing a positive integer n and forming the group $\mathbb{Z}$ mod n, sometimes denoted $\mathbb{Z}_n$. This group has as its set, the integers $\mathcal{G} = \{0, 1, \ldots, n-1\}$, addition modulo n for its binary operation, 0 as the identity, and the additive inverse of each element modulo n for inverses. This group is both Abelian and cyclic and is generated by all $g \in \mathcal{G}$ where $(g, n) = 1$.

Non-negative integers with modular multiplication

Using modular arithmetic again, let n be the modulus of the group. The elements of the group are the set $\{a \in \mathbb{Z}^+ : a < n$ and $(a, n) = 1\}$ (i.e., the positive integers less than n that are coprime to n). This set with multiplication mod n, and identity element 1 forms an Abelian group.

Permutations of a finite set

Let $S = \{a, b, c\}$, a three element set. Let S_3 denote the group of all permutations on S. We will represent the permutations of this set, using the cycle notation defined above. S_3 consists of the elements: $\{\{\}, \{a, b\}, \{a, c\}, \{b, c\}, \{a, b, c\}, \{c, b, a\}\}$. S_3 is one group in a set of groups known as the symmetric group, S_n. The symmetric groups contain enough interesting properties themselves, to serve as a dedicated area of study. See [104] and [364] for more details.

The binary operation on permutations in the symmetric group is that of composition (i.e., executing one permutation followed by another), denoted by $\circ$. Composition is applied from right to left. For example, $\{b, c\} \circ \{a, b, c\} = \{a, c\}$. The identity element is the null permutation, $\{\}$. The inverse of a permutation is the permutation that "undoes" the actions of the first permutation. In this example, the transpositions $\{a, b\}$, $\{b, c\}$, and $\{a, c\}$ serve as their own inverses. The remaining two permutations, $\{a, b, c\}$ and $\{c, b, a\}$ serve as each other's inverse.

The symmetry group of a square

Consider the square with vertices labeled, a, b, c, and d in Figure 2.1 below. The square may be repositioned by a combination of counterclockwise rotations and reflections on the various axes of symmetry. Each of these repositioning moves is an element in the symmetry group of a square, also known as the dihedral group of order 8 or $\mathbb{D}_4$ (the subscript 4 is a convention with

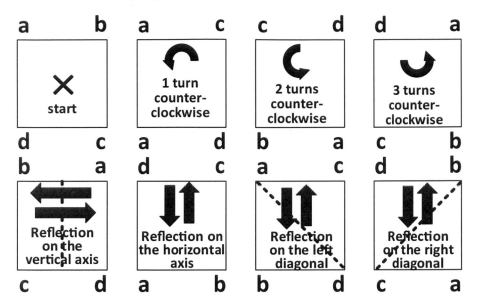

Figure 2.1
The symmetry group of the square, $\mathbb{D}_4$

the dihedral groups wherein the subscript value is one-half of the group order). As it is with the symmetric group, $\mathbb{D}_4$ is only one of a number of dihedral groups, associated with various polygons. In fact, there are a multitude of groups associated with geometric objects in various dimensions along with their properties. These also serve as a subject of study themselves. See [316] and [272]. Again, wherever one finds symmetry, one will find groups.

2.6.3.4 The discrete logarithm problem

Group theory and the use of groups in cryptography finds its greatest value in what is known as the *discrete logarithm problem*. This problem can be easily stated as follows: given a group $\mathcal{G}$, and a generator of the group (or a large subgroup) $g \in \mathcal{G}$, we have the equation:

$$g^x = a \tag{2.20}$$

where g^x represents the application of the group's binary operation to g to itself, x times.

Theorem 2.6.7 *The theorem states that given $\mathcal{G}$ of sufficiently large order, along with g and a, it is computationally infeasible to compute x.*

The existence of the discrete logarithm problem provides a powerful mechanism that can be used in the design of public key cryptographic primitives. The consistency of group properties across different groups make it possible to implement discrete logarithm primitives that are the same in all other respects, except that of the specific group that is being used. As we shall see, a whole class of discrete logarithm-based primitives that utilize the group of integers under multiplication modulo n, can be converted to a more efficient set of primitives based on the group of elliptic curves.

2.6.4 Rings

2.6.4.1 Ring structure and operations

> **Definition 2.6.8** *A ring* $\mathcal{R}$ *consists of a set and two binary operations that we will denote as* $+ : \mathcal{R} \times \mathcal{R} \to \mathcal{R}$ *and* $\times : \mathcal{R} \times \mathcal{R} \to \mathcal{R}$.

In addition, each ring possesses two identity elements, one for each operation that we shall denote as 0 and 1, respectively. All rings have the following properties:

1. The first operation $+ : \mathcal{R} \times \mathcal{R} \to \mathcal{R}$, along with the identity element 0, form an Abelian group.

2. The ring $\mathcal{R}$ is also closed under the second binary operation $\times : \mathcal{R} \times \mathcal{R} \to \mathcal{R}$ in the same manner as the ring is closed under the $+$ operation.

3. The identity element 1 is unique for the ring $\mathcal{R}$ and for all $a \in \mathcal{R}$, $a \times 1 = 1 \times a = a$.

4. The binary operation $\times : \mathcal{R} \times \mathcal{R} \to \mathcal{R}$ is associative.

5. The binary operation $\times : \mathcal{R} \times \mathcal{R} \to \mathcal{R}$ distributes over the binary operation $+ : \mathcal{R} \times \mathcal{R} \to \mathcal{R}$. In other words, for all $a, b, c \in \mathcal{R}$, $a \times (b + c) = (a \times b) + (a \times c)$ and $(b + c) \times a = (b \times a) + (c \times a)$.

2.6.4.2 Examples of rings

The ring of integers $\mathbb{Z}$

The set of all integers, with standard addition and multiplication as the two binary operators, along with 0 as the additive identity and 1 as the multiplicative identity, serves as the archetype for most rings. The ring of integers, $\mathbb{Z}$, also possesses some additional structural properties that distinguish it from many other rings. For example, the second binary operation, $\times$, is not required to be commutative for all rings. However, in the ring of integers, it is.

Also, in the definition of a ring, there is no requirement that for any $a, b \in \mathcal{R}$ (a ring) where $a \neq 0$ and $b \neq 0$, that $a \times b$ cannot equal 0, and many rings do not possess this property. However, $\mathbb{Z}$ does. Elements such as a and b above are called *zero divisors*. Rings such as $\mathbb{Z}$ which contain no zero divisors are called *integral domains*. $\mathbb{Z}$ is also a unique factorization domain, a principal ideal domain, and a Euclidean domain, each with its own well defined properties. See [107] for a comprehensive treatment of the properties of each of these additional algebraic structures which begin with those of a ring.

Integers modulo an integer

For any positive integer $n \in \mathbb{Z}$, with $n > 1$, the integers modulo n form a ring $\mathbb{Z}_n$ using the standard binary operations of addition and multiplication (mod n) and having 0 and 1 as the respective identity values. Consider the ring $\mathbb{Z}_8$. It is comprised of the integers $\{0, 1, 2, 3, 4, 5, 6, 7\}$ with both addition and multiplication performed (mod 8). For example:

$$
\begin{aligned}
2 + 7 &= 1 \\
5 \times 7 &= 3 \\
2 \times 4 &= 0 \text{ (2 and 4 are zero divisors in this ring)} \\
3 \times (5 + 6) &= 3 \times 3 = 1 = (3 \times 5) + (3 \times 6) = 7 + 2 = 1
\end{aligned}
$$

Polynomial rings over $\mathbb{Z}$

Given the ring of integers $\mathbb{Z}$, we can create another ring, denoted $\mathbb{Z}[\mathcal{X}]$, where $\mathcal{X}$ represents one or more indeterminates; for example, $\mathcal{X} = \{x, y, z\}$. $\mathbb{Z}[\mathcal{X}]$ consists of all polynomials with the indeterminates $\mathcal{X}$ and coefficients in $\mathbb{Z}$. The two binary operations are the standard addition and multiplication of polynomials. Just as with $\mathbb{Z}$, 0 and 1 serve as the identity elements.

2.6.4.3 Ideals

An (double-sided) ideal of a ring $\mathcal{R}$ is a subset $\mathcal{I} \subseteq \mathcal{R}$ that is first, a subgroup of $\mathcal{R}$ under the first binary operation. Also, for each $i \in \mathcal{I}$ and $r \in \mathcal{R}$, $i \times r \in \mathcal{I}$ and $r \times i \in \mathcal{I}$. For example, in the ring $\mathbb{Z}$, for any element $n \in \mathbb{Z}$, all $r \in \mathcal{R}$ such that $n | r$ form the ideal denoted $n\mathbb{Z}$. Concretely, consider the case where $n = 5$. The set of all integral multiples of 5, both positive and negative, form the ideal $5\mathbb{Z} \subseteq \mathbb{Z}$.

2.6.5 Vector Spaces

2.6.5.1 Vector space structure and operations

Vector spaces are usually first encountered in a course on linear algebra. However, they continue to make their appearance in other areas of abstract algebra, as we shall soon see. A vector space over $\mathcal{F}$ consists of an Abelian group $\mathcal{G}$ and a field $\mathcal{F}$ that adhere to the following axioms:

1. The elements of $\mathcal{G}$ are called *vectors*, and obey all group properties. Most importantly, two vectors $u, v \in \mathcal{G}$ may be added together to produce another vector $w \in \mathcal{G}$. Vector addition is both associative and commutative.

2. The elements of $\mathcal{F}$ are called *scalars* and may be applied to vectors through an operation called *scalar multiplication*.

   For all scalars $a, b \in \mathcal{F}$, and any vector $v \in \mathcal{G}$, $a(bv) = (ab)v$.

3. The multiplicative identity element of the field $\mathcal{F}$ 1, serves as the identity element for scalar multiplication, i.e., $1v = v$.

4. Scalar multiplication distributes over vector addition, i.e., for $a \in \mathcal{F}$, and $u, v \in \mathcal{G}$, $a(u + v) = au + av$.

5. Scalar addition distributes over vectors, i.e., for $a, b \in \mathcal{F}$, and $v \in \mathcal{G}$, $(a + b)v = av + bv$.

2.6.5.2   Examples of vector spaces

The Euclidean space $\mathbb{R}^3$

The geometric space, $\mathbb{R}^3$, associated with our familiar 3-dimensional space, consists of vectors of the form $\{x, y, z\}$ where x, y, and $z \in \mathbb{R}$ and scalars $a \in \mathbb{R}$. Vector addition is performed elementwise, i.e., $\{x_1, y_1, z_1\} + \{x_2, y_2, z_2\} = \{x_1 + x_2, y_1 + y_2, z_1 + z_2\}$, and the zero vector $\{0, 0, 0\}$ serves as the identity element. Scalar multiplication is just simple multiplication of real numbers.

The field of complex numbers, $\mathbb{C}$

When 18th and 19th century mathematicians were grappling with the question of general solutions to polynomials in various degrees, especially the cubic (i.e., degree 3) and quartic (i.e., degree 4), they were repeatedly confronted with solutions that included a number that was the square root of -1, which today we refer to as i. At the time, the claim of the existence of roots of negative numbers was considered dubious at best. However, things changed when mathematicians such as Caspar Wessel, Jean-Robert Argand, and Abbé Adrien-Quentin Buée gave a geometrical interpretation of the complex numbers. William Rowan Hamilton, however, rejected the geometric approach to explaining complex numbers, and instead proposed a method wherein complex numbers could be represented as pairs of real numbers [302]. Stated differently, he demonstrated how the complex numbers can be represented as a 2-dimensional vector space over $\mathbb{R}$. Specifically:

- Let a complex number $z = a + bi$ be represented as a vector (a, b) where a and $b \in \mathbb{R}$. a represents the "real part" of z and b represents the "imaginary part". For example, the complex number $3 + 5i$ is represented by the vector $(3, 5)$.

- Complex addition is performed component-wise, so for two complex numbers, $z_1 = 2 + 9i$ and $z_2 = 4 + 2i$, $z_1 + z_2 = (2, 9) + (4, 2) = (6, 11) = 6 + 11i$.

- Complex multiplication is a bit more complicated. Given two complex numbers, (a, b) and (c, d), $(a, b) \times (c, d) = (ac - bd) + (ad + bc)i$.

- Scalar multiplication of reals over complex numbers is as expected. Given $r \in \mathbb{R}$ and $(a, b) \in \mathbb{C}$, where $a, b \in \mathbb{R}$, $r(a, b) = (ra, rb)$.

It is a simple matter to verify that Hamilton's representation of the complex numbers as pairs of numbers in $\mathbb{R}$ fulfills all the requirements of a vector space.

2.6.6 Fields

2.6.6.1 Field structure and operations

A field $\mathbb{F}$ is a ring (as well as an integral domain, unique factorization domain, principal ideal domain, and Euclidean domain) that possesses the following additional properties:

1. Each element $f \in \mathbb{F}$, where $f \neq 0$, has a unique inverse element f^{-1} under the multiplicative binary operation $\times : \mathbb{F} \times \mathbb{F} \to \mathbb{F}$, such that $f \times f^{-1} = f^{-1} \times f = 1$.

2. The binary operation, $\times : \mathbb{F} \times \mathbb{F} \to \mathbb{F}$ is commutative, such that for all $f, g \in \mathbb{F}$, $f \times g = g \times f$.

3. For any field $\mathbb{F}$, the *characteristic* of the field is the smallest number of times the multiplicative identity must be added to itself in order to get the additive identity in return. If no such number exists, the characteristic of the field is equal to 0.

2.6.6.2 Examples of fields

The fields most familiar to us are usually infinite. These include the rational numbers $\mathbb{Q}$, the real numbers $\mathbb{R}$, and the complex numbers $\mathbb{C}$. In each case, both a well-formed additive and multiplicative binary operation is defined. Also, 0 and 1 serve as the respective additive and multiplicative identities for each of these fields. In $\mathbb{Q}$ and $\mathbb{R}$, each element of the field possesses a unique multiplicative inverse commonly referred to as the reciprocal. For any $q \in \mathbb{Q}$ or $r \in \mathbb{R}$, the multiplicative inverse is denoted as $\frac{1}{q}$ and $\frac{1}{r}$, respectively. In the case of the complex numbers $\mathbb{C}$, the multiplicative inverse of $z \in \mathbb{C}$ can be calculated by multiplying the numerator and denominator of the reciprocal $\frac{1}{z}$ by the complex conjugate of z and simplifying. The fields $\mathbb{Q}$, $\mathbb{R}$, and $\mathbb{C}$ are all of characteristic 0.

Another somewhat common example of a field is that of the *rational functions* defined as follows. Given a field $\mathbb{E}$, the set $\mathbb{F}$ of equivalence classes of expressions of the form $\frac{p(\mathcal{X})}{q(\mathcal{X})}$, where $p(\mathcal{X})$ and $q(\mathcal{X})$ are polynomials with co-efficients in the field $\mathbb{E}$ and with $q(\mathcal{X}) \neq 0$, form a field. The two binary operators are defined as one would expect, i.e., the schoolbook notion of addition and multiplication of polynomial fractions over the field $\mathbb{E}$ (while keeping in mind that the result is a representative of an equivalence class), and the identity elements are that of the field $\mathbb{E}$.

For an example of a field that is neither common nor completely intuitive, one should consider the p-adic numbers. See [348] or [143] for a thorough treatment of this very interesting class of fields.

2.6.6.3   Finite (Galois) fields

In contrast to the fields outlined above, which are usually of infinite order, we turn now to the finite fields, which are the fields most commonly used in modern cryptography (e.g., in elliptic curve cryptography). For every prime number p there exists a field of order p, comprised of the set of integers $\{0, \ldots, p - 1\}$ along with addition and multiplication modulo p as the two binary field operations, and 0 and 1 as the associated identity elements. These finite fields are often referred to as *prime fields* and are denoted $\mathbb{F}_p$ or $\mathbb{GF}_p$. For any prime number p, one and only one finite field of order p exists, specifically the prime field $\mathbb{F}_p$. Each prime field $\mathbb{F}_p$ has characteristic p.

In addition to the prime fields, for each prime p and positive integer d, there exists one and only one field of order p^d, where d is called the *degree* of the finite field. These fields are also of characteristic p. Prime fields can be considered to be those finite fields of degree 1. Please note that all finite fields are of order p^d for some prime p and some positive integer d. No other finite fields exist.

2.6.6.4   Finite field elements when degree $d > 1$

Finite fields of degree $d > 1$ behave in most respects, just like the prime fields. The same additive binary operation is used, and the same additive and multiplicative identities exist. However, when $d > 1$, multiplication offers a bit of a challenge. In order to recognize the challenge and address it, one must first establish how we represent elements of a finite field of degree greater than 1. Finite fields of degree $d > 1$ are isomorphic to the vector space of degree d over the prime field $\mathbb{F}_p$ and, therefore, its elements may be represented as elements of this isomorphic vector space.

Take for example, the finite field $\mathbb{F}_{125}$, where $125 = 5^3$. This finite field is of characteristic 5 and of degree 3. Its elements can be represented as the elements of the degree 3 vector space over $\mathbb{F}_5$. Each element f is represented as $\{n_0, n_1, n_2\}$, where each n_x is an element of $\mathbb{F}_5$. Addition is defined element-wise modulo 5, and the elements $\{0, 0, 0\}$ and $\{1, 1, 1\}$ serve as the respective additive and multiplicative identity elements.

But what about multiplication? It turns out that finite field multiplication in fields with degree greater than 1, benefits from an alternative representation of the finite field. This is accomplished by representing field elements as polynomials with indeterminate x instead of as elements of a vector space. Using this representation and the example above, each field element $\{n_0, n_1, n_2\}$ is represented instead as $n_0 + n_1 x + n_2 x^2$, with indeterminate x and coefficients n_i in $\mathbb{F}_5$. By utilizing this representation, multiplication becomes simple multiplication of polynomials modulo an irreducible polynomial with coefficients reduced modulo p. An irreducible polynomial is one that cannot be expressed as the product of two or more irreducible polynomials with coefficients in the same field. It is important to note that while a polynomial may be irreducible in one field, it may be reducible in another. For example, over $\mathbb{R}$, the polynomial $x^2 + 1$ is irreducible. However, over $\mathbb{C}$, the same polynomial can be reduced to $(x - i)(x + i)$.

Returning to the finite field $\mathbb{F}_{125}$, we will use the irreducible polynomial, $1 + x^2 + x^3$. Let us consider the elements $a = 1 + 2x + 4x^2$ and $b = 3 + x + 2x^2$. Computing their product ab:

$$\left(1 + 2x + 4x^2\right)\left(3 + x + 2x^2\right) \quad (\text{mod } 1 + x^2 + x^3) = \tag{2.21}$$
$$3 + 7x + 16x^2 + 8x^3 + 8x^4 \quad (\text{mod } 1 + x^2 + x^3)$$

First reducing the coefficients (mod 5) gives $3 + 2x + x^2 + 3x^3 + 3x^4$ (mod $1 + x^2 + x^3$). Finally, reducing by the polynomial $(1 + x^2 + x^3)$ gives the element $3 - x + x^2 = 3 + 4x + x^2$.

2.6.6.5 Finite field use in AES

As already mentioned, finite fields are used throughout modern cryptography. One noteworthy example is in order. Without going into the details of the design of this mainstay primitive, the Advanced Encryption Standard (AES) makes use of two finite fields in its design. First, in the round step known as "SubBytes", known as the S-box, or substitution box, makes use of $\mathbb{F}_{2^8}$ with the polynomial $1 + x + x^3 + x^4 + x^8$ as its irreducible polynomial for multiplication.

The round step known as "MixColumns" represents state columns in what is called the D-box as polynomials over $\mathbb{F}_{2^8}$, each of degree less than four. These polynomials are multiplied by a fixed polynomial and then reduced modulo the polynomial $x^4 + 1$ (which is reducible in $\mathbb{F}_{2^8}$).

In both cases, the fact that $\mathbb{F}_{2^8}$ is a field of characteristic 2 (i.e., a binary field), allows each byte to be represented as an element in this finite field (i.e., a polynomial of degree ≤ 7 with binary coefficients) and vice-versa. This provides many opportunities to optimize the implementation of AES in today's computer hardware and software. The reader is highly encouraged to consult [94] for a much more detailed treatment of this topic.

2.7 Elliptic Curves

2.7.1 Introduction

Definition 2.7.1 *An elliptic curve $\mathcal{E}$ over a field $\mathbb{F}$ is defined by an equation of the form: $y^2 + axy + by = x^3 + cx^2 + dx + e$, where $a, b, c, d, e \in \mathbb{F}$*

An equation in this form is called a *Weierstrass equation*. If $\mathbb{F}$ is not of characteristic 2 or 3, a Weierstrass equation can be simplified considerably. In this case, a *Simplified Weierstrass equation* is defined as:

$$y^2 = x^3 + ax + b$$

Cases of characteristic 2 and 3 can be reduced similarly. However, our focus will be on elliptic curves of characteristic > 3, and therefore, we will restrict our usage to elliptic curves defined using the Simplified Weierstrass format above.

Elliptic curves have been the object of study by mathematicians from the time of Diophantus (3rd century AD) [57]. In 1985, Neal Koblitz [226] and Victor Miller [289] independently proposed how elliptic curves could be used as a basis for cryptosystems. Shortly thereafter, in 1987, Hendrik Lenstra published a paper in the *Annals of Mathematics* that demonstrated an algorithm for utilizing elliptic curves to factor integers [244]. These events marked the emergence of elliptic curves as the foundation of a new type of cryptography that addresses some of the emerging problems with existing cryptographic primitives.

Many of the cryptographic primitives utilized in public key algorithms such as the Diffie-Hellman key exchange, The Digital Signature Algorithm, the various El Gamal primitives rely on the difficulty of the discrete logarithm problem. As stated above, the discrete logarithm problem applies to any group of sufficiently large order. However, as computational power increases, the order of the group used must continue to grow in order to be able to thwart brute force attacks by those possessing sufficient computational resources. As the order of the group increases, the size of keys used in these algorithms grows, and the performance of these algorithms erodes. As we shall see below, the geometry of elliptic curves over a field provides the means to define a group, using the points on the elliptic curves as elements of the group. Using elliptic curve groups in cryptography provides an alternative that can deliver equal levels of security (e.g., resistance to brute force attack), while continuing to use smaller order groups and smaller keys.

2.7.2 Elliptic Curves

In elliptic curve cryptography all elliptic curves are defined over finite fields, both prime and those of higher degree. However, in order to provide a pictorial representation of elliptic curves and their arithmetic, we will begin by looking at an elliptic curve over $\mathbb{R}$. Consider the elliptic curve $y^2 = x^3 - 3x + 1$. Plotting this over $\mathbb{R}$ is shown in Figure 2.2.

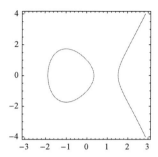

Figure 2.2
The elliptic curve $y^2 = x^3 - 3x + 1$ over $\mathbb{R}$

As stated above, it is possible to construct a group out of the points of an elliptic curve like the one pictured in Figure 2.2, as well as those over other fields, including finite fields. This group is geometric in nature, with the binary operation being a special form of point addition that will be defined below. The additive identity is an additional point called the *point at infinity* represented by ∞. It is defined as a single point one finds by extending a vertical line in either direction from any point on the elliptic curve. For a geometric interpretation, note Figure 2.3.

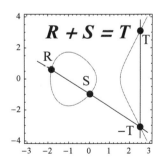

Figure 2.3
The elliptic curve $y^2 = x^3 - 3x + 1$ over $\mathbb{R}$ with geometric point addition

Geometrically, point addition works as follows. Given a point $\mathcal{R}$ and a point $\mathcal{S}$ on the elliptic curve, draw a line $\overline{\mathcal{RS}}$. Except for the case where $\overline{\mathcal{RS}}$ is vertical (i.e., it has no slope), the line will intersect the elliptic curve in a

3rd point called $-\mathcal{T}$. Reflecting $-\mathcal{T}$ across the x axis will give another point, $\mathcal{T}$. This point $\mathcal{T}$ is considered to be the sum of points $\mathcal{R}$ and $\mathcal{S}$ on the elliptic curve (see Figure 2.3). Points $\mathcal{T}$ and $-\mathcal{T}$ are the additive inverses of one other. Adding them geometrically results in a vertical line that extends to the point at infinity. Given that the elliptic curve is a group, this is to be expected as $\mathcal{T} + -\mathcal{T} = \infty$. If a point $\mathcal{P}$ is placed on a tangent of the elliptic curve that does not intersect with any other point on the curve, and a tangent line is drawn, it too will result in a vertical line that extends to the point at infinity. Putting this into group theoretic terms, this tangential point $\mathcal{P}$ is its own additive inverse, i.e., $\mathcal{P} + \mathcal{P} = \infty$. Finally, though the proof is not included here, point addition is associative as is required of all algebraic groups.

2.7.3   Point Representation

As stated above, elliptic curves used in cryptography almost always are those over a finite field $\mathbb{F}$. Therefore, a point on an elliptic curve can be represented as a tuple of the form (a, b), where $a, b \in \mathbb{F}$. This point representation is known as the *affine point representation*. In this representation, the group identity is the point at infinity, represented by ∞, and the additive inverse of a point $\mathcal{P} = (x, y)$ is $-\mathcal{P} = (x, -y)$. As we will see in the next section, point addition using the simplified Weierstrass form of the curve with affine points requires 2 multiplications, 1 squaring, and 1 field inversion [451]. As a rule, field inversion should be avoided when possible, as it is usually much slower than multiplication. Converting the points to an alternative representation like those outlined below as well as converting the point addition and doubling formulae to work with the new point representation can obviate the need for any field inversions, thus providing significant performance improvements during point addition and point doubling. Some algorithms go so far as to mix point representations (e.g., mixing affine and Jacobian representations) in order to gain other performance improvements.

2.7.3.1   Projective coordinates

One of the simplest and most natural conversions is to take the affine coordinates and convert them to *projective coordinates*; computing with the elliptic curve in projective space. The reader is encouraged to consult a text on projective geometry for a more in depth treatment of the topic. When using projective coordinates, points take the form $(x : y : z)$, which is an equivalence class of points (i.e., each point can have multiple representations). The set of projective points represented by $[x : y : 0]$ is the *line at infinity*. Using projective coordinates changes the elliptic curve equation $y^2 = x^3 + ax + b$ to $y^2 z = x^3 + axz^2 + bz^3$. Although changing to projective coordinates increases the number of multiplications and squarings required during adding, it completely removes the need for field inversions.

2.7.3.2 Jacobian coordinates

Changing projective coordinates slightly to what are known as *Jacobian Coordinates* provides faster computation of point doubling (only 3 multiplications and 6 squarings). Also, as with projective coordinates, the use of Jacobian coordinates removes the need for field inversion. With Jacobian coordinates, the point $(x : y : z)$ represents the affine point $\left(\frac{x}{z^2}, \frac{y}{z^3} \right)$. As a result, the elliptic curve $y^2 = x^3 + ax + b$ becomes $y^2 = x^3 + axz^4 + bz^6$. The point at infinity has coordinates $(1 : 1 : 0)$ and the additive inverse of a point $(x : y : z)$ is $(x, -y, z)$.

2.7.3.3 Other coordinate systems

In addition to the two examples outlined above, others exist, including Edwards coordinates, Chudnovsky coordinates, and López-Dahab projective coordinates. Each has been created to facilitate the implementation of optimally performing algorithms for elliptic curve operations. Both [160] and [81] provide a much more detailed treatment of the various coordinate systems and the algorithms that utilize them, than this short gloss on the topic does.

2.7.4 Point Addition and Doubling

Using affine coordinates, we consider the addition of points $\mathcal{R} = (x_r, y_r)$ and $\mathcal{S} = (x_s, y_s)$ to give point $\mathcal{T} = (x_t, y_t)$ over the curve $y^2 = x^3 - ax + b$. When $\mathcal{S} = -\mathcal{R}$, the solution is immediate, as $\mathcal{R} + \mathcal{S}(= -\mathcal{R}) = \infty$, i.e., the point at infinity. In the general case, however, the following formula can be used when $\mathcal{R} \neq \mathcal{S}$:

$$
\begin{aligned}
x_t &= \left(\frac{y_s - y_r}{x_s - x_r} \right)^2 - x_r - x_s \text{ and} \\
y_t &= \left(\frac{y_s - y_r}{x_s - x_r} \right) (x_r - x_t) - y_r
\end{aligned}
\tag{2.22}
$$

Efficient algorithms for computing repeated addition of a point to itself (e.g., $17\mathcal{P} = \underbrace{\mathcal{P} + \mathcal{P} + \mathcal{P} + \ldots + \mathcal{P}}_{17 \text{times}}$) make generous use of point doubling. When a point is located at a vertical tangent, adding it to itself results in the point at infinity, ∞. However, for most cases, the formula below is used to compute $2\mathcal{R} = (x_d, y_d)$ where $\mathcal{R} = (x_r, y_r)$ over the curve $y^2 = x^3 + ax + b$:

$$
\begin{aligned}
x_d &= \left(\frac{3x_r^2 + a}{2y_r} \right)^2 - 2x_r \text{ and} \\
y_d &= \left(\frac{3x_r^2 + a}{2y_r} \right)^2 (x_r - x_d) - y_r
\end{aligned}
\tag{2.23}
$$

2.8 Conclusion

Mathematics has always been a fascinating subject; expansive in its breadth as well as its depth. As to breadth, it is safe to say that two research mathematicians who are both experts in their chosen focus area may find themselves completely unfamiliar with the concepts that their respective colleague uses routinely. As to depth, a simple example from history will suffice. The Oxford professor, Dr. Andrew Wiles, spent over seven years dedicating himself to the sole task of developing a proof to the simple conjecture that for any integer $n > 2$, there do not exist three positive integers a, b, and c such that the equation $a^n + b^n = c^n$ holds. This 360 year old open problem, known as Fermat's Last Theorem, had captured his attention from childhood, and had been studied by multitudes before him, without success. In the opinion of one respected mathematician, only one in one thousand mathematicians will be able to comprehend the entirety of the 129 pages of Wile's proof. Mathematics, in spite of its enormity, has never failed to entice the curious to explore its fathomless dimensions. With the coming of age of computing devices, the need for robust cryptographic solutions that protect these devices has become mandatory. Thankfully, this boundless world of mathematics has become the foundation for these solutions. While at this point, cryptography does not make use of every area of mathematics, it is a mathematical subject. As such, it has been the intent of this author to provide a short introduction to those mathematical topics considered most relevant to the study of cryptographic constructions. In addition, the references found in Appendix B should provide a rich source of additional material to fill in the gaps that this chapter could not. May this chapter serve as a useful "kick start" to the reader as they engage in reading the remainder of this book. The following chapter provides guidance in the design and efficient implementation of block ciphers. What better chapter to follow this one, then one that makes use of probability theory, permutations, modular arithmetic, groups, and finite fields. To the reader, this author simply says, "Enjoy!"

2.9   Appendix A: Symbols and Notation

Symbol	Definition
$\mathbb{C}$	The field of complex numbers
$\mathbb{R}$	The field of real numbers
$\mathbb{Q}$	The field of rational numbers
$\mathbb{Z}$	The ring of integers
$\mathbb{Z}^+$	The set of positive integers
$\in$	Is an element of, i.e., set membership
$|\mathcal{S}|$	The cardinality of a finite set $\mathcal{S}$, i.e., the number of elements in $\mathcal{S}$
$\phi(x)$	Euler's totient function
$\mathcal{S} \le \mathcal{G}$	$\mathcal{S}$ is a subgroup of $\mathcal{G}$
∞	The point at infinity on an elliptic curve
$a\|b$	a divides b
$a \nmid b$	a does not divide b
(a, b)	The GCD of a and b
$\left(\frac{a}{p}\right)$	The Legendre symbol
$(\bmod\ n)$	Reduction modulo n
$\sigma : \mathbb{Z} \to \mathbb{C}$	A function σ with domain $\mathbb{Z}$ and codomain $\mathbb{C}$
$[0, 1]$	The closed (i.e., it contains both endpoints) interval over the real numbers from 0 to 1

2.10   Appendix B: References

See the back of the book for a full list of references. Here we sort some key sources by category.

2.10.1   General

[103] Whitfield Diffie and Martin Hellman. New directions in cryptography. *IEEE Transactions on Information Theory*, IT-22(6):644–654, November 1976.

[138] Oded Goldreich. *Foundations of Cryptography II*. Cambridge University Press, Cambridge, 2004.

[139] Oded Goldreich. *Foundations of Cryptography I*. Cambridge University Press, Cambridge, 2006.

[171] Jeffrey Hoffstein, Jill Pipher, and Joseph Silverman. *Introduction to Mathematical Cryptography*. Springer Verlag, New York, 2008.

[194] David Kahn. *The Codebreakers*. Scribner, New York, 1996.

[201] Jonathan Katz and Yehuda Lindell. *Introduction to Modern Cryptography*. Chapman & Hall / CRC, Boca Raton, 2008.

[227] Neal Koblitz. The uneasy relationship between mathematics and cryptography. *Notices of the American Mathematical Society*, 54(8):972–979, September 2007.

[285] Alfred Menezes, Paul van Oorschot, and Scott Vanstone. *Handbook of Applied Cryptography*. CRC Press, Boca Raton, 1996.

[360] Antonio Lloris Ruiz, Encarnación Castillo Morales, Luis Parrilla Roure, and Antonio García Ríos. *Algebraic Circuits*. Springer Verlag, Berlin, 2014.

[397] Simon Singh. *The Code Book*. Anchor, New York, 2000.

2.10.2 Probability Theory

[42] Rabi Bhattacharya and Edward Waymire. *A Basic Course in Probability Theory*. Springer Verlag, New York, 2007.

[219] Achim Klenke. *Probability Theory*. Springer Verlag, New York, 2013.

[243] Mario Lefebvre. *Basic Probability Theory With Applications*. Springer Verlag, New York, 2009.

2.10.3 Information Theory

[20] Robert Ash. *Information Theory*. Dover Publications, New York, 1991.

[52] Monica Borda. *Fundamentals in Information Theory and Coding*. Springer Verlag, New York, 2011.

[87] Thomas Cover and Joy Thomas. *Elements of Information Theory*. John Wiley & Sons, Hoboken, 2006.

[298] Stefan Moser. *A Student's Guide to Coding and Information Theory*. Cambridge University Press, Cambridge, 2012.

[379] Peter Seibt. *Algorithmic Information Theory: Mathematics of Digital Information Processing*. Springer Verlag, New York, 2006.

[385] Claude Shannon. Communication theory of secrecy systems. *Bell System Technical Journal*, 28:28, 656–715.

[386] Claude Shannon and Warren Weaver. *The Mathematical Theory of Communication.* University of Illinois Press, Urbana, 1998.

2.10.4 Combinatorics

[2] Martin Aigner. *A Course in Enumeration.* Springer Verlag, New York, 2007.

[6] Fatima Amounas. A novel approach for enciphering data-based ECC using catalan numbers. *International Journal of Information & Network Security*, 2(4):339–347, August 2013.

[92] Paul Cull, Mary Flahive, and Robby Robson. *Difference Equations: From Rabbits to Chaos.* Springer Verlag, New York, 2005.

[113] Saber El Aydi. *An Introduction to Difference Equations.* Springer Verlag, New York, 2005.

[118] Graham Everest, Alf van der Poorten, Igor Shparlinski, and Thomas Ward. *Recurrence Sequences.* American Mathematical Society, Providence, 2003.

[177] T. C. Hu and M. T. Shing. *Combinatorial Algorithms.* Dover Publications, Mineola, 2002.

[192] Stasys Jukna. *Extremal Combinatorics.* Springer Verlag, New York, 2011.

[273] George Martin. *Counting: The Art of Enumerative Combinatorics.* Springer Verlag, New York, 2001.

[408] Richard Stanley. *Enumerative Combinatorics: Advances and Frontiers.* Cambridge University Press, Cambridge, 2012.

[415] K. R. Sudha, A. Chandra Sekhar, and Prasad Reddy. Cryptography protection of digital signals using some recurrence relations. *International Journal of Computer Science and Network Security*, 7(5):203–207, May 2007.

[436] J. H. van Lint and R. M. Wilsom. *A Course in Combinatorics.* Cambridge University Press, Cambridge, 2001.

[457] Hans Wussing. *The Genesis of the Abstract Group Concept: A Contribution to the History of the Origin of Abstract Group Theory.* Dover Publications, Mineola, 2007.

2.10.5 Number Theory

[7] James Anderson and James Bell. *Number Theory with Applications.* Prentice Hall, Upper Saddle River, 1997.

[8] Titu Andreescu and Dorin Andrica. *Number Theory—Structure, Examples, and Problems*. Springer Verlag, New York, 2009.

[9] Titu Andreescu, Dorin Andrica, and Ion Cucurezeanu. *An Introduction To Diophantine Equations*. Springer Verlag, New York, 2010.

[18] Benno Artmann. *Euclid: The Creation of Mathematics*. Springer Verlag, New York, 1999.

[27] Maria Welleda Baldoni, Ciro Ciliberto, and Giulia Maria Piacentini Cattaneo. *Elementary Number Theory, Cryptography and Codes*. Springer Verlag, New York, 2008.

[75] Lindsay Childs. *A Concrete Introduction to Higher Algebra*. Springer Verlag, New York, 2009.

[76] Benny Chor and Oded Goldreich. An improved parallel algorithm for integer gcd. *Algorithmica*, 5(1–4):1–10, June 1990.

[78] Henri Cohen. *Number Theory: Volume I: Tools and Diophantine Equations*. Springer Verlag, New York, 2007.

[79] Henri Cohen. *Number Theory: Volume II: Analytic and Modern Tools*. Springer Verlag, New York, 2007.

[80] Henri Cohen. *A Course in Computational Algebraic Number Theory*. Springer Verlag, New York, 2010.

[84] W. A. Coppel. *Number Theory: An Introduction to Mathematics*. Springer Verlag, New York, 2009.

[91] Richard Crandall and Carl Pomerance. *Prime Numbers: A Computational Perspective*. Springer Verlag, New York, 2005.

[96] Abhijit Das. *Computational Number Theory*. CRC Press, Boca Raton, 2013.

[116] Euclid and Thomas Heath. *The Thirteen Books of the Elements Volume 2*. Dover Publications, Mineola, 2012.

[117] G. Everest. *An Introduction to Number Theory*. Springer Verlag, New York, 2006.

[125] Benjamin Fine and Gerhard Rosenberger. *Number Theory: An Introduction Via the Distribution of Primes*. Birkhauser, Boston, 2006.

[143] Fernando Quadros Gouvea. *p-Adic Numbers: An Introduction*. Springer Verlag, New York, 2003.

[161] G. H. Hardy and E. M. Wright. *An Introduction to the Theory of Numbers*. Oxford University Press, New York, 1980.

[189] Gareth Jones and Josephine Jones. *Elementary Number Theory*. Springer Verlag, New York, 1998.

[225] Neal Koblitz. *A Course in Number Theory and Cryptography*. Springer Verlag, New York, 2001.

[270] Yu. I. Manin and Alexei A. Panchishkin. *Introduction to Modern Number Theory: Fundamental Problems, Ideas, and Theories*. Springer Verlag, Berlin, 2005.

[304] Wladyslaw Narkiewicz. *The Development of Prime Number Theory: From Euclid to Hardy and Littlewood*. Springer Verlag, New York, 2010.

[318] Carl Olds, A. Rockett, and P. Szusze. *Continued Fractions*. Mathematical Association of America, Washington, DC, 1992.

[343] Paulo Ribenboim. *The Little Book of Big Primes*. Springer Verlag, New York, 1991.

[344] Paulo Ribenboim. *The New Book of Prime Number Records*. Springer Verlag, New York, 1995.

[345] Hans Riesel. *Prime Numbers and Computer Methods for Factorization*. Springer Verlag, New York, 2011.

[348] Alain Robert. *A Course in p-Adic Analysis*. Springer Verlag, New York, 2010.

[351] Kenneth Rosen. *Elementary Number Theory and Its Applications*. Addison-Wesley, Boston, 2000.

[365] F. Saidak. A new proof of euclid's theorem. *American Mathematical Monthly*, 113(10):937–938, 2006.

[367] József Sándor, Dragoslav S. Mitrinovic, and Borislav Crstici. *Handbook of Number Theory I*. Springer Verlag, New York, 2005.

[368] József Sándor, Dragoslav S. Mitrinovic, and Borislav Crstici. *Handbook of Number Theory II*. Springer Verlag, New York, 2005.

[378] Manfred Schroeder. *Number Theory in Science and Communications: With Applications In Cryptography, Physics, Digital Information, Computing, and Self-Similarity*. Springer Verlag, New York, 2008.

[392] Victor Shoup. *A Computational Introduction to Number Theory and Algebra*. Cambridge University Press, Cambridge, 2005.

[409] William Stein. *Elementary Number Theory: Primes, Congruences, and Secrets—A Computational Approach*. Springer Verlag, New York, 2008.

[424] Gérald Tenenbaum and Michel Mendès France. *The Prime Numbers and Their Distribution*. American Mathematical Society, Providence, 2000.

[444] Sartorius von Waltershausen. Gauss zum Gedächtniss, 1856.

[463] Don Zagier. The first 50 million prime numbers. *The Mathematical Intelligencer*, 1(1 Supplement):7–19, December 1977.

2.10.6   Efficient Modular Integer Arithmetic

[31] Paul Barrett. mplementing the rivest shamir and adleman public key encryption algorithm on a standard digital signal processor. In *Advances in Cryptology—CRYPTO '86 Proceedings*, volume 263 of *Lecture Notes in Computer Science*, pages 311–323, 1987.

[53] Antoon Bosselaers, René Govaerts, and Jose Vandewalle. comparison of three modular reduction functions. In *Advances in Cryptology—CRYPTO '93 Proceedings*, volume 773 of *Lecture Notes in Computer Science*, pages 175–186, 1994.

[197] Anatoli Karatsuba and Yuri Ofman. Multiplication of many-digital numbers by automatic computers. *Proceedings of the USSR Academy of Sciences*, 145:293–294, 1962.

[294] Peter Montgomery. Modular multiplication without trial division. *Mathematics of Computation*, 44(170):519–521, April 1985.

[453] André Weimerskirch and Christof Paar. Generalizations of the karatsuba algorithm for polynomial multiplication. Unpublished—available from International Association for Cryptologic Research (eprint.iacr.org), March 2002.

2.10.7   Abstract Algebra

[17] Michael Artin. *Algebra*. Addison-Wesley, Boston, 2010.

[63] Celine Carstensen, Benjamin Fine, and Gerhard Rosenberger. *Abstract Algebra: Applications to Galois Theory, Algebraic Geometry, and Cryptography*. De Gruyter, Berlin, 2011.

[88] David Cox. *Galois Theory*. John Wiley & Sons, Hoboken, 2004.

[94] Joan Daemen and Vincent Rijmen. *The Design of Rijndael: AES — The Advanced Encryption Standard*. Springer, New York, 2002.

[107] David Dummit and Richard Foote. *Abstract Algebra*. John Wiley & Sons, Hoboken, 2003.

[130] John Fraleigh. *A First Course in Abstract Algebra*. Addison-Wesley, Boston, 2002.

[132] Lisl Gaal. *Classical Galois Theory*. American Mathematical Society, Providence, 1998.

[178] Thomas Hungerford. *Algebra*. Springer Verlag, New York, 1980.

[184] Nathan Jacobson. *Basic Algebra I*. Dover Publications, Mineola, 2009.

[185] Nathan Jacobson. *Basic Algebra II*. Dover Publications, Mineola, 2009.

[220] Anthony Knapp. *Advanced Algebra*. Birkhauser, Boston, 2007.

[221] Anthony Knapp. *Basic Algebra*. Birkhauser, Boston, 2007.

[224] Neal Koblitz. *Algebraic Aspects of Cryptography*. Springer Verlag, Berlin, 1998.

[239] Serge Lang. *Algebra*. Springer Verlag, New York, 2005.

[240] Serge Lang. *Undergraduate Algebra*. Springer Verlag, New York, 2005.

[246] Rudolf Lidl and Gunther Pilz. *Applied Abstract Algebra*. Springer Verlag, New York, 1997.

[284] Alfred Menezes et al. *Applications of Finite Fields*. Springer Verlag, New York, 2010.

[301] Gary Mullen and Daniel Panario. *Handbook of Finite Fields*. Chapman & Hall/CRC, Boca Raton, 2013.

[302] Paul Nahin. *An Imaginary Tale: The Story of $\sqrt{-1}$*. Princeton University Press, Princeton, 1998.

[357] Louis Halle Rowen. *Algebra: Groups, Rings, and Fields*. A K Peters, Wellesley, 1994.

[381] Igor Shafarevich. *Basic Notions of Algebra*. Springer Verlag, New York, 2005.

[411] Ian Stewart. *Galois Theory*. Chapman & Hall / CRC, Boca Raton, 1990.

[434] B. L. van der Waerden. *Algebra: Volume I*. Springer Verlag, New York, 1991.

[435] B. L. van der Waerden. *Algebra: Volume II*. Springer Verlag, New York, 2003.

2.10.8 Group Theory

[4] J. L. Alperin and Rowen Bell. *Groups and Representations*. Springer Verlag, New York, 1995.

[16] Mark Armstrong. *Groups and Symmetry*. Springer Verlag, New York, 2010.

[21] Avner Asher and Robert Gross. *Fearless Symmetry: Exposing the Hidden Patterns of Numbers*. Princeton University Press, Princeton, 2006.

[104] John Dixon and Brian Mortimer. *Permutation Groups*. Springer Verlag, New York, 1996.

[186] Gordon James and Martin Liebeck. *Representations and Characters of Groups*. Cambridge University Press, Cambridge, 2001.

[272] George Martin. *Transformation Geometry: An Introduction to Symmetry*. Springer Verlag, New York, 1996.

[316] Viacheslav Nikulin. *Geometries and Groups*. Springer Verlag, New York, 2002.

[356] Joseph Rotman. *An Introduction to the Theory of Groups*. Springer Verlag, New York, 1999.

[364] Bruce Sagan. *The Symmetric Group: Representations, Combinatorial Algorithms, and Symmetric Functions*. Springer Verlag, New York, 2010.

[410] Benjamin Steinberg. *Representation Theory of Finite Groups*. Springer Verlag, New York, 2011.

2.10.9 Algebraic Geometry

[45] Robert Bix. *Conics and Cubics*. Springer Verlag, New York, 1998.

[89] David Cox, John Little, and Donal O'Shea. *Using Algebraic Geometry*. Springer Verlag, New York, 2005.

[90] David Cox, John Little, and Donal O'Shea. *Ideals, Varieties, and Algorithms: An Introduction to Computational Algebraic Geometry and Commutative Algebra*. Springer Verlag, New York, 2008.

[163] Joe Harris. *Algebraic Geometry: A First Course*. Springer Verlag, New York, 1995.

[165] Robin Hartshorne. *Algebraic Geometry*. Springer Verlag, New York, 1997.

[170] James William Peter Hirschfeld, G. Korchmaros, and F. Torres. *Algebraic Curves over a Finite Field*. Princeton University Press, Princeton, 2008.

[174] Audun Holme. *A Royal Road to Algebraic Geometry*. Springer Verlag, New York, 2011.

[255] San Ling, Huaxiong Wang, and Chaoping Xing. *Algebraic Curves in Cryptography*. CRC Press, Boca Raton, 2011.

[257] Qing Liu. *Algebraic Geometry and Algebraic Curves*. Oxford University Press, Oxford, 2006.

[327] Daniel Perrin. *Algebraic Geometry: An Introduction*. Springer Verlag, New York, 2007.

[342] Miles Reid. *Undergraduate Algebraic Geometry*. Cambridge University Press, Cambridge, 1989.

[371] Hal Schenck. *Computational Algebraic Geometry*. Cambridge University Press, Cambridge, 2003.

[382] Igor Shafarevich. *Basic Algebraic Geometry 1*. Springer Verlag, New York, 2013.

[383] Igor Shafarevich. *Basic Algebraic Geometry 2*. Springer Verlag, New York, 2013.

[403] Karen Smith, Lauri Kahanpää, Pekka Kekäläinen, and William Traves. *An Invitation to Algebraic Geometry*. Springer Verlag, Berlin, 2010.

2.10.10 Elliptic Curves

[19] Avner Ash and Robert Gross. *Elliptic Tales: Curves, Counting, and Number Theory*. Princeton University Press, Princeton, 2012.

[57] Ezra Brown and Bruce Myers. Elliptic curves from mordell to diophantus and back. *American Math. Monthly*, 109(7):639–649, August 2002.

[81] Henri Cohen, Gerhard Frey, Roberto Avanzi, Christophe Doche, Tanja Lange, Kim Nguyen, and Frederik Vercauteren. *Handbook of Elliptic and Hyperelliptic Curve Cryptography*. Chapman & Hall/CRC, Boca Raton, 2005.

[160] Darrel Hankerson, Alfred Menezes, and Scott Vanstone. *Guide to Elliptic Curve Cryptography*. Springer Verlag, New York, 2004.

[168] Haruzo Hida. *Elliptic Curves and Arithmetic Invariants*. Springer Verlag, New York, 2013.

[179] Dale Husemöller. *Elliptic Curves*. Springer Verlag, New York, 2003.

[226] Neal Koblitz. Elliptic curve cryptosystems. *Mathematics of Computation*, 48(177):203–209, January 1987.

[238] Serge Lang. *Elliptic Curves: Diophantine Analysis*. Springer Verlag, New York, 1978.

[244] Hendrik Lenstra. Factoring integers with elliptic curves. *Annals of Mathematics*, 126(3):649–673, November 1987.

[283] Alfred Menezes. *Elliptic Curve Public Key Cryptosystems*. Kluwer Academic Publishers, Boston, 1993.

[290] Victor Miller. Use of elliptic curves in cryptography. In *Advances in Cryptology—CRYPTO '85 Proceedings*, volume 218 of *Lecture Notes in Computer Science*, pages 417–426, 1986.

[354] Michael Rosing. *Implementing Elliptic Curve Cryptography*. Manning Publications, Greenwich, 1998.

[393] Joseph Silverman. *Advanced Topics in the Arithmetic of Elliptic Curves*. Springer Verlag, New York, 1994.

[394] Joseph Silverman. *The Arithmetic of Elliptic Curves*. Springer Verlag, New York, 2009.

[395] Joseph Silverman and John Tate. *Rational Points on Elliptic Curves*. Springer Verlag, New York, 2010.

[451] Lawrence Washington. *Elliptic Curves, Number Theory, and Cryptography*. Chapman & Hall/CRC, Boca Raton, 2008.

2.10.11 Lattice Theory

[64] J. W. S. Cassels. *An Introduction to the Geometry of Numbers*. Springer Verlag, New York, 1997.

[83] John Conway and Neal Sloane. *Sphere Packings, Lattices, and Groups*. Springer Verlag, New York, 1999.

[110] Wolfgang Ebeling. *Lattices and Codes*. Springer Spektrum, Berlin, 2013.

[156] Harris Hancock. *Development of the Minkowski Geometry of Numbers: Volume 1*. Dover Phoenix, Mineola, 2005.

[157] Harris Hancock. *Development of the Minkowski Geometry of Numbers: Volume 2*. Dover Phonenix, Mineola, 2005.

[198] Oleg Karpenkov. *Geometry of Continued Fractions*. Springer Verlag, New York, 2013.

[274] Jacques Martinet. *Perfect Lattices in Euclidean Space*. Springer Verlag, New York, 2010.

[277] Jiri Matousek. *Lectures on Discrete Geometry*. Springer Verlag, New York, 2002.

[288] Daniele Micciancio and Shafi Goldwasser. *Complexity of Lattice Problems: A Cryptographic Perspective*. Kluwer Academic Publishers, Boston, 2002.

[313] Phong Nguyen and Brigitte Vallée. *The LLL Algorithm: Survey and Applications*. Springer Verlag, New York, 2009.

2.10.12 Computational Complexity

[140] Oded Goldreich. *Computational Complexity: A Conceptual Perspective*. Cambridge University Press, Cambridge, 2008.

[141] Oded Goldreich. *Studies in Complexity and Cryptography*. Springer Verlag, New York, 2011.

[193] Stasys Jukna. *Boolean Function Complexity: Advances and Frontiers*. Springer Verlag, New York, 2012.

[314] Andre Nies. *Computability and Randomness*. Oxford University Press, Oxford, 2009.

[355] Jorg Rothe. *Complexity Theory and Cryptology: An Introduction to Cryptocomplexity*. Springer Verlag, Berlin, 2005.

[421] John Talbot and Dominic Welsh. *Complexity and Cryptography: An Introduction*. Cambridge University Press, Cambridge, 2006.

3

Block Ciphers

Deniz Toz

K. U. Leuven

Josep Balasch

K. U. Leuven

Farhana Sheikh

Intel Corporation

CONTENTS

3.1   Introduction

Public and private key cryptography can provide secure and reliable transmission of digital information over either a wired or wireless channel. In order to operate efficiently in both wired and wireless real-time environments, cryptography systems must deliver high throughput. For wireless communication, low-power is an additional necessary requirement. Specialized hardware to meet these needs as (wired or wireless) general-purpose programmable devices with software implementations cannot deliver the required energy-efficiency and throughput. This chapter presents an introduction to block ciphers typically used in private key cryptography systems and the most common block cipher used today is the Advanced Encryption Standard also known as AES. Its predecessor was the Data Encryption Standard. The next section presents the mathematical definition of a block cipher followed by a historical review of block ciphers. The remainder of the chapter then focuses on AES and its implementations in both hardware and software.

3.2   Block Cipher Definition

A block cipher is a mathematical function that acts on fixed-length input values and returns output values of the same length by using a secret key. The process of transforming the input, known as *plaintext*, is called *encryption*. The resulting output is called *ciphertext* and is unaccessible to anyone without the knowledge of the secret key. The original message can be revealed by an inverse process called *decryption*. More formally we define a block cipher as follows:

Definition 3.2.1 *A block cipher is a function $E : \{0,1\}^n \times \{0,1\}^k \to \{0,1\}^n$, $E(P, K) = C$ such that for each $K \in \{0,1\}^k$, E is a bijection (i.e., invertible mapping) from $\{0,1\}^n$ to $\{0,1\}^n$, where P is the plaintext, C is the ciphertext, and K is the secret key.*

The plaintext is partitioned into blocks of fixed size n and each block is encrypted separately using the same secret key. The ciphertext is then obtained by combining the outputs of all encryptions. The exact method by which the outputs are combined are specified by the *mode of operation* of the block cipher. A brief summary of these modes will be provided later in this chapter.

Most block ciphers use an iterative round function (based on Shannon's product cipher [387]) as the building block. The main idea is to combine two or more simple operations such as modular arithmetic, substitution, and permutation to obtain a more secure cipher (than either of its components). Almost all of the algorithms used today are based on this concept. Each round function is key-dependent; hence, the initial key (known as the *master key*) is expanded into the round keys by the key scheduling algorithm.

3.3 Historical Background

Even though computers were initially only for governmental and military use, in the 1960s they became affordable and powerful enough also for the private sector. Consequently, the need of a common system for communicating with the other companies in addition to internal communication arose. Lucifer [120] was developed by IBM to fulfill this need.

With the beginning of the information age in the 1970s, the exchange of digital information became an essential part of our society. In 1973, the National Bureau of Standards (NBS) made a call for a candidate symmetric-key encryption algorithm for the protection of sensitive but unclassified information. Unfortunately, none of the proposals were found viable and a second call was issued in 1974. After the evaluation phase, the submission of IBM which was heavily influenced by Lucifer, was chosen to become the Data Encryption Standard (DES) after some modifications.

At the end of 1980s and the beginning of 1990s new block ciphers were designed as an alternative to DES. Some examples include RC5, IDEA, FEAL, Blowfish, and CAST. Meanwhile the cryptographers made a great effort in analyzing the security of DES. Differential cryptanalysis [44] and linear cryptanalysis [278] which are the core of many other cryptanalysis techniques that are used today, were introduced in this period. The DESCHALL Project, which consisted of thousands of volunteers connected over the Internet, was the first to break DES (by using exhaustive key search) in public in 1997. Only two years later, it was possible to perform an exhaustive key search for DES in less than a day.

Obviously the short key-length of DES was no longer sufficient for sensitive applications. Moreover, being initially designed for hardware performance, DES was not as efficient in software as the new block ciphers. In 1997, NIST (National Institute of Standards and Technology) made a public call for a new algorithm "capable of protecting sensitive government information well into the next century" [306] to replace the DES.

As a result, fifteen submissions were submitted to the competition and after two years evaluation five of them were chosen as finalists. Finally in 2001 Rijndael [94], designed by Rijmen and Daemen, was chosen as the Advanced

Encryption Standard (AES). Unsurprisingly, in the past 10 years many attacks have been published against AES, yet none of them is considered a practical threat to its security. All the attacks were either against reduced round AES or they worked under some special conditions until the recent work of [50] which reduces the complexity of exhaustive search by a factor of four.

3.4 DES

DES [305] is a block cipher designed by IBM. It was chosen as the Data Encryption Standard in 1977 by the National Bureau of Standards (NBS) for the protection of sensitive, unclassified governmental data. DES accepts a 64-bit plaintext P and a 128-bit user key as inputs and is composed of 16 rounds. The bits of the input block are first shuffled by using an initial permutation (IP), and then the permuted input is divided into two branches L_0 and R_0 where each branch has 32 bits. The two branches are updated by using a round function f as follows:

$$L_i = R_{i-1}$$
$$R_i = L_{i-1} \oplus f(R_{i-1}, RK_i)$$

Here, RK_i is the round key generated from the 56-bit secret key by the key schedule. The number of rounds for DES is 16 and in the last round the swap operation is omitted. Finally, the ciphertext value (C) obtained after the inverse initial permutation (IP^{-1}).

3.4.1 The Round Function

In each round, the 32-bit input block is first expanded to 48 bits by a key expansion E. The expansion permutation simply duplicates half of the bits as follows: let the output of the key expansion be considered as eight groups of 6 bits each, then the j-th group is composed of the bits $(4j - 4, 4j - 3, \ldots 4j + 1)$.

The expanded input block is then XORed with the 48-bit round key and the result passes the substitution layer. DES uses eight 6×4-bit S-boxes $S_1, S_2, \ldots, S_3$ for substitution. This step is the only non-linear operation of DES. Finally all words are permuted by the permutation P. The sketch of DES is given in Figure 3.1.

3.4.2 The Key Schedule

The key schedule derives the sixteen 48-bit round keys from the master key. DES has a very simple key schedule: The 56-bit key is divided into two

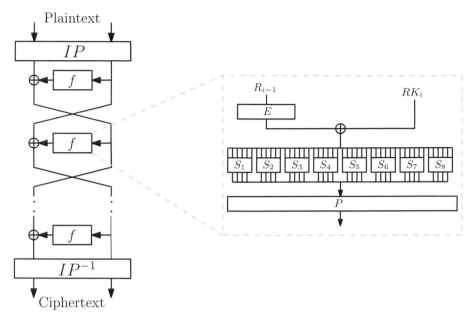

Plaintext

Ciphertext

Figure 3.1
The Feistel structure and the round function of DES

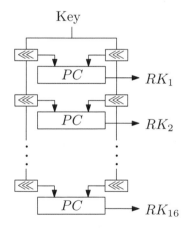

Key

Figure 3.2
The key schedule of DES

branches of 28 bits and each branch is cyclically shifted left by one or two bits (the number of shifts is specified for each round). Then 48-bit subkey is selected by Permuted Choice (PC) and is composed of 24 bits from the left half, and 24 from the right half. The key schedule of DES is depicted in Figure 3.2.

3.4.3 Decryption

The decryption function is identical to the encryption function except the round keys are used in the reserve order. Permutations IP and IP^{-1} cancel out each other and due to symmetry of the Feistel structure the same inputs enter the round function f. The same key schedule is used for decryption.

3.5 AES

Rijndael [94] is a block cipher designed by Daemen and Rijmen and is a substitution-permutation network following the wide-trail strategy. Both the block length and the key length can be any multiple of 32 bits, with a minimum of 128 bits and a maximum of 256 bits, independently of each other with key size greater than or equal to block size. The 128-bit block variant of Rijndael has been chosen as the Advanced Encryption Standard (AES).

In AES, each data block (plaintext, ciphertext, subkey, or intermediate step) is represented by a 4×4 `state matrix` of bytes. The bytes are considered as elements of the finite field $GF(2^8)$ (i.e., a polynomial with coefficients in $GF(2)$). The irreducible polynomial $m(x) = x^8 + x^4 + x^3 + x + 1$ is used to construct $GF(2^8)$; hence, the multiplication is done and the multiplicative inverse is defined accordingly. Here, addition is done in $GF(2)$ which is the XOR operation.

3.5.1 The Round Transformation

The state is initialized with the plaintext block, and then it is then transformed by iterating a round function. The final state gives the ciphertext block. The round function is composed of the four transformations `SubBytes` (SB), `ShiftRows` (SR), `MixColumns` (MC), and `AddRoundKey` (AK). (See Figure 3.3.) Before the first round, there exists a whitening layer consisting of `AddRoundKey` only, and in the last round the `MixColumns` operation is omitted.

The number of rounds N_r for AES varies with the key length N_k; $N_r = 10$ for 128-bit keys ($N_k = 4$), $N_r = 12$ for 192-bit keys ($N_k = 6$) and $N_r = 14$ for 256-bit keys ($N_k = 8$).

3.5.1.1 The `SubBytes` step

The `SubBytes` step is a non-linear byte substitution (8×8-bit S-box) that acts on every byte of the state. The S-box used in AES is denoted by S_{RD} and is composed of two maps, f and g.

$$S_{RD}(x) = f \circ g(x) = f(g(x))$$

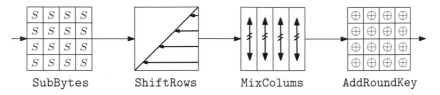

Figure 3.3
The round transformation of AES

Both maps are defined in $GF(2^8)$ and have simple algebraic expressions. Let $x \in GF(2^8)$, then g maps x to its multiplicative inverse x^{-1}. f is an invertible affine transformation given by $f(x) = Ax + b$.

$$f(x) := \begin{bmatrix} 1 & 1 & 1 & 1 & 1 & 0 & 0 & 0 \\ 0 & 1 & 1 & 1 & 1 & 1 & 0 & 0 \\ 0 & 0 & 1 & 1 & 1 & 1 & 1 & 0 \\ 0 & 0 & 0 & 1 & 1 & 1 & 1 & 1 \\ 1 & 0 & 0 & 0 & 1 & 1 & 1 & 1 \\ 1 & 1 & 0 & 0 & 0 & 1 & 1 & 1 \\ 1 & 1 & 1 & 0 & 0 & 0 & 1 & 1 \\ 1 & 1 & 1 & 1 & 0 & 0 & 0 & 1 \end{bmatrix} \cdot \begin{bmatrix} x_7 \\ x_6 \\ x_5 \\ x_4 \\ x_3 \\ x_2 \\ x_1 \\ x_0 \end{bmatrix} + \begin{bmatrix} 0 \\ 1 \\ 1 \\ 0 \\ 0 \\ 0 \\ 1 \\ 1 \end{bmatrix}$$

3.5.1.2   The `ShiftRows` step

The `ShiftRows` (SR) step is a cyclic shift of bytes in a row that acts individually on each of the last three rows of the state. Row i is shifted i bytes to the left so that the byte at position j is row moves to position $(j - i) \bmod 4$. It aims to provide the optimal diffusion. Similarly, the inverse operation is a cyclic shift of the 3 bottom rows to the right by the same amount.

3.5.1.3   The `MixColumns` step

The `MixColumns` step is a linear transformation that acts independently on every column of the state. The columns of the state are considered as polynomials over $GF(2^8)$ and multiplied with a fixed polynomial $c(x) = 03 \cdot x^3 + 01 \cdot x^2 + 01 \cdot x + 02$ modulo $(x^4 + 1)$. Alternatively, the modular multiplication can be written as a matrix multiplication. Let $b(x) = c(x) \cdot a(x)$, then:

$$\begin{bmatrix} b_0 \\ b_1 \\ b_2 \\ b_3 \end{bmatrix} = \begin{bmatrix} 02 & 03 & 01 & 01 \\ 01 & 02 & 03 & 01 \\ 01 & 01 & 02 & 03 \\ 03 & 01 & 01 & 02 \end{bmatrix} \times \begin{bmatrix} a_0 \\ a_1 \\ a_2 \\ a_3 \end{bmatrix}$$

The inverse operation is similar to `MixColumns`: each column is multiplied with a fixed polynomial $d(x)$ where $c(x) \cdot d(x) \equiv 1 \bmod (x^4 + 1)$. It is computed as $d(x) = 0B \cdot x^3 + 0D \cdot x^2 + 09 \cdot x + 0E$.

3.5.1.4   The `AddRoundKey` step

The `AddRoundKey` step is the bitwise exclusive-or (XOR) of the round key with the intermediate state. To invert the `AddRoundKey`, one simply performs the XOR operation with the same round key. Therefore, `AddRoundKey` is its own inverse. The round key of the i-th round is denoted by `ExpandedKey`$[i]$ and is derived from the master key by using a key schedule.

3.5.2   The Key Schedule

The key schedule derives $(N_r + 1)$ 128-bit round keys `ExpandedKey`$[\cdot]$ from the master key. It consists of a linear array of 4-byte words denoted by $W[i]$ for $0 \leq i \leq 4 \cdot (N_r + 1)$. The round key of the i-th round is given by the words $4i$ to $4i + 3$ of W.

There are two versions of the key expansion function, depending on the key size N_k. For both versions, the first N_k words $W[0], W[1], \cdots, W[N_k - 1]$ are directly initialized with the words of the master key. The remaining key words are generated recursively in terms of the previous key words. For $N_k \leq 6$

$$W[i] = \begin{cases} W[i - N_k] \oplus S_{RD}(W[i - 1] \lll 8) \oplus RC[i/N_k] & \text{if } i \equiv 0 \bmod N_k \\ W[i - N_k] \oplus W[i - 1] & \text{otherwise} \end{cases}$$

For $N_k = 8$, we have

$$W[i] = \begin{cases} W[i - N_k] \oplus S_{RD}(W[i - 1] \lll 8) \oplus RC[i/N_k] & \text{if } i \equiv 0 \bmod N \\ W[i - N_k] \oplus S_{RD}(W[i - 1]) & \text{if } i \equiv 4 \bmod N \\ W[i - N_k] \oplus W[i - 1] & \text{else} \end{cases}$$

In the above equations $\lll$ denotes the rotation of the word to the left, $RC[\cdot]$ are the fixed round constants, and they are independent of N_k. The round key RK_i is given by the words $W[N_b \cdot i]$ to $W[N_b \cdot (i + 1)]$.

3.5.3   Decryption

The ciphertext can be decrypted in a straightforward way by using the inverse functions `InvSubBytes`, `InverseShiftRows`, `InverseMixColumns`, and `AddRoundKey` in a reverse order. There exists also an equivalent algorithm for decryption in which the order of the (inverse) steps is the same as in the encryption with a modified key schedule (i.e., `InverseMixColumns` is applied to each `ExpandedKey`$[i]$). This equivalent algorithm is more convenient for software implementation purposes.

3.6 Modes of Operation

In block ciphers, every message block is processed in a similar way called the *mode of operation*. To obtain the ciphertext C, the plaintext P is first padded such that the length of the input is a multiple of the block length. Then it is divided into t blocks such that $P||pad = P_1||P_2||...||P_t$. Up to now many modes of operation have been proposed. Some common modes are given below.

- **Electronic Code Book (ECB).** This is the most straightforward mode of operation in which each plaintext block is encrypted independently by using the same key. Its simplicity and suitability for parallelization are among the advantages of this mode. However, it does not hide patterns in the plaintext, i.e., identical plaintext blocks are encrypted into identical ciphertext blocks. Therefore, the use of this mode is not recommended for cryptographic applications. The scheme is depicted in Figure 3.4.

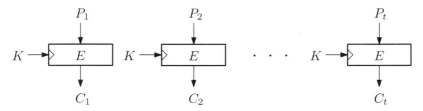

Figure 3.4
Electronic Code Book (ECB) mode of operation

- **Cipher Block Chaining (CBC).** It is the most widely used mode of operation in which each plaintext block is XORed with the previous ciphertext block (an initialization vector (IV) is used for the first plaintext block) before passing through the block cipher. This guarantees that each ciphertext block depends on all previous plaintext blocks, avoiding repetitions and providing randomness. The encryption operation is no longer parallelizable whereas for decryption it is sufficient to have two consecutive ciphertext blocks, and it can still be performed independently. The scheme is depicted in Figure 3.5.

- **Cipher Feedback (CFB).** In CFB mode the previous ciphertext block is encrypted and then XORed with the current plaintext block to obtain the ciphertext block. As in CBC mode, for the first block an initialization vector is used. Hence, this mode of operation converts the block cipher into a self-synchronizing stream cipher. Again, the feedback prevents the parallelization of encryption whereas decryption can be done independently provided that two consecutive blocks are available. This mode was initially

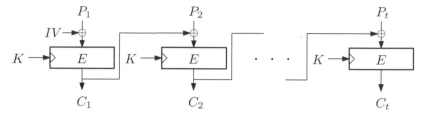

Figure 3.5
Cipher Block Chaining (CBC) mode of operation

designed to encrypt r-bit message blocks and to transmit without delay where $1 \leq r \leq n$ (typically $r = 1$ or $r = 8$). In this case, the r leftmost bits of the output block ($E(S, K)$) is XORed with the plaintext bits and returned as ciphertext bits. The state S is shifted to left by r bits and previous cipher-text bits are inserted before each iteration as feedback. This is depicted in Figure 3.6.

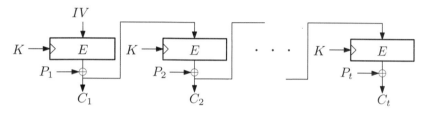

Figure 3.6
Cipher Feedback (CFB) mode of operation

- **Counter (CTR).** In this mode of operation the counter is encrypted and then the result is XORed with the plaintext block to obtain the ciphertext block. The counter incremented before it is used for the next block. Although a function can be used for incrementation, incrementing by one is the easiest and the most common application. The use of nonce (a random or pseudo-random number) is suggested and the same IV and key combination must not be used more than once for protection against attacks. CTR mode allows parallelization. This is depicted in Figure 3.7.

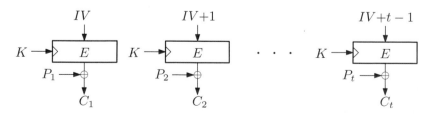

Figure 3.7
Counter (CTR) mode of operation

3.7 Composite Field Representation of S-box

Galois Field $GF(2^k)$ is a finite field containing 2^k elements and any k linearly independent elements form a basis. For instance, the set

$$B_1 = \{1, \alpha, \alpha^2, \ldots, \alpha^{k-1}\}$$

is a basis for $GF(2^k)$ provided that the elements of B are linearly independent. Each element $a \in GF(2^k)$ can be represented as a polynomial of degree $d < k$ as

$$a = \sum_{i=0}^{k-1} a_i \alpha_i$$

where the coefficients $a_i \in GF(2)$ and B_1 are called the polynomial basis. Various types of bases have been studied extensively; standard, normal, and dual basis can be given among the most common representations. Different basis representations can be used to simplify the implementation of arithmetic operations.

If k is a composite integer (i.e., $k = mn$) then the fields $GF(2^k)$ and $GF((2^m)^n)$ are isomorphic to each other, and it is possible to represent the elements of $GF(2^k)$ as polynomials of degree $d < n$ with coefficients in $GF(2^m)$. In this case, $G(2^k)$ is a *composite field*, and $GF(2^m)$ is called the *ground field*. Now, let B_2 be a polynomial basis for $GF((2^m)^n)$:

$$B_2 = \{1, \beta, \beta^2, \ldots, \beta^{n-1}\}$$

Then $a \in GF(2^k)$ can be represented as:

$$a = \sum_{i=0}^{n-1} a_i' \beta_i$$

where the coefficients $a_i' \in GF(2^m)$. The composite field representation provides relatively efficient implementations when the arithmetic operations (such as multiplication, inversion, and exponentiation) rely on table lookups.

The AES S-box uses the mapping $g : x \to x^{-1}$ where x^{-1} is the multiplicative inverse in $GF(2^8)$. This inversion is the most costly step of the S-box transformation in hardware implementation of AES. The simplest method for finding the inverse is to use a lookup table; however, this method requires $2^m \times m$ bits of memory which increases exponentially with m. An algorithm based on Euclid's algorithm whose area requirement is proportional to m and requires $2m$ cycles per inversion has been presented by Brunner et al. in [58].

The first efficient hardware implementation of the multiplicative inversion in $GF(2^8)$ has been proposed by Rijmen [346] and first implemented by Rudra et al. [358] and Wolkerstorfer et al. [455]. The main idea is to decompose the elements of $GF(2^8)$ into linear polynomials over the subfield $GF(2^4)$. In this case, an element $a \in GF(2^8)$ can be represented as:

$$a \cong a_1 x + a_0$$

where $a_1, a_0 \in GF(2^4)$. The multiplication is performed modulo an irreducible polynomial with degree two to ensure that the result is a linear polynomial. Let the irreducible polynomial be denoted as $x^2 + \alpha x + \beta$, then the multiplicative inverse of a can be computed as:

$$a^{-1} \cong a_1 (a_1^2 \beta + a_1 a_0 \alpha + a_0^2)^{-1} x + (a_0 + a_1 \alpha)(a_1^2 \beta + a_1 a_0 \alpha + a_0^2)^{-1}$$

As a result, the inversion in $GF(2^8)$ can be expressed in terms of multiplications, squarings, additions, and inversion in $GF(2^4)$. Moreover, the elements of $GF(2^4)$ can be further decomposed into polynomials over the $GF(2^2)$ before implementing the final operations in $GF(2)$.

$$GF(2^8) \cong GF((2^4)^2) \cong GF(((2^2)^2)^2) \cong GF((((2)^2)^2)^2)$$

This method is known as *tower field decomposition*. Satoh et al. [370] and Mentens et al. [286] follow this approach in their implementations. In all the above implementations the field elements are represented by using polynomial basis.

3.8   Software Implementations of AES

Due to its structural and algebraic simplicity, the AES algorithm is well suited for implementations on a wide variety of processors. Generic software implementations of AES are present in most general purpose cryptographic libraries and in a wide variety of programming languages, such as OpenSSL and OpenPGP (in C) or BouncyCastle (in Java). More specific, target-optimized implementations are found across the literature for a full spectrum of devices and architectures, ranging from ultra-constrained 4-bit CPUs [183] to multicore graphics processors (GPUs) [164].

In the rest of this section we give a brief review of state-of-the-art implementations targeted to embedded processors (8-bit and 32-bit) and to general purpose processors (32-bit and 64-bit). Optimizations often balance the common tradeoff speed vs. memory usage (RAM and/or ROM), and can be performed either at high-level or at low-level, i.e., by describing alternative reformulations of the AES building blocks or by exploiting microarchitectural features specific to the target CPU. The former techniques can typically be ported to a wide range of devices; whereas, the latter improvements are only valid for particular architectures.

Speed reports vary significantly depending on the benchmarking method employed and on the choice of mode of operation. Direct comparison between proposals is thus often impossible, even if one considers only speed and not other parameters such as ROM or RAM requirements. In order to partially tackle this issue, the public eSTREAM benchmarking interface brings forward a common framework for comparison of software cryptographic implementation. For AES in particular, a detailed overview (in terms of speed) of existing proposals when considering counter mode of operation (AES-CTR) can be found in [37].

3.8.1   8-Bit Processors

This type of embedded processor is often found as the central element of smart cards, and is thus one of the most common targets for cryptographic implementations. Smart cards are generally constrained in program memory (flash or ROM), RAM, clock speed, or even arithmetic capabilities, and implementers have to balance and adapt these parameters according to their design goals. As the wordsize of 8-bit processors perfectly matches the basic unit for processing in the AES algorithm, namely a byte, a one-to-one mapping from the AES specification to its implementation is possible.

The non-linear transformation of the AES or SubBytes step is typically implemented as a lookup table (S-box), thus avoiding complex operations at the cost of storing a 256-byte array in memory. Aligning this table in a 256-byte memory boundary, e.g., starting at memory address 0x100, 0x200, etc., often results in cycle savings when updating the address pointer. The AddRoundKey step can be straightforwardly implemented as most 8-bit processors support additions in $GF(2^8)$ via the XOR (exclusive-or) instruction, while the ShiftRows step requires only to transpose the AES state bytes.

The matrix multiplication in the MixColumns step is often rewritten in a small series of instructions [94] that employ only field additions and field multiplications by 02:

$$t = a[0] \oplus a[1] \oplus a[2] \oplus a[3]; \quad \text{a is a 4-byte column}$$
$$u = a[0];$$
$$v = a[0] \oplus a[1]; \quad v = xtime(v); \quad a[0] = a[0] \oplus v \oplus t;$$
$$v = a[1] \oplus a[2]; \quad v = xtime(v); \quad a[1] = a[1] \oplus v \oplus t;$$
$$v = a[2] \oplus a[3]; \quad v = xtime(v); \quad a[2] = a[2] \oplus v \oplus t;$$
$$v = a[3] \oplus u; \quad v = xtime(v); \quad a[3] = a[3] \oplus v \oplus t;$$

Multiplications by 02 in $GF(2^8)$, denoted as *xtime*, can be carried out in two different ways. One option is to execute a left-shift operation of the input value v and, if carry occurs, perform an additional XOR operation with the value $1B$, which corresponds to the AES irreducible polynomial. A second option consists in implementing *xtime* as a lookup table, thus avoiding conditional executions that might enable timing attacks.

Regarding the decryption process, an efficient method to compute the more complex InvMixColumns step making only use of *xtime* can be also found in [94].

3.8.2   32-Bit Processors

A well-known technique to speed up AES implementations on 32-bit processors is the use of the so-called T-tables [94]. In a nutshell, this optimization allows execution of all AES round transformations with only four table lookups and four XOR operations per column. It requires 4 kB of storage space, which can be reduced to 1 kB at the cost of some extra rotations. The same technique can be efficiently applied for the decryption process, albeit using different lookup tables.

Further implementations on 32-bit processors include the work of Bertoni et al. [41]. They propose to restructure the AES algorithm in a different way than its standard formulation in order to allow a better exploitation in 32-bit processors. The proposed modifications result in considerable performance improvements in decryption on various 32-bit platforms such as ST22 smart cards, ARM7TDMI and ARM9TDMI processors, and Intel PentiumIII general purpose processors. Atasu et al. [22] exploit peculiarities of the 32-bit ARM instruction set architecture and propose a new implementation for the AES encryption linear mixing layers. Darnall and Kuhlman [95] explore various implementation tradeoffs on ARM7TDMI platforms focusing on three typical critical counts for embedded devices: execution time, ROM, and RAM.

3.8.3   64-Bit Processors

One of the first implementations of AES on the Intel x64 architecture was due to Matsui and Nakajima [279]. They reported a constant-time implementation with a throughput of 9.2 cycles/byte on an Intel Core 2, conditioned to the use of 2048-byte input data sizes previously transposed to a bitsliced format. Around 1 cycle/byte is required for bitsliced format conversion. A downside of this proposal is that smaller input sizes need to be padded to a 2048-byte boundary, thus creating significant overheads. The semi-bitsliced implementation due to Könighofer [232] reduces the minimum input data size to 64 bytes, but the throughput drops to 19.81 cycles/byte on an Athlon 64. Bernstein and Schwabe [38] report 10.57 cycles/byte on an Intel Core 2 and 10.43 cycles/byte on an Athlon 64.

A more recent bitsliced AES implementation due to Käsper and Schwabe [199] achieves 7.59 cycles/byte on an Intel Core 2 and 6.92 cycles/byte on Intel Core i7. The minimum input data size is 128-bytes which is stored into eight 128-bit XMM registers. This implementation takes advantage of Intel's microarchitectural SSSE3 instruction set extensions to improve results by a factor 30% compared to other implementations.

3.9 Hardware Implementations of AES

The computational complexity of mapping modular Galois-field arithmetic onto general-purpose processors create power and performance bottlenecks, especially when high-throughput real-time operation is required. Special-purpose hardware AES accelerators address the need for energy-efficiency real-time AES encryption/decryption. Special-purpose hardware implementations of the AES algorithm started to appear in the literature at around the same time that the AES algorithm was being chosen as the standard by NIST in 2001 [441]. The efficiency of hardware implementations of AES is determined largely by how well the S-box inverse and MixColumn steps are implemented. The hardware implementation is routing-limited so an additional parameter to consider in the hardware implementation is the wire routing strategy. For example, AES instructions on Intel's Haswell platform requires 0.63 and 4.44 cycles/byte for CTR and CBC-encryption respectively, which when converted to Gb/Joule is many orders of magnitude more energy-efficient than generic assembly or C-compiled code [440]. The most challenging components of implementing AES using hardware are computing the multiplicative inverse in the SBox and routing the computational elements of the SBox and Mixed Columns. In this section we review three key AES hardware implementations and discuss future directions.

3.9.1 AES SBox Optimization for Hardware

The first efficient hardware implementation of AES was reported in [455] which describes optimizations to the AES Sbox inverse operation. Inversion in the finite field $GF(2^8)$ is calculated using combinational logic in the finite field $GF(2^4)$ which reduces the complexity of the inverse computation significantly. This implementation is contrasted with previous implementations that rely on look-up tables. The work presented a hardware implementation of the SBox which results in an area of $0.108mm^2$ in a fairly old CMOS 0.6μm process.

The AES hardware implementation area and power is largely determined by the implementation of the MixColumn operation and the SBox operation. The remaining operations are trivially implemented using XOR gates, shifts, and clever routing techniques. This work shows that the implementation significantly reduces area over a ROM implementation of the SBox inversion step. The basic idea that is employed in this work and later on leveraged in subsequent publications relies in representing the finite field $GF(2^8)$ AES generator polynomial in a composite field $GF((2^4)^2)$. The authors show that this can be done because the finite field $GF(2^8)$ is isomorphic to the finite field $GF((2^4)^2)$, where for each element in $GF(2^8)$ there exists exactly one element in $GF((2^4)^2)$. The representation and mapping are explained mathematically in the above referred paper and will not be repeated here. The implementation requires simple gates such as inverters, XORs, and multiplexers.

3.9.2   AES for Microprocessors

The second noteworthy implementation that we review was reported in 2010 [276] where the modular inversion is also carried out in the composite $GF(2^4)^2$ field which enables an efficient circuit implementation at the cost of mapping/inverse-mapping overhead. However, in this implementation the mapping overhead is amortized over all the AES rounds by moving the transformation outside of the iteration loop. This is accomplished by computing the entire AES round in the composite field $GF(2^4)^2$ in contrast to most conventional implementations that compute only the SBox in the composite field as presented in the previous section. The implementation of the SBox and the modular inverse is computed by combinational logic as in [455] with further optimizations in the logic resulting in a smaller gate count. The other features of this implementation include a reconfigurable datapath for encrypt/decrypt, optimized ground and composite-field polynomials, integration of bypass multiplexers into affine transforms, difference-matrix computation for the InvMixColumn step, and a folded ShiftRow datapath organization. Careful consideration has been given to wire routing to optimize layout area.

The 53Gbps AES encrypt/decrypt hardware accelerator is manufactured in 45nm CMOS process resulting in 67% reduction in interconnect length and 20% reduction in total area. The accelerator is reconfigurable to support different modes of AES ECB: AES-128, AES-192, and AES-256. The implementation consumes 125mW total power measured at 1.1V. This state-of-the-art implementation at the time was accomplished by amortizing the mapping the entire AES round to the composite field $GF(2^4)^2$ and amortizing this mapping over the multiple iterations required. The authors implemented a fused MixColumns/InverseMixColumns circuit to save area and folded the SBox datapath to reduce wire routing overhead. The reader is referred to the authors' detailed paper in [276] to fully understand the mechanisms and design techniques required to achieve high-throughput required for high-performance microprocessors.

3.9.3   AES for Ultra-Low Power Mobile Platforms

In contrast to the above implementation targeted towards high-performance computing, there is an increasing requirement for encrypted data storage and communication across low-power devices such as RFID tags, Internet-of-Things (IOT) compute devices, and even mobile phones and wearables. AES has become the defacto block cipher for content protection, memory encryption, and network security. However, the large area cost and high energy consumption for parallel, multiple round implementation is prohibitive for small embedded devices. The first application of hardware accelerated AES was in fact for RFID applications where encryption was necessary to secure and protect data. In [121] the authors report an AES hardware implementation which encrypts 128-bit block of data within 1000 clock cycles and consumes below 9μA on 0.35μm CMOS process. Ingrid Verbauwhede is famous for implementing low-cost cryptographic processing hardware solutions starting from AES for RFIDs, with the first low-power implementation presented in 2002 [237]. The most recent attempt at implementing an ultra-low power AES hardware accelerator was presented in 2014 [275].

In this most recent implementation of AES targeted towards ultra-low power encryption/decryption, the authors present 2090-gate AES-128 encrypt/decrypt engine fabricated in 22nm CMOS consuming 13mW at 1.1GHz operational frequency at 0.9V. Optimum energy efficiency is achieved at a supply voltage of 430mV which results in energy-efficiency of 289Gbps/W. The key to implementing a low-power version of AES is to serialize the computation of the SBox. That is, only implement one SBox and reuse it over time. The authors in [275] have also mathematically optimized the implementation by comprehensive exploration of the $GF(2^4)^2)$ polynomial space to select non-symmetric ground and extension-field polynomials for encrypt and decrypt that minimize layout area by up to 9% area reduction over previous work.

3.10 Future Directions

There is continuing work to develop extreme low-power implementations of the now defacto AES standard used for encryption of media content, memory and disk storage, and network communication. Body Area Networks (BAN) are now emerging as the next phase in personal data communication over wireless networks where it is essential to protect privacy and secure data communication. However, in this space, extremely low energy solutions are necessary. Whether it is AES that can be implemented in a "nano" regime or whether it is other ultra-low cost block ciphers, there will be continuing efforts to develop new algorithms and implementations to meet the growing demand for secure data storage and transmission.

4

Secure Hashing: SHA-1, SHA-2, and SHA-3

Ricardo Chaves

University of Lisbon

Leonel Sousa

University of Lisbon

Nicolas Sklavos

University of Patras

Apostolos P. Fournaris

T.E.I. Western Greece

Georgina Kalogeridou

University of Cambridge

Paris Kitsos

T.E.I. Western Greece

Farhana Sheikh

Intel Corporation

CONTENTS

4.1 Introduction

A hash function is a cryptographic operation that converts a variable length input message to a fixed output called message hash or message digest. Hash functions can mostly be found in authentication schemes, random number generators, data integrity mechanisms, and digital signatures.

A modern cryptographic hash function can compute the hash value for any initial message. One of the basic hash function properties is the difficulty of finding two different messages with the same hash value (collision resistance). Due to this property, hash functions are widely used in popular security protocols like the Transport Layer Security or the Internet Protocol Security.

4.2 SHA-1 and SHA-2 Hash Functions

In 1993, the Secure Hash Standard (SHA) was first published by the NIST. In 1995, this algorithm was revised [309] in order to eliminate some of the initial weakness. The revised algorithm is usually referenced as SHA-1. In 2001, the SHA-2 hashing algorithm was proposed. The revised SHA-2 algorithm considers larger Digest Messages (DM), making it more resistant to possible attacks and allowing it to be used with larger data inputs, up to 2^{128} bits in the case of SHA512. The SHA-2 hashing algorithm is the same for the SHA224, SHA256, SHA384, and SHA512 hashing functions, differing in the size of the operands, the initialization vectors, and the size of the final DM. The next sections start by describing the round computation of the SHA-1 and SHA-2 algorithms followed by the description of the needed input data padding and expansion.

4.2.1 SHA-1 Hash Function

The SHA-1 produces a single output 160-bit message digest (the output hash value) from an input message. This input message is composed of multiple blocks. The input block, of 512 bits, is split into 80×32-bit words, denoted as W_t, one 32-bit word for each computational round of the SHA-1 algorithm, as depicted in Figure 4.1.

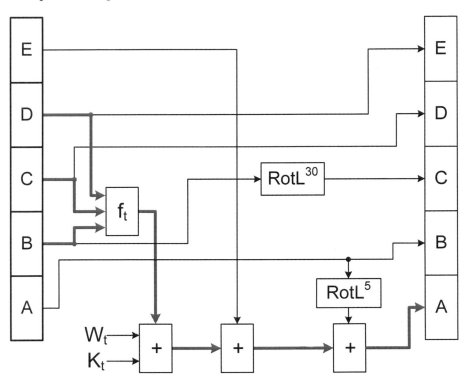

Figure 4.1
SHA-1 round calculation

Each round comprises additions and logical operations, such as bitwise logical operations (f_t) and bitwise rotations to the left ($RotL^i$). The calculation of f_t depends on the round (t) being executed, as well as the value of the round dependant constant K_t. The SHA-1 80 rounds are divided into four groups of 20 rounds, each with different values for K_t and the applied logical functions (f_t) [71].

Table 4.1 presents the values of K_t and the logical function executed, according to the round. The symbol $\wedge$ represents the bitwise *AND* operation and $\oplus$ represents the bitwise *XOR* operation.

The initial values of the A to E variables in the beginning of each data block calculation correspond to the value of the current 160-bit hash value

Table 4.1
SHA-1 logical functions

Rounds	Function	K_t
0 to 19	$(B \wedge C) \oplus (\bar{B} \wedge D)$	0x5A827999
20 to 39	$B \oplus C \oplus D$	0x6ED9EBA1
40 to 59	$(B \wedge C) \oplus (B \wedge D) \oplus (C \wedge D)$	0x8F1BBCDC
60 to 79	$B \oplus C \oplus D$	0xCA62C1D6

or Digest Message (DM). After the 80 rounds have been computed, the A to E 32-bit values are added to the current DM. The Initialization Vector (IV) or the DM for the first block is a predefined constant value. The output value is the final DM, after all the data blocks have been computed. In some higher level protocols such as the keyed-Hash Message Authentication Code (HMAC) [310], or when a message is fragmented, the IV may differ from the constant specified in [309].

To better illustrate the algorithm a pseudo code representation is depicted in Figure 4.2.

$$
\begin{array}{l}
\text{DM} = DM_0 \text{ to } DM_4 = \text{IV} \\
\textbf{for } \text{for each data\_block } \textbf{do} \\[4pt]
\quad W_t = expand(\text{data\_block}) \\
\quad A = DM_0 \text{ ; } B = DM_1 \text{ ; } C = DM_2 \text{ ; } D = DM_3 \text{ ; } E = DM_4 \\[4pt]
\quad \textbf{for } t= 0, t\leq 79, t{=}t{+}1 \textbf{ do} \\
\quad\quad \text{Temp} = RotL^5(A) + f_t(B,C,D) + E + K_t + W_t \\
\quad\quad E = D \\
\quad\quad D = C \\
\quad\quad C = RotL^{30}(B) \\
\quad\quad B = A \\
\quad\quad A = \text{Temp} \\
\quad \textbf{end for} \\[4pt]
\quad DM_0 = A + DM_0 \text{ ; } DM_1 = B + DM_1 \text{ ; } DM_2 = C + DM_2 \\
\quad DM_3 = D + DM_3 \text{ ; } DM_4 = E + DM_4 \\
\textbf{end for}
\end{array}
$$

Figure 4.2
Pseudo code for SHA-1 function

4.2.2   SHA256 Hash Function

In the SHA256 hash function, a final DM of 256 bits is produced. Each 512-bit input block is expanded and fed to the 64 rounds of the SHA256 function in words of 32 bits each (denoted by W_t). Like in the SHA-1, the data scrambling is performed according to the computational structure depicted in Figure 4.3

by additions and logical operations, such as bitwise logical operations and bitwise rotations. The several input data blocks are mixed with the current state and the 32-bit round dependent constant (K_t).

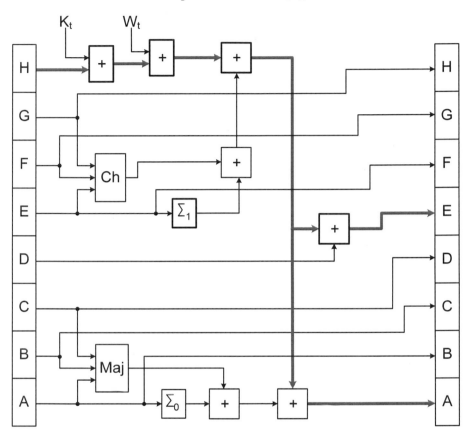

Figure 4.3
SHA-2 round calculation

The 32-bit values of the A to H variables are updated in each round and the new values are used in the following round. The IV for these variables is given by the 256-bit constant value specified in [309], being set only for the first data block. The consecutive data blocks use the partial DM computed for the previous data block. Each SHA256 data block is processed in 64 rounds, after which the values of the variables A to H are added to the previous DM in order to obtain the new value of the DM. Comparing the SHA round computation depicted in Figure 4.1 and Figure 4.3, it is noticeable a higher computational complexity of the SHA-2 algorithm in regard to the SHA-1 algorithm. To better illustrate this algorithm a pseudo code representation is depicted in Figure 4.4. The final Digest Message (DM) for a given data stream is given by the final DM value, obtained after the last data block is processed.

```
DM = DM₀ to DM₇ = IV
for each data_block i do

    W = expand(data_block)
    A = DM₀ ; B = DM₁ ; C = DM₂ ; D = DM₃
    E = DM₄ ; F = DM₅ ; G = DM₆ ; H = DM₇

    for t= 0, t≤ 63 {79}, t=t+1 do
        T₁ = H + Σ₁(E) + Ch(E,F,G) + Kₜ + Wₜ
        T₂ = Σ₀(A) + Maj(A,B,C)
        H = G ; G = F ; F = E ;
        E = D + T₁
        D = C ; C = B ; B = A
        A = T₁ + T₂
    end for

    DM₀ = A + DM₀ ; DM₁ = B + DM₁
    DM₂ = C + DM₂ ; DM₃ = D + DM₃
    DM₄ = E + DM₄ ; DM₅ = C + DM₅
    DM₆ = D + DM₆ ; DM₇ = E + DM₇
end for
```

Figure 4.4
Pseudo code for SHA-2 algorithm

4.2.3 SHA512 Hash Function

The SHA512 hash function computation is identical to that of the SHA256 hash function, differing in the size of the operands, 64 bits instead of 32 bits as for the SHA256. The DM has twice the width, 512 bits, and different logical functions are used [309]. The values W_t and K_t are 64 bits wide and each data block is composed of 16×64-bit words, having in total 1024 bits.

Table 4.2 details the logical operations performed in both the SHA256 and SHA512 algorithms, namely Ch, Maj, Σ_i, and σ_i, where $ROTR^n(x)$ represents the right rotation operation by n bits, and $SHR^n(x)$ the right shift operation by n bits.

4.2.4 Data Block Expansion for the SHA Function

The SHA-1 round computation, described in Figure 4.1, is performed 80 times, and in each round a 32-bit word, obtained from the current input data block (W_t), is used. Since each input data block only has 16×32-bit words (512 bits), the remaining 64×32-bit words are derived from the data expansion operation. This data expansion is performed by computing (4.1), where $M_t^{(i)}$ denotes the first 16×32-bit words of the i-th data block.

$$W_t = \begin{cases} M_t^{(i)} & , 0 \leq t \leq 15 \\ \\ RotL^1(W_{t-3} \oplus W_{t-8} \oplus W_{t-14} \oplus W_{t-16}) & , 16 \leq t \leq 79 \end{cases} \qquad (4.1)$$

Table 4.2
SHA256 and SHA512 logical functions

Designation	Function
Maj(x,y,z)	$(x \wedge y) \oplus (x \wedge z) \oplus (y \wedge z)$
Ch(x,y,z)	$(x \wedge y) \oplus (\overline{x} \wedge z)$
$\Sigma_0^{\{256\}}(x)$	$ROTR^2(x) \oplus ROTR^{13}(x) \oplus ROTR^{22}(x)$
$\Sigma_1^{\{256\}}(x)$	$ROTR^{14}(x) \oplus ROTR^{18}(x) \oplus ROTR^{41}(x)$
$\sigma_0^{\{256\}}(x)$	$ROTR^7(x) \oplus ROTR^{18}(x) \oplus SHR^3(x)$
$\sigma_1^{\{256\}}(x)$	$ROTR^{17}(x) \oplus ROTR^{19}(x) \oplus SHR^{10}(x)$
$\Sigma_0^{\{512\}}(x)$	$ROTR^{28}(x) \oplus ROTR^{34}(x) \oplus ROTR^{39}(x)$
$\Sigma_1^{\{512\}}(x)$	$ROTR^{14}(x) \oplus ROTR^{18}(x) \oplus ROTR^{41}(x)$
$\sigma_0^{\{512\}}(x)$	$ROTR^1(x) \oplus ROTR^8(x) \oplus SHR^7(x)$
$\sigma_1^{\{512\}}(x)$	$ROTR^{19}(x) \oplus ROTR^{61}(x) \oplus SHR^6(x)$

For the SHA-2 algorithm, the round computation, depicted in Figure 4.3, are performed for 64 rounds (80 rounds for the SHA512). In each round, a 32-bit word (or a 64-bit word for the SHA512) from the current data input block is used. Once again, the input data block only has 16 words, resulting in the need to expand the initial data block to obtain the remaining W_t words.

This expansion is performed by computing (4.2), where $M_t^{(i)}$ denotes the first 16 words of the i-th data block and the operator $+$ describes the arithmetic addition operation.

$$W_t = \begin{cases} M_t^{(i)} & , 0 \leq t \leq 15 \\ \sigma_1(W_{t-2}) + W_{t-7} + \sigma_0(W_{t-15}) + W_{t-16} & , 16 \leq t \leq 63 \; \{\text{or } 79\} \end{cases} \quad (4.2)$$

4.2.5 Message Padding

In order to assure that the input data are a multiple of 512 bits, as required by the SHA-1 and SHA256 specification, the original message has to be padded. For the SHA512 algorithm the input data must be a multiple of 1024 bits.

The padding procedure for a 512-bit input data block is as follows: for an original message composed of n bits, the bit "1" is appended at the end of the message (the $n + 1$ bit), followed by k zero bits, where k is the smallest solution to the equation $n + 1 + k \equiv 448 \mod 512$. The last 64 bits of the padded message are filled with the binary representation of n, i.e., the original message size. This operation is better illustrated in Figure 4.5 for a message with 536 bits (010 0001 1000 in binary representation).

$$\overbrace{}^{24bits} \quad \overbrace{}^{423bits} \quad \overbrace{}^{64bits}$$

$$\underbrace{10\cdots\cdots01}_{\text{first data block}} \quad \underbrace{\overbrace{10\cdots01}\,1\,\overbrace{00\cdots00}\,0\cdots0\,01000011000}_{\text{last data block}}$$

Figure 4.5
Message padding for 512-bit data blocks

For the SHA512 message padding, 1024-bit data blocks are considered with the last 128 bits, not 64 bits, reserved for the binary representation of the original message.

4.2.6   Hash Value Initialization

In the SHA-1 and SHA-2 standards, the initial value of the DM is a constant value, that is loaded at the beginning of the computation. The DM value can be loaded into the SHA state registers by using set/reset signals. However, if the SHA-2 algorithm is to be used in a wider set of applications and in the computation of fragmented messages, the initial DM is no longer a constant value. In these cases, the initial value is given by an Initialization Vector (IV) that has to be loaded into the A to E variables, in the SHA-1 case, and in the A to H variables, in the SHA-2 case.

4.3   SHA-1 and SHA-2 Implementations

This section presents the several techniques that have been proposed in the related state of the art allowing to achieve SHA-1/2 implementations with significant improvements in terms of area and throughput. Given the similarities between data paths of the SHA-1 and SHA-2 algorithms, the following details the proposed techniques with a particular focus on the SHA-1 algorithm. This allows us to simplify the description of the proposed improvements, given the simpler data path of the SHA-1 algorithm. Nevertheless, given their similitude, the presented techniques can also be exploited for the SHA-2 algorithm, as detailed in the presented state of the art. Regarding the SHA256 and SHA512 implementations, the main difference lays in the length of the data path. Sklavos [401] explores this to design a computational structure capable of computing both the SHA256 and the SHA512 hash functions with a negligible area and delay increase.

In its canonical form, SHA implementations are bound by the data dependency between rounds. The performance of the SHA implementation can be improved by registering the output of the data expansion block so it does not impact in the performance of the SHA compression structure. Carry Save

Adders (CSA) can also be used to perform the additions of intermediate values, only using one full adder, saving area resources, and reducing the computation delay [144, 400].

Independently of these optimizations to further improve the design of the SHA algorithm, data dependencies must be taken into account.

Dadda [93] improves the SHA computation by balancing the delay in the computation of the data expansion. This improvement starts by identifying the critical path in the structure performing the expansion of the input data blocks into 32-bit words, in particular for the SHA256 algorithm. The authors reduce the critical path by dividing the computation of the output block (W_t) into stages separated by a register. Allowing to double the operating frequency and consequently the throughput of the data expansion computation. This approach requires additional area resources for the pipeline registers and control logic and imposes an additional clock cycle of latency.

The computational structures of the SHA compressions are relatively simple. However, in order to compute the values of one round the values from the previous round are required. This data dependency imposes a sequentiality in the processing, preventing the parallel computation between rounds. Dadda [93] proposes the use of a quasi-pipeline technique in this compression stage, allowing for a fast pipelined SHA architecture, using registers to break the long critical path within the SHA core. However, given the high data dependency of the SHA algorithm, additional control logic is required in order to properly control the partial pipeline registers. Nevertheless, the proposed quasi-pipelined design achieves a shorter critical path, allowing to achieve higher data throughputs.

Chaves [70, 71] further improves the pipeline usage in the SHA implementation by analyzing the existing data dependencies, proposing the functional rescheduling of the SHA arithmetic operations. As depicted in Figures 4.2 and 4.4, the bulk of the SHA-1 and SHA-2 round computation is oriented for the computation of the A value (and E for the SHA-2 algorithm). The remaining values do not require any particular computation.

For the particular case of the SHA-1 algorithm the value of A is calculated with the addition of the previous value of A along with the other internal values. Nevertheless, since only the parcel $RotL^5(A_t)$ depends on the variable A_t, the value A_{t+1} can be pre-computed using the remaining values that do not require computation on that cycle, producing the intermediate carry and save values, as depicted in Figure 4.6. By splitting computation of the value A and rescheduling it to different computational cycles, the computation can be optimized using an additional pipeline stage. With this, the critical path is further reduced with a minimum area overhead. In this particular case the critical path is restricted to a bit-wise rotation, with as a negligible impact, a CSA, and a final addition, as depicted in gray in Figure 4.6. An identical rescheduling is performed for the SHA-2 algorithm, considering its more complex data path [70].

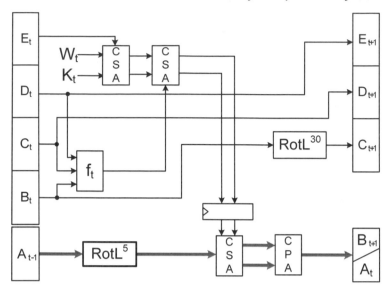

Figure 4.6
SHA-1 round rescheduling

Chaves [72] also considers hardware reuse in the addition of the final round value with the current Digest Message (DM) value. Rather than using one adder for the addition of each internal variable, A to H in the case of the SHA-2, the data dependencies are once more explored to minimize the needed hardware. The proposed DM addition also allows the loading of different IV at the cost of a few selection logic units. This hardware reuse allows to further reduce the footprint of the SHA implementations.

Going in the opposite direction, Lien [247] proposes the use of unrolling techniques to improve the maximum achievable throughput. Even though the data dependencies exist, only part of the computation is affected by this computation. As depicted in Figure 4.7 for the SHA-1 algorithm, by unrolling the computation twice, it is possible to compute in parallel two SHA rounds with a minimal increase in delay, regarding the canonical implementation, as illustrated by the gray lines in Figures 4.1 and 4.7.

The authors in [247], propose to unroll the loop up to five times, resulting in the parallel computation of 5 rounds with a similar computational delay. As expected, this loop unrolling results in an increase in area requirements, proportional to the considered loop unrolling. To further improve the delay the computation of the first logical function (f_t), depicted with a shaded box in Figure 4.7, can be rescheduled to the previous clock cycle. The authors propose the same approach to the SHA-2 algorithm. Regardless of its more complex data path, identical improvements are achieved for SHA-2.

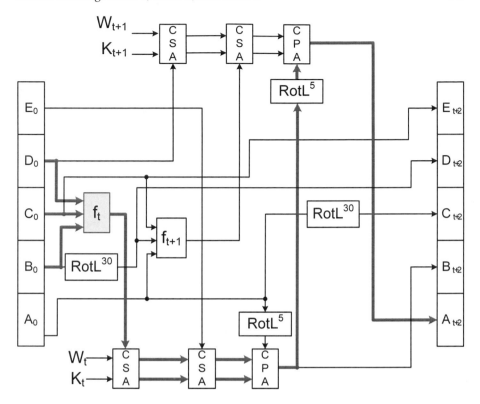

Figure 4.7
SHA-1 2× loop unrolling

Given the concurrent computation of more than one round, the hardware performing the block expansion also need to be modified in order to output several round input values (W_t) for the compression stage [281]. McEvoy [281] further improves the SHA-2 throughput by combining the use of unrolling techniques with the quasi-pipeline solution proposed by Dadda [93].

All of the above proposed improvements to the SHA-1 and SHA-2 computation play with the data dependencies in order to improve the resulting structures, in area and/or throughput [242, 390]. Nevertheless, these optimizations are limited by the existing data dependencies. Lee [242] proposes the Iterative Bound Analysis method to identify the architectural performance limit imposed by the data dependencies of these algorithms. With this approach, the authors once more used pipeline and unrolling techniques to further improve the performance metrics of the obtained implementations.

Table 4.3 presents the resulting performance metrics considering the several implementation technologies considered in the related SHA-1 and SHA-2 state of the art.

Table 4.3
SHA-1 and SHA-2 implementations

Algorithm	Technology [ASIC/FPGA]	Area [Gates/Slices]	Throughput [Mbps]	Efficiency [Throughput/Area]
SHA-1 Gremb. [144]	Virtex	730	462	0.63
SHA-1 Lien [247]-basic	Virtex-E	484	659	1.36
SHA-1 Lien [247]-unrolled	Virtex-E	1484	1160	0.78
SHA-1 Chaves [71]	Virtex-E	388	840	2.16
SHA-1 Chaves [71]	Virtex-2P	565	1420	2.51
SHA256 Sklavos [400]	Virtex	1060	326	0.31
SHA256 McEvoy [281]	Virtex-2	1373	1009	0.74
SHA256 Chaves [70]	Virtex-2	797	1184	1.49
SHA256 Chaves [70,133]	Virtex-5	433	1630	3.76
SHA256 Dadda [93]	$0.13\mu m$	n.a.	7420	n.a.
SHA256 [242]	$0.13\mu m$	22025	5975	0.27
SHA256 [369]	$0.13\mu m$	15329	2370	0.15
SHA512 Sklavos [400]	Virtex	2237	480	0.21
SHA256/512 Sklavos [401]	Virtex	2384	291/467	0.12/0.20
SHA512 Gremb. [144]	Virtex	1400	616	0.44
SHA512 Lien [247]-basic	Virtex-E	2384	717	0.30
SHA512 Lien [247]-unrolled	Virtex-E	3521	929	0.26
SHA512 Chaves [70]	Virtex-E	1680	889	0.53
SHA512 McEvoy [281]	Virtex-2	2726	1329	0.49
SHA512 Chaves [70]	Virtex-2	1666	1534	0.92
SHA512 [242]	$0.13\mu m$	43330	9096	0.21
SHA512 [369]	$0.13\mu m$	27297	2909	0.11

From the presented results it is clear that significant improvements to the SHA-1 and SHA-2 implementations can be achieved if the data dependencies are properly explored and hardware re-usage is considered. However, it is also clear that a tradeoff has to be made between less area demanding and higher throughput structures. The results suggest that adequate compromises can be achieved in order to achieve a higher throughput per used area resources.

4.4 SHA-3

One of the widely accepted hash functions of past years is SHA-1, which was introduced by the National Institute of Standards and Technology (NIST) in 1995 [311]. SHA-1 processes 512-bit message blocks, operating at 80 rounds, and outputs a 160-bit message digest as described in earlier sections of this chapter. For many years, SHA-1 appeared to be collision resistant, until several researchers have proven that it is possible, under certain conditions, in a computationally efficient way to find message collisions, thus reducing the SHA-1 security level to that of an 80-bit block cipher. Due to this issue, in 2001 NIST announced the follower of SHA-1, denoted as SHA-2. The SHA-2 hash function standard employs four hash functions with different and longer message digests at 224, 256, 384, and 512 bits. Unfortunately, this standard did not meet NIST's high expectations, so in October 2008 the institute announced an open competition for a new cryptographic hash function called SHA-3, aimed at replacing SHA-1 and SHA-2 [308].

The competition for SHA-3 had at first 64 algorithm submissions. After a thorough study, 51 out of the 64 candidates passed to round 1. NIST selected 14 out of 51 algorithms for round 2 and concluded at five algorithms for the final competition round (round 3). Those finalist algorithms were BLAKE, Grøstl, JH, Keccak, and Skein and met all the required SHA-3 criteria specified by NIST. The most noticeable such criteria were high security, diversity, analysis, and performance. Moreover, NIST had set the requirement that the function to be chosen as the SHA-3 standard will have to be an open-source hash algorithm, be available worldwide for public use, and be suitable for a wide range of software and hardware platforms. Furthermore, the SHA-3 winner algorithm had to support a message hash of 224, 256, 384, and 512 bits and a maximum length message of at least $(2^{64} - 1)$ bits. Finally, NIST specified additional characteristics to be evaluated for the new SHA-3 choice, like simplicity, flexibility, computational efficiency, memory use, and licensing requirements. On October 2012 the competition ended and NIST announced the SHA-3 winner. Out of the five excellent algorithms that reached round 3, Keccak was chosen as the new SHA-3 standard.

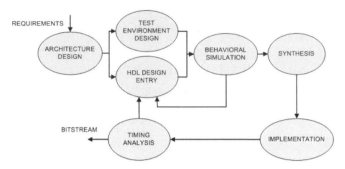

Figure 4.8
FPGA design flow

In this section, we describe and evaluate the 5 SHA-3 finalists and analyze their efficiency from hardware perspective in order to justify NIST choice for Keccak. To achieve that, we implement the 5 finalists in FPGA technology and extract performance measurements. FPGAs were chosen as the best mean implementing the various operations and specification of the five finalist cryptographic hash functions since they are reprogrammable and allow rapid prototyping. Similar approach for SHA-3 candidates evaluation was followed by others research groups where compact implementations were used for comparisons [209] or 0.13 μm standard-cell CMOS technology [150] or embedded FPGA Resources (like DSPs and BRAMs) [388]. However, this chapter's approach is entirely different since all the functions have been implemented in FPGA under the same design philosophy thus ensuring the accuracy of the compared results and providing fairness in comparisons.

4.4.1 FPGA Background

Field Programmable Gate Arrays (FPGA) are integrated circuits that can be configured by a VLSI designed after fabrication. This configuration is defined using a hardware description language (HDL), like VHDL and Verilog, or a schematic design, and it can be used in order to implement any logical function. The HDLs are mostly used in large structures, but schematic entries are preferred for easier visual image design. The design flow employed by the most FPGA designers can be seen in Figure 4.8.

During architecture design, the requirements and problems of the target system functionality are analyzed and a document is provided as output presenting the interfaces and the operations of the system's structural blocks. During environment design, the system's test environment and suitable behavioral models are specified and realized. The outcome of environmental design is used in the behavioral simulation stage where the outputs of the HDL model are compared with the provided behavioral model. During synthesis stage, the behaviorally verified HDL code is converted to a digital circuit schematic, denoted as netlist. The netlist is used in the implementation stage to designate the FPGA components needed for the realized system. The output of this stage is a file called bitstream generator that is used for timing analysis operations in order to verify if a design follows the timing rules, set by the developer [280].

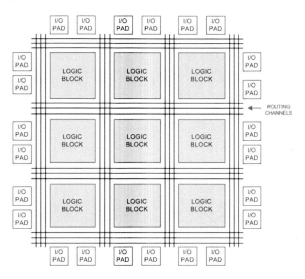

Figure 4.9
FPGA structure

The architecture of an FPGA consists of an array of logic blocks along with appropriate routing channels and I/O pads, as described in Figure 4.9 [228]. Each logic block contains lookup tables (LUT), flip-flops and special purpose hardware structures like full adders, etc. A typical logic block is shown in Figure 4.10. Up-to-date FPGA devices may also include DSP elements and RAM blocks (depending on the FPGA manufacturer and family). Due to the logic block array, the FPGA can compute any complex combinational function or simple logic gate configuration.

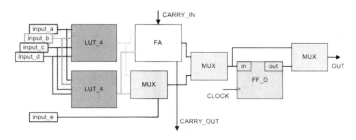

Figure 4.10
The FPGA logical cell

FPGAs apart from digital operations can also support analog functions. The most common such function is an FPGA characteristic capable of making pin signals propagate faster and stronger on high speed channels. Furthermore, the connection to differential signaling channels and the mixed signal FPGAs that let blocks operate as a system-on-chip, make FPGAs advantageous for many applications.

Originally, FPGAs were applied in telecommunications and networking letting microprocessors handle all other computation work. However, the computation application demands for low area, small delay, high speed, and parallelism, over the years could not be matched by microprocessors. FPGAs, having the ability to fix into the silicon higher level functionality, reduce the required area, and provide high speed to every function, are ideal candidates to address the above issue. So, nowadays, digital signal processing, computer vision and medical imaging, aerospace, cryptography and defense systems, computer hardware, and ASIC prototyping are gradually been migrated on FPGA technology replacing other implementation platforms like ASICs, general or specialized microprocessors.

4.4.2 SHA-3 Finalist Hash Function Algorithms

4.4.2.1 Blake

BLAKE hash function follows the iteration mode of HAIFA (Framework for Iterative Hash Functions). BLAKE internal structure is based on the LAKE hash function (which is a version of stream cipher ChaCha) and its associated compression algorithm. The main benefits of the BLAKE is its strong security, its high performance, and its parallelism. BLAKE hash family consists of four different function members: BLAKE-28, BLAKE-32, BLAKE-48, BLAKE-64. The characteristics of each family member are basically different in the word length they use for the data transformation, in the applied message block, the produced message digest, as well as in the used salt. The BLAKE members specifications are described in Table 4.4.

Regardless of the differences that each BLAKE family member may have, the architecture of each member is practically the same. The main hardware components and functions that are used in BLAKE architectures are described in Figure 4.11. The BLAKE algorithm consists of two parts: the com-

Table 4.4
BLAKE Hash family

Hash Algorithm	Word	Message	Block	Digest	Salt
BLAKE-28	32bits	$< 2^{64}$	512	224	128
BLAKE-32	32bits	$< 2^{64}$	512	256	128
BLAKE-48	64bits	$< 2^{128}$	1024	384	256
BLAKE-64	64bits	$< 2^{128}$	1024	512	256

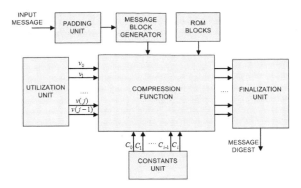

Figure 4.11
Generic BLAKE architecture

pression function and the iteration mode. The compression function needs four inputs: the chain value h, the block message m, the salt value s, and the counter value t. The BLAKE hashing process follows three stages: the initialization process, the round function, and the finalization process. The first process uses h,s,t to create a 4×4 matrix that initializes a 16-word value v to come up with v_i $(i = 0, \ldots, 15)$ different initial states.

These states are inputted to the round function, where the compression function G_i is computed in r partially parallel rounds for $i = 0, \ldots 7$. The output of this stage is a new state value v that is used for generating the chain value $h' = h'_i$ for $i = 0, \ldots 7$ during finalization. In the finalization stage, values h, s, v are XORed in order to produce a new chain value $(h'_i = h_i \oplus s_{i \bmod 4} \oplus v_i \oplus v_{i+8})$. BLAKE-32 needs $r = 14$ rounds of its compression function to come up with a result while BLAKE-64 needs $r = 16$ rounds [23], [398].

The message in BLAKE is extended through padding so that its length is congruent to 447 modulo 512 (BLAKE-28/BLAKE-32) or 895 modulo 1024 (BLAKE-48/BLAKE-64). The padded bits that will be added are from 1 to 512. The first bit in this sequence is 1, followed by zeros. Finally, for BLAKE-32 and 64, at the padding's end a bit 1 is added followed by a 64-bit unsigned big-endian representation of the message's bit length l. Though the BLAKE iteration hashing the padded message is divided into 16-word blocks, if the last block contains none of the bits present in the first block, the counter is set to zero and the final message digest is outputted.

4.4.2.2 Grøstl

The quality of analysis, the strong security features, and the well defined design principles are some of the most significant advantages of the Grøstl hash function algorithm. The hash function of this algorithm is based on round iterations of a compression function f with inputs original message blocks and the previous round's f output (Figure 4.12). The initial message is divided in m_i blocks, each of l bits [136]. Value l for Grøstl-224 and -256 is equal to 512, using 10 rounds while for Grøstl-384 and -512, is 1024 with 14 rounds.

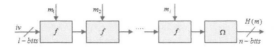

Figure 4.12
The Grøstl cryptographic hash function

The compression function f, as depicted in Figure 4.13, is computed following the equation $f(h, m) = P(h \oplus m) \oplus Q(m) \oplus h$ where Q and P are two permutation operations.

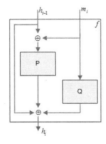

Figure 4.13
The Grøstl compression function

Permutations P and Q are based on the Rijndael block cipher algorithm (AES) and their design consists of four round transformations: AddRoundConstant, SubBytes, ShiftBytes, and MixBytes [136], [28]. AddRoundConstant XORs a round dependent constant $C[i]$ (where i is the Grøstl round) with every byte of each l bits input state A. The output of this function is $A \longleftarrow A \oplus C[i]$. SubBytes transformation replaces all bytes in the state matrix with other values taken from the Grøstl S-boxes (similarly to the AES algorithm's Sboxes). ShiftBytes makes a cyclic shift of all bytes in a row of the matrix A according to a vector $\sigma = [\sigma_0, \sigma_1, \sigma_2, ...\sigma_7]$ where σ_i represents the number of bits to be shifted in row i. Shiftbytes in Q permutation uses the vector: $\sigma = [1, 3, 5, 7, 0, 2, 4, 6]$ and in P the vector $\sigma = [0, 1, 2, 3, 4, 5, 6, 7]$. In functions where larger permutations are required, the ShiftBytesWide transformation is used in Q_{1024} with vector $\sigma = [1, 3, 5, 11, 0, 2, 4, 6]$ and P_{1024} with $\sigma = [0, 1, 2, 3, 4, 5, 6, 11]$. Finally, MixBytes transforms the bytes of each column in the state matrix A to elements of the finite field F_{256}. This means that each column of a matrix A is multiplied with an 8×8 matrix B in F_{256}. Each row of matrix B is left-rotated related to each previous row of B. The output message is equal to $A \longleftarrow B \times A$.

4.4.2.3 JH

The JH hash function family has four members, JH-224, JH-256, JH-384, JH-512. The main characteristics of JH are high parallelism in computing, use of the same hardware components for all four JH family members, and easy implementation. JH uses a generalized AES design methodology and in combination with the JH compression function provides strong security.

JH operates using a hash block (H) of 1024 bits and message blocks (M_i) of 512 bits as presented in Figure 4.14 [456]. The final output bits are compressed using the following function $H_i = F_d(H_{i-1}, M_i)$.

The JH compression function F_d consists of several steps as can be seen in Figure 4.14. Initially, the left half of the hash block H_{i-1} is XORed with the message block M_i. The generated message goes into a bijective function E_d, which is a block cipher with constant key. This function consists of a set of operations that are iterated for 35 rounds. The two 4-bit Sboxes contained in E_d are updated in every round based on a round constant vector, consisting of 256 bits. This constant vector is computed in mostly parallel and its update is based on the SBox chosen to be used in each round. The final operation of E_d is a linear transformation and permutation, where, at least, one 4-bit Sbox is used. Finally, the left part of the value generated from the bijective function E_d, becomes the hash value of the initial message's leftmost part.

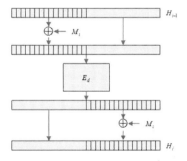

Figure 4.14
The JH compression function

Similarly, the right half of H_{i-1} is also inputted to the E_d function without XORing with M_i. This part passes through the same compression procedure like the left part and then is XORed with M_i to become the hash value of the initial message's rightmost part.

4.4.2.4 Keccak

Keccak is the algorithm chosen by NIST as the new SHA-3 standard. This algorithm is based on the sponge construction described in Figure 4.15, taking advantage of its double resistance to attacks and collisions [39]. Moreover, this construction allows multiple lengths for input and output messages and

Table 4.5
Keccak bitrate r values

Hash bits	Bitrate r
224	1152 bits
256	1088 bits
384	832 bits
512	576 bits

leads to the fastest implementations compared to all the other four candidates. Keccak is a family of hash functions, each member of which is characterized by two values, the bitrate r and the capacity c (Keccak[r,c]). The bitrate is dependent on the output hash size, according to Table 4.5, and the factor c, the capacity of the hash function, is equal to $c = 1600 - r$ state bits [40]. The input and output size of the messages at every round of Keccak is a 5×5 matrix, with entries of 64-bit words, (a total of 1600 bits). A complete permutation of the Keccak f function needs 24 rounds, each of those rounds consisting of five steps $(\theta, \rho, \pi, \chi, \iota)$.

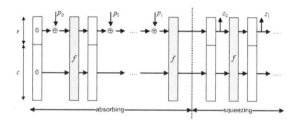

Figure 4.15
The Keccak sponge construction

The basic functionality of Keccak can be summarized in the following steps:

- initialization of the message state to 0 at every round,

- padding of the input message and dividing it into r bits,

- XORing of these bits with the initial bits of the state and performing an f permutation operation called absorbing,

- apply the block permutation squeezing, at the same rate as in the previous steps, and output the required bits from the hashed outputs z_i.

4.4.2.5 Skein

Skein is characterized by speed, simplicity, security, parallelism, and flexibility. The hash algorithm is based on three basic components: the Threefish block cipher, the Unique Block Iteration (UBI) chaining, and an argument system, which contains a configuration block and optional arguments [124], [426]. The general idea of this algorithm is similar to Keccak sponge function, providing Skein with the ability to generate arbitrary hash values from fixed size initial messages of 256, 512, or 1024 bits. The difference from other approaches is relevant to the state of capacity and security, because Skein tweaks the bits in a way that is unique for its block. This operation is possible through the use of Threefish tweakable block cipher.

Threefish cipher uses XOR, addition, and rotation in order to come up with a ciphertext result. It uses an N-bit encryption key, which is a power of 2 equal or greater to 256 and a 128-bit tweak in order to encrypt an N-bit plaintext block. Furthermore, Treefish needs 72 rounds for Threefish-256, -512 and 80 rounds for Threefish-1024 to come up with a result. Each round is characterized by a specific number of nonlinear mixing functions, denoted as MIX [124], which are 2 for Threefish-256, 4 for -512, and 8 for -1024. MIX uses as input two 64-bit words. After every 4 rounds, an N-bit subkey is added. The Skein one round operations are presented in Figure 4.16.

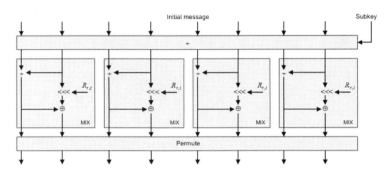

Figure 4.16
One round of Skein

The UBI chaining mode uses the Threefish algorithm in order to build the Skein compression function in order to map an arbitrary input size message to a fixed size output. Skein uses the following arguments:

- a key, to choose if Skein will use the MAC or the KDF function,

- a configuration block,

- the personalization string, which is used to create different functions for different uses,

- a hashed public key,

- the key derivation identifier,

- the nonce value, used in stream cipher mode,

- the message that is going to be hashed, and

- the generated output.

Skein was implemented in a version suitable for HW acceleration in a system-on-chip platform in [445]. In this publication, the authors present an area-efficient, energy-efficient Skein-512 implementation which offers 58Gbps throughput at a total latency of only 20 clock cycles. The core data-path is unrolled to accommodate 8 rounds of Threefish which are pipelined. This allows parallel computation of two independent hashes, but requires two tweak generators and two key schedulers to independently supply two subkeys to keep the hardware pipeline filled with two independent messages during each cycle. This increases latency from 10 cycles to 20 cycles. The total area for the implementation is 60.4Kgate equivalents.

4.5   Reconfigurable Hardware Architecture Approaches

One of the most important criteria that NIST set in the SHA-3 competition was performance efficiency. As such, all the candidate algorithms were implemented in various platforms in order to provide accurate, detailed, and comprehensive performance measurements for the evaluation process. Hardware implementation is a vital part of this process since it provides an insight of how the candidate algorithms behave in restricted hardware resources systems [218], [399]. Especially during round 3 of the competition, where all algorithms offered very strong security, the performance criteria played a very important role in determining the SHA-3 winner.

Hardware implementation for testing the SHA-3 finalists is mainly focused on FPGA technology. FPGA design offers flexibility, high performance, reconfigurability, and is a widely used implementation tool for testing but also as a commercial application solution. The design approach to be used for implementing the five SHA-3 finalists can be derived from the algorithms' basic functionality that is based on iterations (rounds). As such, iterative design, parallelism, and pipelining techniques can be very successful at improving the hash function implementation efficiency in terms of speed and/or required hardware resources. In this section, we describe the hardware approaches used in this chapter's implementations of the five SHA-3 finalist hardware architectures [350].

4.5.1   Iterative Design

Iterative design approach is very fitting for implementations where iterative operations are needed. Thus, it is ideal for hash functions and is presented in Figure 4.17. In this design approach, the initial message is configured in two separated steps. Firstly, the message is padded following the specifications of each hash algorithm in order to be expanded to the algorithm's appropriate bit length. Secondly, the message scheduler produces a message sub-block provided to every round or step of the algorithm. The data produced in each round of the algorithm, can be saved in a memory block, such as ROM. An iteration block implements the hash function functionality for one round (including the compression function), and this block is reused as many times as the number of rounds with appropriate inputs provided from the rest of the architecture's components. The various operations of each hash function round that are completed in several steps are synchronized using the hash finite state machine.

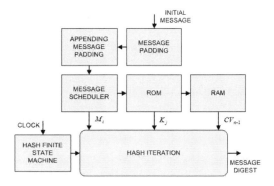

Figure 4.17
Hash function iterative design approach

Finally, the output block of hash iteration is XORed with the initial hash values, which are stored for this function until the end of any algorithm iteration. Block RAMs can be used to store parts of the output at the end of any iteration, in order to become inputs for the next hash function round. It is obvious that depending on the characteristics of the implemented hash algorithm, the iterative design is adapted, varied, or qualified accordingly.

4.5.2 Pipeline Design

The pipelining design approach can be used to process a large amount of data in a faster way, because through pipeline registers the critical path of the datapath can be considerably reduced. The main disadvantage of this technique is the high chip covered area due to the additional registers.

Using the above technique, inner-round hash function pipeline design can be achieved by dividing round operations into pipeline stages separated by registers as depicted in Figure 4.18.

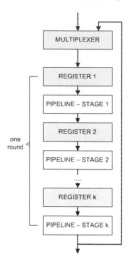

Figure 4.18
Inner-round pipelining

4.5.3   Results

Following the previous section's design approaches we have implemented all five SHA-3 finalist hash function algorithms using the VHDL hardware description language and the XILINX FPGA Virtex 6 device in an effort to identify the performance characteristics that led NIST into choosing Keccak as the new SHA-3. All implementations follow similar design principles in order to provide comparison accuracy and fairness.

We use throughput, area, and the throughput-per-area metrics in order to determine the performance of the implemented hash function algorithms. For the area metric the number of FPGA slices was used rather than dedicated FPGA resources such as block RAMs or DSPs [334] for the sake of fairness in comparisons. We excluded the use of block RAMs since they can be too big for the hash function's actual needs. The obtained results are for 256-bit message digest and are presented in Table 4.6 for the iterative design approach and in Table 4.7 for the pipeline design approach.

Table 4.6
Iterative design implementation results

Implementation	Throughput (Mbps)	Area (Slices)	Throughout/Area
BLAKE	2460	1790	1.37
Grøstl	10310	2800	3.68
JH	6010	922	6.51
Skein	1506	1200	1.25
Keccak	**11991**	**1210**	**9.90**

Table 4.7
Pipeline design implementation results

Implementation	Throughput (Mbps)	Area (Slices)	Throughout/Area
BLAKE	5319	2048	2.60
Grøstl	19764	3248	6.08
JH	7410	1503	4.93
Skein	2258	1406	1.60
Keccak	**17429**	**1507**	**11.56**

As can be observed from Tables 4.6 and 4.7 the SHA-3 finalist hash function implementations achieved remarkable time performance and very good area results. However, the performance efficiency, more accurately measured by the *throughput/area* ratio, of the SHA-3 winner (Keccak) is considerably better than the ration of the other finalist implementations.

To further verify this remark, comparisons where made of the proposed implementations with previous works in terms of the synthesis and time performance, provided in Table 4.8. Note that iterative architectures are symbolized as $NAME_{1R}$ while the pipeline architectures are symbolized as $NAME_P$. In [209] compact implementations using 64-bit bus internal datapath were proposed for the five finalists using a common methodology, which allows fair performance comparisons. The implementations of [209] do not employ any dedicated FPGA resources such as block RAMS or DSPs. Both of the proposed design techniques (iterative and inner-pipeline designs) are much better in terms of time performance compared to the implementations in [209] although they are worse in terms of area resources. However, the *throughput/area* performance ratio of the proposed implementations is far better than the designs in [209].

For the five implementations in [150] a 0.13 μm IBM process using standard-cell CMOS technology was used. One round of the hash function algorithms was selected as design methodology. We provide the results of [150] in Table 4.8 for the sake of completeness since comparisons between ASIC and FPGA designs are not compatible (lack of fairness).

Finally, in [388] implementations, dedicated FPGA resources such as block RAMs and DSPs were employed for a one round design approach similar to [150]. The proposed implementations outperform the identical one presented in [388].

Table 4.8 results of previous works indicate that the best time performance is achieved by Grøstl hash function while the second better by Keccak hash function. In terms of area resources BLAKE hash function implementation is the most compact implementation while Keccak is the biggest one. However, the implementations proposed in this book chapter based on the iterative and pipelining approach, providing in general better results than the other similar works, highlight the considerable *throughput/area* perfor-

Table 4.8
SHA-3 finalists comparison results

Implementations	Throughput (Mbps)	Area (Slices)	Throughput/Area
BLAKE [209]	132	117	0.75
BLAKE [150]	2.13 Gbps	34.15 kGEs	62.47
BLAKE [388]	1534	662	2.31
$BLAKE_{1R}$ Proposed	2460	1790	1.37
$BLAKE_P$ Proposed	5319	2048	2.60
Grøstl [209]	960	260	3.27
Grøstl [150]	9.31 Gbps	124.34 kGEs	85.58
Grøstl [388]	8057	1627	4.95
$Grøstl_{1R}$ Proposed	10310	2800	3.68
$Grøstl_P$ Proposed	19764	3248	6.08
JH [209]	222	240	0.73
JH [150]	3.05 Gbps	49.29 kGEs	34.08
JH [388]	3120	1066	4.95
JH_{1R} Proposed	6010	922	6.51
JH_P Proposed	7410	1503	4.93
Keccak [209]	145	144	0.77
Keccak [150]	10.67 Gbps	42.49 kGEs	14.7
Keccak [388]	11252	1338	8.41
$Keccak_{1R}$ Proposed	11991	1210	9.90
$Keccak_P$ Proposed	17429	1507	11.56
Skein [209]	223	240	0.77
Skein [150]	3.05 Gbps	66.36 kGEs	39.71
Skein [388]	2359	1264	1.86
$Skein_{1R}$ Proposed	1506	1200	1.25
$Skein_P$ Proposed	2258	1406	1.60

mance ratio of Keccak in comparison with the other implementations. This remark provides a very good justification of NIST's choice for Keccak as the new SHA-3 standard since from overall performance perspective this algorithm has the best potential.

4.6 Conclusions

NIST SHA-3 hash competition ended in October 2012 selecting Keccak among 5 candidates as the successor of SHA-2. In this chapter we described SHA-1, SHA-2, and SHA-3 finalists, suggested design approaches applicable to all those finalists and provided implementations based on these approaches in order to evaluate the finalists hash functions in an effort to justify NIST's choice. All finalist algorithms are highly secure and have their advantages. BLAKE is very strong in security, performance and parallelism. Grøstl has significant quality of analysis, security features and well defined design values. JH favors parallel computing, easy implementation. Skein can provide outcomes fast, in a simple manner and favors parallelism and flexibility. However, Keccak having the privilege of the sponge construction, that takes advantage of its double resistance to attacks and collisions, is the most strong among the other finalist due to its elegant design, flexibility and excellent efficiency in hardware implementations. This efficiency was highlighted in our work by the provided FPGA hardware implementation results that clearly indicated Keccak's advantage in throughput per area performance metrics, suggesting a well balanced, from implementation perspective, algorithm worthy of being SHA-3. In 2014, NIST published the draft FIPS 202 "SHA-3 Standard: Permutation-Based Hash and Extendable-Output Functions" and the standardization process is in progress as of April 2015.

5

Public Key Cryptography

Nathaniel Pinckney
University of Michigan, Ann Arbor

David Money Harris
Harvey Mudd College

Nan Jiang
Harvey Mudd College

Kyle Kelley
Harvey Mudd College

Samuel Antao
University of Lisbon

Leonel Sousa
University of Lisbon

CONTENTS

5.1　Introduction

Almost all forms of data communication in the modern day require security transmission and storage. This can be due to requirements to secure data residing on the application platform or due to the support of communication protocols. In particular, private key and public key cryptography are used to guarantee secure communication of information. Private key cryptography uses a secret key distributed to sender and receiver prior to the communication to cipher and decipher data transmitted over an unsecured channel. The most widely used private key ciphering algorithm is the Advanced Encryption Standard (AES) [307]. The AES algorithm is discussed in detail in Chapter 3. In contrast, public key cryptography is based on no prior knowledge of a secret key by the sender and receiver before data transmission which requires complex mathematical functions that can be efficiently computed but are difficult to invert. Public key cryptography is based on a distributed method to create a secure communication channel using a mathematically computed secret. Public key cryptography algorithms support digital signatures that can be used to validate data from a given signature generated by the sender. Two popular public key cryptography algorithms in wide use today are the Rivest-Shamir-Adleman (RSA) algorithm and the Elliptic Curve Cryptography (ECC) algorithm. Efficient implementations of public key cryptography algorithms are continuously being explored and implemented largely as high-performance hardware accelerators to support the large computational demands. These accelerators typically take advantage of efficient parallelization and programmability to improve flexibility and enable configuration.

This chapter addresses the efficient implementations of public key cryptography algorithms by first presenting efficient implementation of modular multiplication which is the key work-horse kernel. This is then followed by a presentation of a methodology to implement parallel versions of the algorithms that are also configurable. The mathematical foundation of public key cryptography algorithms is presented in Chapter 2.

5.2 Very Large Scalable Montgomery Multiplier Implementations

5.2.1 Introduction to Montgomery Multiplication

The efficiency of a public key cryptosystem is highly dependent on the performance of modular multiplication. This operation can be implemented either as an interleaved method, where multiplication and reduction steps are alternately performed, or as a full multiplication followed by the reduction of the product. Whereas the first approach fits hardware implementations better, due to its high regularity and memory efficiency, the latter enables the use of sub-quadratic multiplication approaches, such as the Karatsuba [196], Toom-Cook [427], or Schonhage-Strassen [375] algorithms.

Many public key cryptosystems are based on modular exponentiation. For example, the RSA algorithm involves a public key e, a private key d, and an odd modulus M that are n-bit integers satisfying the property $x^{de} \bmod M = x \; \forall \; x$. Alice publishes her public key and keeps her private key secret. Bob securely sends a message x by computing and transmitting $y = x^e \bmod M$. Alice then computes $y^d \bmod M = x^{de} \bmod M = x$, recovering the message. An eavesdropper, Eve, who intercepts y is unable to compute x because d is secret. Guessing d from e is as hard as factoring M. Because there is no known method to efficiently factor large numbers, guessing d and recovering x is intractable as long as n is large enough (typically 256-4096 bits).

Modular exponentiation is performed with repeated modular multiplication. Conventional modular multiplication consists of multiplication followed by a division to determine the remainder; the division is slow and costly. Montgomery's algorithm [293] transforms the problem to eliminate the division.

The Montgomery multiplication of two n-bit words X and Y produces an n-bit product Z using Algorithm 5.1. The remainder of this introduction provides a high-level overview of the hardware implementation choices.

Algorithm 5.1 Montgomery multiplication algorithm.

Multiply: $Z = X \times Y$
Reduce: $Q = Z \times M \bmod R$
$\qquad\quad Z = [Z + Q \times M'] / R$
Normalize:
if $Z \geq M$ **then**
$\quad Z = Z - M$
end if

Early hardware implementations of Montgomery's algorithm were customized to specific word lengths, n [36,115,319,413,460]. Tenca and Koç [423] introduced a *scalable* Montgomery multiplier using multiple passes through one or more w-bit processing elements (PE) to perform the multiplication for reconfigurable word lengths. Their radix-2 PE considers w bits of the multiplicand and one bit of the multiplier at a time. Scalable designs generalize to higher radix. A radix-2^v PE considers w bits of the multiplicand and v bits of the multiplier in each step. The third dimension of parallelism is to cascade p PEs to form a systolic array.

The primary metric for cost and throughput of Montgomery multiplication is $m = pwv$, the number of partial product bits processed in each cycle by the systolic array. Like an ordinary multiplier, an n-bit Montgomery multiplier has to perform n^2 operations to combine all n bits of the multiplicand with all n bits of the multiplier. The cycle time T_c grows with w and especially v as the hardware in each PE becomes more complex. Using a systolic array of PEs, the multiplication time is $O(T_c n^2 / m)$. The hardware area is $O(m)$. The area-time product is thus $O(T_c n^2)$.

For a given amount of hardware, m, the choices of w, v, and p have several impacts. Without loss of generality, define $v \leq w$; otherwise, the multiplier and multiplicand could be swapped to satisfy this definition. As was just mentioned, w and v influence the cycle time. Furthermore, each PE must perform two multiplications ($X \times Y$ and $Z \times M'$) and a shift. The arrangement of these operations within the PE influences the cycle time and the latency l for information to move between successive PEs in the systolic array. For an n-bit multiplicand, at most $p_{max} = n/wl$ processing elements can be gainfully employed simultaneously.

This section investigates the hardware implementation choices for scalable Montgomery multipliers. Section 5.2.2 explains the Montgomery multiplication algorithm and how it simplifies finite field arithmetic for cryptographic applications. Section 5.2.3 explains the scalable radix-2 algorithm, the hardware in basic PEs, and the arrangement of PEs into a systolic array. Section 5.2.4 generalizes the algorithm to radix-2^v high radix designs. To reduce the latency, cycle time, and hardware, the PE can employ several architectural approaches including parallelism, quotient pipelining, and left shifting, described in Section 5.2.5. Section 5.2.6 analyzes the cycle count for an n-bit Montgomery multiplication. Section 5.2.7 illustrates a variety of logic-level implementations. These implementations can generally be classified as low-radix, in which the running total Z is produced in carry-save redundant format using carry-save adders, and high-radix, in which Z is produced in standard binary format using $w \times v$-bit multipliers. Section 5.2.8 compares post-layout area, timing, and power of the various design results in an industrial 45 nm process. Finally, Section 5.2.9 provides a summary of tradeoffs when designing Montgomery multipliers.

5.2.2 Montgomery Multiplication

Conventional modular multiplication of n-bit integers $Z = X \times Y \bmod M$ involves an ordinary multiplication of X and Y to produce a $2n$-bit result, followed by a costly division by M to produce an n-bit remainder Z. Montgomery's algorithm transforms the integers into M-residues, operates on these residues, then transforms the M-residues back into integers. Montgomery multiplication of residues involves three n-bit multiplications and a shift, which is less costly than traditional modular multiplication. For modular exponentiation, which requires repeated modular multiplication, the simplified Montgomery multiplication step justifies the cost of the transformations at the beginning and end.

Specifically, let the modulus M be an odd n-bit number such that $2^{n-1} < M < 2^n$. $R = 2^n$ and R^{-1} is the multiplicative inverse satisfying $RR^{-1} \equiv 1 \bmod M$. The M-residue of an integer $A < M$ is $\overline{A} \equiv AR \bmod M$. There is a one-to-one correspondence between integers and M-residues for $0 < A < M\text{-}1$. The Montgomery product of two M-residues is

$$\overline{Z} = \text{MM}(\overline{X}, \overline{Y}) = \overline{X}\overline{Y}R^{-1} \bmod M$$

This is verified by observing $\overline{X}\,\overline{Y}R^{-1} \bmod M = (XR)(YR)R^{-1} \bmod M = XYR \bmod M = ZR \bmod M$.

Montgomery multiplication uses a precomputed term M' satisfying RR^{-1} – $M'M = 1$ and the steps shown in Algorithm 5.1. Observe that Montgomery multiplication requires three integer multiplications and a right shift by n. To avoid discarding information, the reduction step adds a multiple of M to Z so that the sum is evenly divisible by R (i.e., the lower n bits of Z are zero). Adding multiples of M does not change the result (modulo M).[1] The result of the reduction step is in the range $0 \leq Z < 2M$. The normalization step subtracts M if necessary to bring the result into the range $0 \leq Z < M$.

Montgomery multiplication can also be used to transform an integer to or from an M-residue:

- $\text{MM}(A, R^2) = AR^2R^{-1} \bmod M = \overline{A}$

- $\text{MM}(\overline{A}, 1) = (AR)R^{-1} \bmod M = A$

It has been observed [49,152,319,446] that if $R > 4M$ and $X, Y < 2M$, then $Z < 2M$. Hence, the normalization step is unnecessary between successive Montgomery multiplications in an exponentiation operation if the operand size is $n_1 = n + 1$ bits. Moreover, [446] also observed that if $\overline{A} < 2M$, the final transformation $A = \text{MM}(\overline{A}, 1)$ will produce a result $A < M$. Hence, normalization is never needed during modular exponentiation. Hence, through the rest of this work, we will eliminate the normalization step by increasing R. Specifically, let $R = 2^{n_2} > 4M$, where, for example, $n_2 = n + 2$. The tradeoff is that a slightly larger multiplication is needed.

5.2.3   Scalable Radix-2 Montgomery Multiplication

The algorithm in Algorithm 5.1 is impractical for hardware implementation with large n (e.g., 1024 bit) because it requires an $n_2 \times n_1$-bit multiplier. A radix-2 Montgomery multiplication algorithm scans the multiplier X one bit at a time, as shown in Algorithm 5.2. X_i is the i-th bit of the multiplier. The partial product Z is initialized to 0. On each step, the multiplicand Y is added to the partial product if the corresponding bit of the multiplier X_i is 1. If the result is odd, the modulus M is added so that the result becomes evenly divisible by 2. Thus, Q_i is simply the least significant bit of Z in the radix-2 algorithm [2]. After n_2 steps, the entire multiplier has been handled and the result has been shifted right to divide by $R = 2^{n_2}$. Note that the normalization step is intentionally omitted.

---

[1]To prove that $Z + Q \times M$ is divisible by R, note that $[Z + Q \times M] \bmod R = [Z + (Z \times M' \bmod R) \times M] \bmod R = [Z + Z \times M' \times M] \bmod R = [Z + Z(RR^{-1} - 1)] \bmod R = ZRR^{-1} \bmod R = 0 \bmod R$.

[2]Note that only one bit of Q is needed because the partial product is only shifted by one bit in each step. Because $M' \bmod 2 = 1$, computing Q becomes trivial for radix 2. The radix-2 algorithm thus involves only two multiplications and a shift at each step. In general, when shifting right by v bits, v bits of Q are needed [108].

Algorithm 5.2 Radix-2 Montgomery multiplication algorithm.

$Z = 0$
for $i = 0$ to n_2-1 **do**
    $Z = Z + X_i \times Y$
    $Q_i = Z \bmod 2$
    $Z = Z + Q_i \times M$
    $Z = Z / 2$
end for

The algorithm in Algorithm 5.2 is still limited because it uses a fixed and slow n_1-bit adder. A better hardware implementation would support variable values of n. The Tenca-Koç *scalable* radix-2 Montgomery multiplication algorithm [423] shown in Algorithm 5.3 solves this problem by breaking Y, M, and Z into e words of w-bit and scanning them one word at a time in the inner loop, where $e = \lceil \frac{n_1}{w} \rceil$. The word size w is commonly in the range of 8 to 64 bits so that addition can be performed in a reasonable cycle time. In this algorithm, Y^j refers to the j-th word of Y. C_a and C_b are running carries between words of the partial product. The right shift is performed by concatenating the least significant bit of the current Z^j word onto the $w - 1$ most significant bits of the previous Z^{j-1} word.

Algorithm 5.3 Tenca-Koç scalable radix-2 Montgomery multiplication algorithm.

$Z = 0$
for $i = 0$ to n_2-1 **do**
    $(C_a, Z^0) = Z^0 + X_i \times Y^0$
    $Q_i = Z \bmod 2$
    $(C_b, Z^0) = Z^0 + Q_i \times M^0$
    for $j = 1$ to e **do**
        $(C_a, Z^j) = Z^j + C_a + X_i \times Y^j$
        $(C_b, Z^j) = Z^j + C_b + Q_i \times M^j$
        $Z^{j-1} = (Z^j{}_0, Z^{j-1}{}_{w-1:1})$
    end for
end for

5.2.3.1 Hardware organization

As shown in Figure 5.1, scalable Montgomery multipliers are organized as a systolic array of p processing elements (PEs), each of which performs the operations in the inner loop. Using multiple PEs can be viewed as unrolling the outer loop. In the radix-2 version, each PE is used for $e + 1$ successive clock cycles to process all of the e words of Y and M and to produce one more word that can be right-shifted to produce Z^{e-1}. Each PE handles one of the bits of X, so p bits of the multiplier are considered in parallel. The PE latency

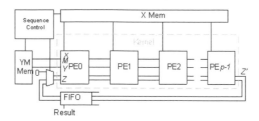

Figure 5.1
Scalable Montgomery multiplier organized as a systolic array

for a right-shifting radix-2 design is $l = 2$ cycles as Z^j must be computed before Z^{j-1} can be shifted. Two memories deliver the bits of X and the words of Y and M to the processing elements. The results of the final PE are fed back through a FIFO to be used in the first PE when it has finished all of the words. The FIFO has a minimum latency of b cycles to bypass the results back even when it is not storing data; this latency typically consists of wire delay and the input multiplexer, so b is usually 1.

The organization of a processing element in the radix-2 Tenca-Koç design is illustrated in Figure 5.2. The processing element keeps Z in carry-save redundant form so that carry-save adders (CSAs) can be used in place of carry-propagate adders (CPAs). Additionally, AND gates are used to compute $X_i \times Y^j$ and $Q_i \times M^j$. The multiplexer captures the least significant bit of Z and holds it as Q_i for the remaining words. The PE also shifts the partial Z right by one bit each cycle, so that the entire result is divided by $R = 2^{n_2}$ after all loop iterations have completed.

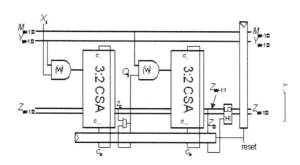

Figure 5.2
Radix-2 Tenca-Koç processing element

5.2.4 Scalable Radix-2^v Montgomery Multiplication

The scalable radix-2 algorithm can be extended to higher radices by processing more bits of X within each PE per cycle, thus reducing cycle count at the cost of increased area and PE complexity. Specifically, a radix-2^v PE handles w-bit words of Y, M, and Z, and v-bit words of X, where v is typically up to 32 bits. Thus, the outer loop requires $f = \lceil \frac{n_2}{v} \rceil$ iterations rather than n_2 iterations used in a radix-2 design. For large v the PE generates partial products using $w \times v$-bit multipliers instead of the w-bit AND gates, while radix-4 and radix-8 architectures use booth encoding and w-bit multiplexers. It also requires a truncated $v \times v$-bit multiplication to calculate Q, which is now v bits rather than just 1. Radix-4 [328, 329] and radix-8 [5] architectures process $v = 2$ and $v = 4$ bits, respectively. The outer i loop iterates over the f v-bit words of X while the inner j loop iterates approximately e w-bit words of Y and M. C^A and C^B are v-bit carry words propagated between iterations of the inner loop. For shifting to work properly, the algorithm requires $v \leq w$, as was assumed earlier.

Algorithm 5.4 Scalable radix-2^v Montgomery multiplication algorithm.

$Z = 0$
for $i = 0$ to f-1 **do**
  $(C^A, Z^0) = Z^0 + X^i \times Y^0$
  $Q_i = (M^{0\prime} \times Z^0) \bmod 2^v$
  $(C^B, Z^0) = Z^0 + Q^i \times M^0$
  for $j = 1$ to e **do**
    $(C^A, Z^j) = Z^j + C^A + X^i \times Y^j$
    $(C^B, Z^j) = Z^j + C^B + Q^i \times M^j$
    $Z^{j-1} = (Z^j_0, Z^{j-1}_{w-1:1})$
  end for
end for

Figure 5.3 shows a simple scalable very high radix processing element [207] that uses dedicated $w \times v$-bit multipliers. The processing element has a latency $l = 4$ cycles, as compared to $l = 2$ cycles for radix-2 designs, because $X^i \times Y^j$ and $Q^i \times M^j$ are pipelined into different stages to reduce the critical path. Carry-propagate adders instead of carry-save adders are used to avoid added area requirements for storing Z in redundant form, since the critical path is dominated by multipliers. Radix-4 and radix-8 implementations do have four cycle latencies since they do not require large multipliers.

5.2.5 Processing Element Architectural Optimizations

A good processing element (PE) design has small area, low latency, and high throughput. Several architectural optimizations can be made to pursue these goals. These include parallelism, quotient pipelining, and left shifting.

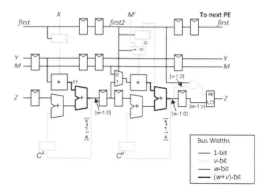

Figure 5.3
Scalable very high radix processing element

5.2.5.1 Parallelism and quotient pipelining

The least significant word of Z faces a long critical path, as is apparent in Algorithm 5.4. The path involves a multiply and accumulate with $X \times Y$, then a multiply with $M^{0'}$ to obtain Q, and finally a multiply with M^0 and accumulate to obtain Z. *Parallel* [5, 188, 206, 328, 329, 384, 447] and *quotient pipelining* [187, 319] algorithms eliminate one of the multiplications and permits the other two multiplications to occur in parallel, reducing the critical path length.

Algorithm 5.5 presents the parallel and quotient pipelining algorithms alongside the standard one. Conceptually, the optimized algorithms are based on unrolling the loop so that it can be reordered to eliminate the dependency, then introducing a precomputed $\hat{M}$ and redefining Q to eliminate a multiplication. The cost is that the operand lengths increase by v bits. Specifically, $R > 4\tilde{M}$ and $X, Y, Z < 2\tilde{M}$, where $\tilde{M} = (M' \mod 2^v) \times M$. Hence, we must choose $n_1 = n + v + 1$ and $n_2 \geq n + v + 2$. The number of inner and outer loop steps, e and f, increase with n_1 and n_2 accordingly. The algorithms use another n-bit precomputed term $\hat{M} = \frac{\tilde{M}+1}{2^v}$; this term can be reused for many consecutive multiplications based on the same modulus so precomputation overhead can be amortized over many multiplications. Note that the number of iterations of the inner loop decreases by 1 because Z is right-shifted at the beginning of the inner loop; in contrast, the standard algorithm right-shifts at the end of the loop and requires an extra iteration to produce bits to shift into the most significant word of Z.

In the parallel algorithm, Q is dependent on the previous value of Z, so one step must finish before the next begins. In the quotient pipelining algorithm, Q is pipelined such that it can be used in a multiplication before the addition in the previous iteration completes. This can eliminate an adder from the critical path at the cost of extra registers to hold intermediate results.

Algorithm 5.5 Scalable radix-2^v Montgomery multiplication algorithms.

$Z = 0$
for $i = 0$ to f-1 **do**
  $(C^A, Z^0) = Z^0 + X^i \times Y^0$
  $Q^i = (M^{0\prime} \times Z^0) \bmod 2^v$
  $(C^B, Z^0) = Z^0 + Q^i \times M^0$
  for $j = 1$ to e **do**
    $(C^A, Z^j) = Z^j + C^A + X^i \times Y^j$
    $(C^B, Z^j) = Z^j + C^B + Q^i \times M^j$
    $Z^{j-1} = (Z^j_{v-1:0}, Z^{j-1}_{w-1:v})$
  end for
end for
$Z = 0$
for $i = 0$ to f-1 **do**
  $Q^i = Z^0 \bmod 2^v$
  $C = 0$
  for $j = 0$ to e-1 **do**
    $(C, Z^j) = (Z^{j+1}_{v-1:0}, Z^j_{w-1:v}) + C + Q^i \times \hat{M}^j + X^i \times Y^j$
  end for
end for
$Z = 0$
$Q^{-1} = 0$
for $i = 0$ to f-1 **do**
  $Q^i = Z^0 \bmod 2^v$
  $C = 0$
  for $j = 0$ to e-1 **do**
    $(C, Z^j) = (Z^{j+1}_{v-1:0}, Z^j_{w-1:v}) + C + Q^{i-1} \times \hat{M}^j + X^i \times Y^j$
  end for
end for
$Z = Z \times 2^v + Q^{f-1}$

The modification requires an extra register to hold Q and an extra iteration to accommodate the delayed Q signal. The computation length increases by v more bits: $\tilde{M} = (M' \bmod 2^{2v}) \times M$, $n_1 = n + 2v + 1$, and $n_2 \geq n + 2v + 2$. However, $\hat{M}$ remains n bits: $\hat{M} = \frac{\tilde{M}+1}{2^{2v}}$.

5.2.5.2 Left shifting

A drawback of many right-shifting algorithms is extra latency waiting to compute the next Z^j word before the current Z^{j-1} word can be shifted right. The processing element latency for most right-shifting algorithms is $l=2$ cycles, where a full cycle is wasted waiting for the next word to be computed. An alternative [5, 162, 188, 328, 329] is to left shift Y and M rather than right-shifting Z. This keeps the words aligned and the next PE can immediately begin processing based on the current PE's output, saving a cycle of latency. Thus, the processing element latency for most left-shifting designs shortens to $l=1$ cycles. However, on each step, v 0's accumulate in the bottom of Y, M, and Z. Each w/v cycles, the entire word of these variables becomes 0. The PE should then stall for one cycle and discard the content-free word, which increases the average PE latency by v/w. By left-shifting PE, the effective latency of the PE drops from 2 to $1 + v/w$ on account of the stall. Left shifting also reduces hardware cost by removing a set of registers to hold Y, M, and Z in each processing element.

5.2.5.3 Unified Montgomery multiplication

All of our attention thus far has been on modular arithmetic in the Galois prime field GF(p) with p being the prime modulus M, suitable for the RSA algorithm (see Chapter 2). The ECC algorithm operates in the Galois binary extension field GF(2^n) modulo some irreducible polynomial $f(x)$ of degree n. Montgomery multiplication can be performed in just the same way except that carries are not propagated in the adders. Therefore, addition reduces to the XOR operation. Tenca and Koç [423] observe that a control signal can be provided to the CSAs to conditionally kill the carries, resulting in a *unified* Montgomery multiplier capable of handling both fields. The hardware cost is low. However, the results in this work assume ordinary adders.

5.2.6 Cycle Count Analysis

The time to perform an n-bit Montgomery multiplication and the number of processing elements p that can be usefully applied to the operation depend on the size of $Z/Y/M$ words w, size of X words v, the processing element cycle latency l, and the FIFO cycle latency b. This section illustrates the flow of data through the systolic array and computes the number of cycles required for Montgomery multiplication.

Figure 5.4 shows the timing of a scalable Montgomery multiplier systolic array with an $l = 2$ cycle latency between successive PEs. Each PE requires

e cycles to process all the words of the operand. The entire kernel of p PEs takes lp cycles before producing the first word of the result. The FIFO takes an additional b cycles to bypass the result back. There are two cases to consider. Figure 5.4(a) shows when $e \geq lp + b$. In this case, the first PE is still busy when the last PE produces its first word. The results of the last PE must be buffered in a FIFO until the first PE is ready to process them. However, all PEs are fully utilized during the multiplication, and the pipeline is saturated. Figure 5.4(b) shows where $e < lp + b$. In this case, the first PE will have finished processing all e words before the last PE produces its first word. The first PE will stall until new data is available. The FIFO is not used, but the performance ceases to improve with added hardware because the PEs will spend more time stalled. The length of each pipeline cycle is bounded by the two cases, and thus is max(e, $lp + b$) clock cycles. If we define the clock period as Tc then the total time per pipeline cycle is $T_c[\max(e, lp + b)]$.

It is most convenient if the final result is produced by the last PE. This requires that the number of outer loop iterations f be divisible by p. Let $f = kp$, where k is defined as the number of pipeline cycles, i.e., the number of times that the kernel must be used to handle all f iterations of the outer loop. Then $n_2 = kpv$ and $R = 2^{kpv} > 4M$ or $4\tilde{M}$. As Tenca [423] observes, using a larger R is fine so long as it is used consistently throughout the calculation. In particular, the initial conversion to residues and the calculations of M' and $\hat{M}$ depend on this value of R. Note that the system could be slightly faster if the result can be tapped out from an earlier PE, but that this requires the ability to select the result from any PE, which adds a result bus and tristate drivers to all of the PEs.

To complete an n-bit multiplication the Montgomery multiplier requires $k = f/p$ successive pipeline, so the total multiplication time is $T_{mul} = k[\max(e, lp + b)]T_c$. Note that several extra cycles are needed for a multiplication to completely finish, but that dependent multiplies can start after T_{mul}.

5.2.7 PE Logic Design

This section presents a variety of processing element logic designs that implement optimizations described in Section 5.2.5. It begins with a scalable radix-2 processing element architecture [162] that improves upon the scalable radix-2 Tenca-Koç, and then describes radix-4 designs [5, 328, 329, 447] that process twice as many bits per processing element. Lastly, it shows four very high radix implementations [187, 206, 207] with dedicated $w \times v$-bit multipliers.

The scalable improved radix-2 [162] shown in Figure 5.5 improves upon the Tenca-Koç PE design [423] (Figure 5.2) by left shifting instead of right shifting every cycle, reducing the processing element latency from $l = 2$ cycles to $l = 1+v/w$ cycles. Because multiplication time is $T_{mul} = k[\max(e, lp + b)]T_c$, reducing l improves performance for small n where $e < lp + b$, the case when the first PE within the systolic array must stall for the next pipeline cycle

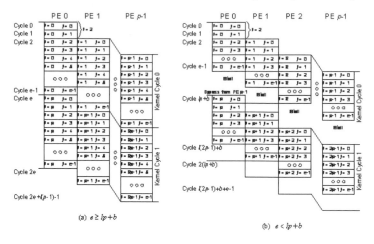

Figure 5.4
Systolic array timing

iteration. However, performance is not improved when $e \geq lp + b$, i.e., when n is large. The critical path of this radix-2 architecture was further reduced in [188] by parallelizing the multiplications, at the cost of a slight increase in cycle count.

The parallelized scalable radix-4 architectures can be implemented with pre-computed operands [328] or with booth-encoding [329, 447]. Within booth-encoded architectures, the intermediate result can be right or left shifted to reduce critical path or reduce latency, respectively.

Figure 5.6 shows a non-booth encoded radix-4 architecture [328] that extends the radix-2 architecture from [188]. Partial product computation is achieved using multiplexers and precomputed $3Y/3\hat{M}$, instead of AND gates

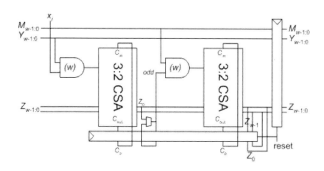

Figure 5.5
Scalable improved radix-2 processing element architecture

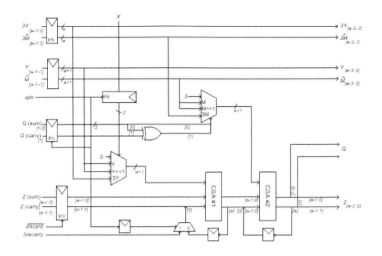

Figure 5.6
Scalable parallelized radix-4 non-booth processing element

in the case of radix-2. As with the radix-2 designs, 3:2 CSAs are used to sum the partial products, and the result is stored in redundant form to reduce clock frequency. Two CSAs also have the advantage of compressing two lower bits of the result into non-redundant form before they shifted left by $v=2$ bits after each cycle.

The radix-4 booth-encoded designs [329] are shown in Figures 5.7 and 5.8, improving upon radix-4 non-booth by booth encoding to remove dependence on precomputed $3Y/3\hat{M}$ in [328] at the cost of additional hardware complexity. Booth encoders and selectors are added to each processing element for partial product computation. The right-shifting booth-encoded design removes booth encoders from the critical path at the cost of higher PE latency l and more registers, while the left-shifting design has a slightly longer critical path but fewer registers. An improved booth-ended radix-4 is given in [447] that maintains right shifting, thus without v/w cycle count penalty, and with an $l=1$ cycle latency. A scalable parallelized booth-encoded radix-8 [5] design further extended radix-4 but was found to offer no benefits over radix-2/4 or very high radix designs.

Figure 5.9 compares the four versions of radix-2^v processing elements. All of the PEs take w bits of Y and Z and v bits of X. Figure 5.9(a) shows the standard very high radix implementation. The PE contains two multiply-accumulate (MAC) units arranged in series. The first MAC computes $Z^j = Z^j + C^A + X^i \times Y^j$ on cycle j. The second multiplier is idle on cycle 0. On cycle 1, it computes $Z^0 \times M'$ and stores the v least significant bits of the result as Q^i. On cycle 2, it computes $Z^0 = Z^0 + C^B + Q^i \times M^0$. On cycle 3, it computes Z^1

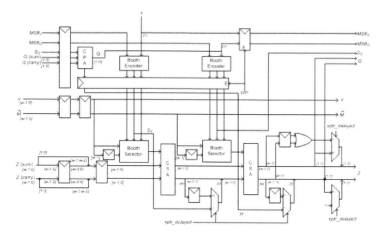

Figure 5.7
Scalable parallelized radix-4 booth-encoded right-shifting processing element

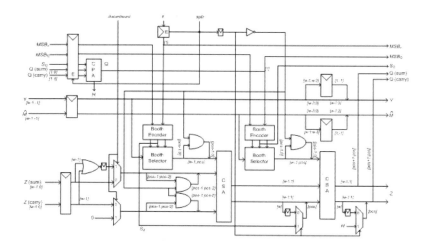

Figure 5.8
Scalable parallelized radix-4 booth-encoded left-shifting processing element

$= Z^1 + C^B + Q^i \times M^1$, then right shifts $\{Z^1, Z^0\}$ by v bits to produce a new Z^0 for the next PE. In general, on cycle $j + 3$, the second multiplier produces Z^j for the next PE. In summary, there is a four-cycle latency between subsequent PEs.

Figure 5.9(b) shows the parallel very high radix implementation that shortens the PE latency to two cycles. Note that the PE receives a precomputed $\hat{M}$ instead of M. On the initial cycle, an enabled register saves Q^i

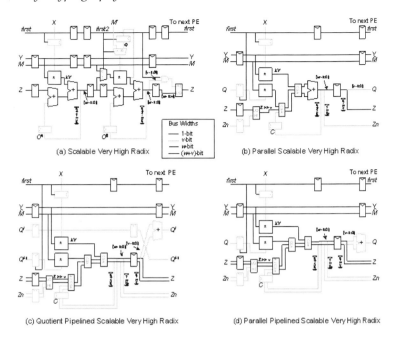

(a) Scalable Very High Radix

(b) Parallel Scalable Very High Radix

(c) Quotient Pipelined Scalable Very High Radix

(d) Parallel Pipelined Scalable Very High Radix

Figure 5.9
Scalable very high radix processing elements

$= Z^0{}_{v-1:0}$. In this design, the PE computes $X^i \times Y^j$ and $Q^i \times M^j$ concurrently. Meanwhile, the most significant bits of Z^j (called $Z_{w-1:v}$) and the least significant bits of Z^{j+1} (called $Zn_{v-1:0}$) are right-shifted by v. The products and shifted Z are added to the running carry using two CSAs and a carry-propagate adder (CPA). On cycle 0, the PE produces $(C, Z^0) = (Z^1{}_{v-1:0}, Z^0{}_{w-1:v}) + Q^i \times \hat{M}^0 + X^i \times Y^0$. On cycle 1, the PE produces $(C, Z^1) = (Z^2{}_{v-1:0}, Z^1{}_{w-1:v}) + C + Q^i \times \hat{M}^1 + X^i \times Y^1$. Now, these two words of Z can be passed to the next PE, so the latency between PEs is two cycles.

Figure 5.9(c) shows the quotient pipelined very high radix implementation that removes the CPA from the critical path. Instead, Z is kept in carry-save redundant form. The addition is only performed to compute Q, which is delayed by a stage before it is used.

Figure 5.9(d) shows the parallel pipelined very high radix implementation. It is an alternative implementation of Algorithm 5.2, but the CPA is moved off the critical path into the second pipeline stage of the PE. Consequently, Z is kept in carry-save redundant form. Q is available at the end of this cycle, so there is no need to delay it by a stage as was done in Figure 5.5(c). Hence, the implementation achieves a critical path comparable to the quotient pipelined design but at lower hardware cost. It also requires fewer cycles because n_1 and n_2 are smaller by v.

5.2.8 Hardware Comparison

The four very high radix, three radix-4, and improved scalable radix-2 Montgomery multiplier described in the previous section were coded in Verilog and simulated with Synopsys VCS to confirm functionality. Systolic arrays of processing elements were synthesized, placed, and routed in an industrial 45nm SOI CMOS process. Prior work compared performance and cost for Xilinx FPGA implementations, but had not compared the designs in a CMOS process appropriate for ASIC implementations which are more flexible with hardware. For example, FPGAs have fixed-size hardware multipliers, such as 16×16, while ASIC synthesis can implement multipliers of arbitrary size. This section presents results on the hardware requirements, cycle count, exponentiation time, and overall performance of the various designs implemented in 45nm CMOS.

Table 5.1 summarizes the post-route results of 16-bit processing elements for each architecture. The very high radix architectures each use two 16×16 multipliers per PE, while the small radix designs use 16-bit CSAs and AND or MUX gates. Table 5.1 also examines $w = v = 8$ and 32 for the parallel architecture, since that architecture had the lowest energy-delay product (EDP) for large n-bit multiplies. Each of these architectures also requires approximately $5n$ bits of SRAM to store X, Y, Z, M, and the FIFO contents, and a small sequence controller, which are not included in the results.

The improved scalable radix-2 architecture has the highest clock speed at 4.78 GHz, since it has the shortest critical path, along with the smallest hardware requirements. By comparison, the scalable very high radix architecture has the lowest clock speed at 1.64 GHz. The quotient pipeline improved on prior work by removing an adder from its critical path, and subsequently has the quickest clock frequency among very high radix designs at 1.76 GHz. As expected, the area requirements per processing element vary dramatically as the radix-2 design uses AND gates to compute partial products, while the very high radix designs use $w \times v$-bit multipliers. The parallel scalable very high radix architecture runs at roughly the same clock speed as other very high radix designs, and operates on the same number of bits per cycle, but consumes roughly half the area per processing element since it has half the CSAs and fewer registers as other very high radix designs. Doubling the w/v word sizes for parallel scalable very high radix from 8 to 16 and from 16 to 32 decreases clock speed by 23% and 41%, respectively. Similarly, in parallel very high radix the area increases by $2.7\times$ from $w=v=8$ to $w=v=16$ and by $3.1\times$ from $w=v=16$ to $w=v=32$.

Table 5.2 summarizes the time required to do 256- and 1024-bit RSA decryptions using a single exponentiation in each of the above-mentioned architectures, along with cycle count. It also compares the implementation technology, hardware cost, and clock frequency. RSA decryption for n-bit keys can be performed using one n-bit modular exponentiation computing y^d mod M. A full n-bit modular exponentiation involves up to $2n+2$ $n_1 \times n_2$-bit

Table 5.1
Processing element clock speed, area, and power in a 45nm CMOS process

Architecture	Reference	w	v	Clock Speed (GHz)	Area / PE (μm^2)	Power / PE (mW)	Logical Critical Path
Parallel Pipelined Scalable Very High Radix	This Work	16	16	1.71	88,650	127	MUL + CSA + REG
Quotient Pipelined Scalable Very High Radix	[187]	16	16	1.76	92,485	132	MUL + CSA + REG
Parallel Scalable Very High Radix	[206]	8	8	2.27	17,622	30	MUL + CSA + CPA + REG
		16	16	1.66	47,176	68	
		32	32	0.98	145,408	139	
Scalable Very High Radix	[207]	16	16	1.64	90,045	130	MUL + CPA + MUX + REG
Scalable Radix 4 Non-Booth	[328]	16	2	3.62	17,600	29	2CSA + BUF + MUX + REG
Scalable Radix 4 Booth Left-Shifting	[329]	16	2	3.06	16,458	25	ENC + INV + MUX5 + 2CSA + AND + 2MUX2 + REG
Scalable Radix 4 Booth Right-Shifting	[329]	16	2	3.05	17,390	27	INV + MUX5 + 2CSA + MUX2 + REG
Improved Scalable Radix 2	[162]	16	1	4.78	4,367	9	2AND + 2CSA + BUF + MUX + REG

Montgomery multiplications to perform the multiplications and squarings, and to convert to and from residue form. Although the average case is better, we use the worst case to prevent timing side-channel attacks. As [446] demonstrates, the result will be correctly reduced to n bits for ordinary scalable Montgomery multipliers. Thus, the total decryption time can be calculated as $T_{decrypt} = (2n + 2)T_{mul}$, where the multiplication time T_{mul} was described in Section 5.2.6. Parallel and quotient pipelined architectures produce an extra v or $2v$ bits that must be reduced; we assume this can be done in software at negligible extra time and energy cost.

When large amounts of hardware are available, very high radix designs perform better ($T_{decrypt}$) because they have a larger v. The quotient pipelined has the quickest $T_{decrypt}$ performance for large n-bit multiplies for a word size $w=16$ because it has a slightly faster clock frequency than the other very high radix architectures. This comes at the cost of highest area and power requirements. The parallel scalable very high radix architecture achieves the lowest energy-delay product (EDP) because it consumes far less area and power, while achieving similar cycle count and clock speed as other very high radix designs. The remaining very high radix designs have an EDP of $1.75\times$ to $1.9\times$ as compared to the parallel scalable architecture for large n-bit multiplies. For parallel very high radix architectures, increasing w/v subsequently reduces EDP as more bits are processed within each processing element, and the overhead in pipelining the algorithm reduces. However, area-efficiency measured using an array-delay product (ADP) reduces as the design becomes larger.

Among very high radix designs, parallel very high radix features the lowest EDP for short $n = 256$ bit multiplies, while the simple scalable very high radix design has nearly $4.5\times$ higher EDP because of its four-cycle latency rather than two-cycle latency. The improved radix-2 design has the lowest overall EDP for a short $n = 256$ bit multiplies since its pipeline can be fully utilized. Increasing w/v for very large radix designs only further decreases efficiency as the performance is directly impacted by pipeline latency— adding more hardware increases latency and power. However, the critical path of the lower radix systems radix-2 implementation is much shorter than that of a typical digital system, while the critical path of a very high radix implementation tends to be better matched. Therefore, synchronous single-clock systems may not be able to take advantage of the faster cycle times of radix-2 designs. The parallel scalable very high radix architecture achieves the best area-delay product (ADP), with the improved radix-2 design increasing ADP by only 11% for long multiplies.

The RSA decryption time $T_{decrypt}$ for a 256-bit and 1024-bit modular exponentiation versus energy are shown in Figure 5.10. Of the very high radix designs, parallel performs modular exponentiation with 47% less energy as quotient, with similar $T_{decrypt}$, for small $n = 256$ bit decryptions. Radix-4 and radix-2 use progressively less energy per decryption, but at increased delay. For a large 1024-bit modular exponentiation parallel very high radix beats other very high radix designs in energy, again with similar exponentiation

Table 5.2
RSA exponentiation clock speed, area, and power in a 45nm CMOS process

Architecture	Reference	w	v	p	Clock Speed (GHz)	Area (μm^2)	Power (mW)	n	Cycles	$T_{decrypt}$ (μs)	EDP (μJ-ms)	ADP (mm^2-μs)
Parallel Pipelined Scalable Very High Radix	This Work	16	16	16	1.71	1,418,400	2,033	256	66	20	0.80	28
								1024	325	390	309	552
Quotient Pipelined Scalable Very High Radix	[187]	16	16	16	1.76	1,479,762	2,114	256	54	16	0.53	23
								1024	331	386	314	571
	[206]	8	8	16	2.27	281,952	559	256	103	23	0.26	6.6
Parallel Scalable Very High Radix								1024	1,171	1058	539	298
		16	16	16	1.66	754,820	1,083	256	54	17	0.30	13
								1024	331	409	181	309
		32	32	16	0.98	2,326,528	3,769	256	46	24	1.27	56
								1024	103	216	103	502
Scalable Very High Radix	[207]	16	16	16	1.64	1,440,725	2,080	256	81	25	1.34	37
								1024	330	413	354	594
Scalable Radix 4 Non-Booth	[328]	16	2	16	3.62	281,596	471	256	171	24	0.28	6.84
								1024	2,178	1,233	717	347
Scalable Radix 4 Booth Left-Shifting	[329]	16	2	16	3.06	263,332	405	256	171	29	0.33	7.56
								1024	2,178	1,459	862	384
Scalable Radix 4 Booth Right-Shifting	[329]	16	2	16	3.50	278,244	440	256	297	44	0.84	12
								1024	2,145	1,256	695	350
Improved Scalable Radix 2	[162]	16	1	16	4.78	69,872	143	256	303	32	0.15	2.3
								1024	4,239	1,818	474	127

time, and similarly beats radix-4 designs. For large exponentiation, the radix-4 designs are sub-optimal and consume more power than the scalable parallel architecture but with much longer decryption times. The scalable radix-2 design consumes the least amount of energy, but has a high latency requirement.

A comparison of decryption time versus area is shown in Figure 5.11. Parallel continues to perform the best with smaller area requirements than other high radix architecture, though the lower radix designs consume far less area at the cost of increased latency. Radix-4 booth-encoded right-shifting and improved scalable radix-2 is best when area limited for both large and small key sizes.

Let $m = pwv$ be a measure of the amount of hardware dedicated to a Montgomery multiplier; specifically, it describes how many partial product bits of XY and QM can be consumed on each cycle. Altogether, approximately n^2 partial product bits must be processed for n-bit Montgomery multiplication. Dropping the lower order terms and rewriting T_{mul} from Section 5.2.6, the number of cycles for a Montgomery multiplication is approximately

$$
\begin{array}{ll}
\frac{n^2}{m} & n \geq lwp \\[2mm]
\frac{nl}{v} & n < lwp
\end{array}
$$

If wp is modest compared to the operand size n, the pipeline is fully utilized and the cycle count decreases linearly with the amount of hardware, m, applied to the problem. If wp is too large, the pipeline begins to stall and the cycle count saturates at a level independent of w and p. In such a case, the cycle count is directly proportional to l, indicating that the latency between PEs limits performance.

Figure 5.12 shows the RSA decryption time $T_{decrypt}$ across multiple designs with 16-bit word sizes for $n = 1024$ as the number of processing elements p is swept. For each curve, w and v are held constant and p is increased. As expected, the cycle count is $O(n^2/m)$ for small m. As more hardware becomes available, the cycle count reaches an inflection point and bottoms out at $O(nl/v)$, then gradually increases because of inefficiencies in very deep pipelines. The inflection point occurs at different m for different architectures, depending on the radix of the design and the latency of each pipeline stage. Past the inflection point, increasing the number of pipeline stages does not improve exponentiation time and performance may deteriorate as the overall pipeline latency increases.

Radix-2 initially performs best for a small number of processing elements for a given area. As more hardware area is available, parallelized very high radix achieves better decryption times because it can process more bits per cycle. The other architectures have similar exponentiation times for small p but perform differentially as the pipeline is under-utilized. The plot for 256-bit Montgomery multiplication is similar, but is shifted down and left because the multiplications are faster but cannot exploit as much hardware.

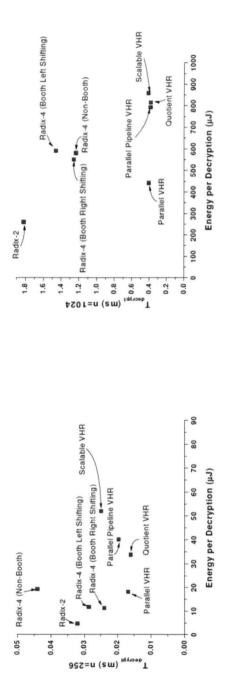

Figure 5.10
RSA exponentiation time ($T_{decrypt}$) vs. energy for different PE architectures

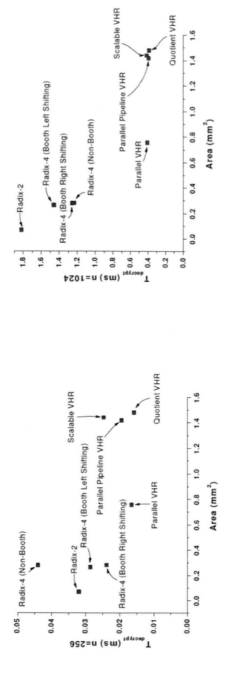

Figure 5.11

RSA exponentiation time $(T_{decrypt})$ vs. area for different PE architectures

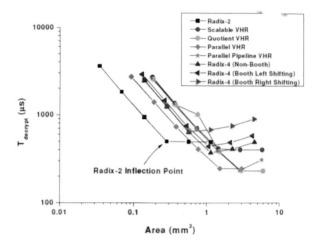

Figure 5.12
1024-bit decryption time ($T_{decrypt}$) vs. area for architectures with 16-bit word sizes as number of processing elements p is swept

5.2.9 Summary

Scalable Montgomery multipliers can be classified by the number of bits of the multiplier X handled by each PE. Radix-2 multipliers process $v = 1$ bit of X. In radix-2, the determination of Q is trivial (it is simply the least significant bit of Z), the partial products are computed with AND gates, and the cycle time is short, but the number of PEs and/or kernel cycles required is large. Radix-4 designs processing $v = 2$ bits of X, increasing the complexity of partial product computation by either pre-computing and selecting $3Y/3\hat{M}$ at the cost of more hardware registers, or by booth encoding which increases critical path length. Very high radix multipliers process many ($v = 8 - 64$) bits of X at a time, and each processing element contains multipliers to compute partial products. Q^i must be determined to v bits of precision. The cycle time is longer and the PE contains more hardware, but fewer PEs and/or kernel cycles are needed.

The radix-2 Montgomery multiplier consumes the least amount of energy and area if the multiplier latency requirement is not stringent, but because of its short critical path requires a much higher clock frequency that may be ill-suited if frequency-matched to a core. With the lowest energy-delay product, the scalable parallel very-high radix achieves the best performance per energy and area across all designs.

5.3 Public Key Cryptography: Programmable and Configurable Processors

5.3.1 Introduction

Most applications today have security as a key feature. This can be either because of the requirement to secure data residing in the platform used to run the applications or due to the support of communication protocols. In particular for the communications, private and public key cryptography is the security guarantee. Private key cryptography uses a secret key distributed prior to the communication that can be used by both entities to cipher and decipher the data to be transmitted over an unsecured channel. The most widely-used private key ciphering algorithm is the AES [307]. The AES algorithm, as most private key cryptographic algorithms, tends to be light-weight and optimized to streaming computing so that high data throughput can be obtained. These optimizations are accomplished by the selection of the best algorithm's parameters so that the implementation better fits the hardware resources. Moreover, some recent general purpose processors include support to dedicated instructions specially tailored to deal with stream ciphering, in particular AES [15, 146].

On the other hand, public key cryptography is based on more complex algorithms and its security relies in the mathematical properties of the underlying functions, which can be efficiently computed but are very difficult to invert. The main feature of the public key cryptography regarding its private key counterpart is that no previous secret information must be known by both parties so as to create a secure communication. Public key cryptography is based on a distributed method to create a common secret that can be used afterwards to securely communicate data. Another important feature of public key cryptography is that it provides methods to digital sign data supporting the validation of data from a given signature generated by the sender. Two main public key cryptography algorithms have been used: the Rivest-Shamir-Adleman (RSA) [347] and Elliptic Curve (EC) cryptography [222, 289].

Due to the interesting features of public key cryptography, efficient implementation of its supporting algorithms have been continuously prospected [11, 13, 419]. Given that, unlike the AES, there is no native support for public key cryptography in general purpose processing platforms, high-performance and efficient cryptographic accelerators are used to support the computational demands of public key cryptography. Given the current trends of the computing platforms, high-performance is usually equivalent to efficient parallelization. Also, there is a constant pressure to develop accelerators which are flexible enough to keep up with the trends and needs of the cryptographic algorithms. Therefore, features such has programmability and reconfiguration are very important.

This section addresses precisely the efficient implementation of public key cryptographic algorithms by presenting a methodology to implement parallel versions of these cryptographic algorithms that can benefit from the programmability and reconfiguration capabilities of the supporting devices. This methodology takes advantage of the Residue Number System (RNS) [291, 405] capabilities to extract parallelism of the public key cryptographic algorithms that typically contain several data dependencies. The utilization of the RNS enables the support of a scalable and flexible implementation of these algorithms. Proof-of-concept implementations of the modular exponentiation (main operation used in the RSA) and of the EC point multiplication (the core operation of the EC cryptography) are also presented for both programmable and reconfigurable devices, namely GPU and FPGA. The methodology presented in this chapter can be easily extended to any other programmable and/or reconfigurable platform that presents support for parallel computation, in particular for data-level parallelism.

This section is organized as follows. Section 5.3.2 presents the details of the public key cryptographic algorithms addressed in this chapter.

Section 5.3.2.6 introduces the RNS and the efficient ways to map the arithmetic operations that underlie the cryptographic algorithms, i.e., the MA, to an RNS version.

Section 5.3.6 describes the methodology to create an RNS version of an algorithm based on modular arithmetic (MA), in particular the cryptographic ones, and Section 5.3.7 presents the implementation of this algorithm with GPU and FPGA.

Finally, Section 5.3.8 presents experimental results and Section 5.3.9 summarizes the work.

5.3.2   Public key Cryptographic Algorithms

In the following subsections the details of the *RSA* and *EC*-based cryptographic algorithms are described, preceded by the properties of the finite field arithmetic that underlie each of these cryptographic algorithms. Nowadays, these two algorithms are used to support most of the public key features required by applications and protocols.

5.3.2.1   The Rivest-Shamir-Aldleman (RSA) algorithm

The *RSA* is a public key cryptographic algorithm proposed in 1978 by Rivest et al. [347]. Nowadays, the *RSA* is the most used public key algorithm with applications in data ciphering and signature generation. Although more efficient alternatives to *RSA* exist [see Section 5.3.2.2], it still is adopted in new applications either because backward compatibility requirements or the users' confidence that it enjoys given that since its proposal it has resisted all tampering efforts — the vulnerabilities exploited by proposed cryptanalysis algorithms have been overcome with bigger, thus more secure, private keys.

In the *RSA* algorithm, a user A sets his public key $K_A = (n_A, e_A)$ and private key $k_A = (n_A, d_A)$, where n_A, e_A, and d_A are integers. With K_A the user can compute a *one-way trapdoor permutation* of an integer x in the range $[0, n_A - 1]$ as:

$$y = f_A(x) = x^{e_A} \bmod n_A \tag{5.1}$$

Definition 5.3.1 (One-way trapdoor permutation) *An operation is called a one-way trapdoor permutation if it corresponds to a function that can be efficiently computed but can only be inverted if extra information is provided. In the case of (5.1), that extra information is the private key of the user A.*

Only the user A knows his private key k_A, thus, he is the only one capable of inverting the permutation in (5.1) and obtain x from y as:

$$x = f_A^{-1}(y) = y^{d_A} \bmod n_A. \tag{5.2}$$

In order for the permutation to be secure, the keys must comply with some constraints. To compute his keys, the user A chooses two prime numbers p_A and q_A, computing $n_A = p_A q_A$. Methods for obtaining these two prime numbers are published with the *RSA*-based digital signature standards [312], which corresponds to a complex task but that only has to be computed once (the user A can use the same keys for several transactions). Factorizing n_A is known to be a computational hard problem, hence any user that only knows n_A cannot compute its factors. To obtain his public key K_A, the user A sets an integer $e_A < n_A$. The value of e_A can be fixed for the algorithm, being often set to $2^{16} + 1$ [406]. To compute the private key, user A computes $d_A = e_A^{-1} \bmod \phi(n_i)$ obtaining k_A, being $\phi(.)$ the Euler's totient function. Given that n_A results from the product of two prime numbers p_A and q_A, $\phi(n_A) = (p_A - 1)(q_A - 1)$. Note that $\#_{[p_A]} \approx \#_{[q_A]} \approx \#_{[n_A]}/2$, where $\#_{[x]}$ stands for the bit-size of x. The value $\phi(n_A)$ cannot be computed except by user A, because no one else knows the primes or can efficiently factorize n_A.

As suggested by (5.1) and (5.2), the main operation to be computed in the *RSA* algorithm is the modular exponentiation. This operation can be accomplished by a square-and-multiply algorithm controlled by the exponent as presented in Algorithm 5.6. Although n does not define a finite field (it is not a prime number), the modular operations in Algorithm 5.6 can be accomplished with algorithms similar to the ones one would use for a finite field (see Chapter 2). However, since n is the product of two random primes, it cannot be picked so that the performance of the arithmetic can be improved. In other words, it is not possible to choose an n with a small Hamming weight neither in its binary representation or *NAF*. Therefore, other alternative algorithms are used to optimize the modular exponentiation. These algorithms, which can either rely in the Montgomery modular multiplication or in the *RNS*, are presented in Section 5.3.2.6.

Algorithm 5.6 k-bit modular exponentiation.

Require: a a k-bit integer, the base;
Require: b a k-bit integer, the exponent;
Require: n a k-bit integer that is the product of two primes ($n = p \times q$);
Ensure: $r = a^b \bmod n$;

  1: $r = a$;
  2: **for** $i = k - 2; i \geq 0; i - -$ **do**
  3:    $r = r \times r \bmod n$;
  4:    **if** $b_i \neq 0$ /* if the i-th bit of b is different from 0 */ **then**
  5:      $r = r \times a \bmod n$;
  6:    **end if**
  7: **end for**
  8: **return** r.

5.3.2.2 Elliptic curve cryptography

Elliptic curve (EC) cryptography was proposed simultaneously by Koblitz [222] and Miller [289] in 1985 to support public key protocols. Since then, *EC* cryptography has been confirmed as a competing alternative to the widely used *RSA* protocol due to its increased efficiency [13, 74, 245]. At the time of its proposals, *EC* cryptography suggested computing efficiency improvements of about 20% regarding the *RSA* protocol, which were expected to increase with the development of the computing capabilities of the target devices. More recently, Koblitz et al. revisited the comparative efficiency of the *EC* cryptography regarding the *RSA* protocol [223], estimating that a 2048-bit public key cipher for the *RSA* protocol can be tampered with 3×10^{20} *MIPS* years whereas a 234-bit public key cipher for an *EC*-based protocol requires 1.2×10^{23} *MIPS* years in order to be compromised. In other words, with *EC* cryptography it is possible to provide 3 orders of magnitude more security with 8 times smaller keys regarding the *RSA* alternative. This analysis properly depicts the advantages of *EC* cryptography motivating the inclusion of this cryptosystem in the current standards for public key ciphering published by recognized institutions such as the *NIST* [312], the *IEEE* [180, 181], or the Certicom [65, 66]. These advantages and the wide acceptance of the *EC* cryptography as a secure and efficient solution for public key protocols is the main motivation for addressing the cryptosystem in this chapter.

The bottom layer of the hierarchy is the underlying finite field that supports addition, multiplication, and the respective additive and multiplicative inverses and identities. A reduction operation is also included in this layer to assure that the output of each of the aforementioned operations remains in the finite field. The middle layer of the hierarchy is composed of the *EC* group basic operations, namely the addition and doubling. Each of the operations in this layer is computed as a sequence of operations of the bottom layer. The top layer is a composite of the elementary operations in the middle layer and is the most important operation in *EC* cryptography, granting the security of the system. The following subsections introduce the type of *EC* and the correspondent arithmetic used in the cryptographic algorithms addressed in this chapter. The underlying finite field arithmetic of the *EC* over $GF(p)$ is discussed in Chapter 2.

5.3.2.3 Elliptic curve over $GF(p)$

Different types of curves have been used for cryptographic purposes over $GF(p)$. Examples of such curves are the Twisted Edwards' Curves that are known to reduce the required computation arithmetic [259]. Herein, without losing generality, a standard *EC* as presented in the standards [312] is addressed, which is also known as short Weierstrass curve.

Definition 5.3.2 (Elliptic Curve over $GF(p)$) *A short Weierstrass EC over $GF(p)$, referred as $E(a, b, GF(p))$, consists of the set of two-coordinated points $P_i = (x_i, y_i) \in GF(p) \times GF(p)$ complying with:*

$$y_i^2 = x_i^3 + ax_i + b, \ a, b \in GF(p), \tag{5.3}$$

together with a point at infinity $\mathcal{O} = (x_i, 0)$, and $4a^3 + 27b^2 \neq 0$.

An additive group with identity $\mathcal{O}$ is defined with this curve. This additive group consists of all *EC* points and an operation referred to as *EC* addition.

Definition 5.3.3 (Elliptic Curve addition over $GF(p)$) *Consider two points $P_1 = (x_1, y_1)$ and $P_2 = (x_2, y_2)$ in $E(a, b, GF(p))$ with p prime. The addition in $E(a, b, GF(p))$ is commutative and consists of the following field operations modulo p [350]:*

- *For $P_1 \neq \mathcal{O}$, $P_2 \neq \mathcal{O}$, and $P_1 \neq P_2$, the point $P_3 = (x_3, y_3) = P_1 + P_2$ (addition) is obtained as:*

$$x_3 = \lambda^2 - x_2 - x_1; \ y_3 = -y_2 + \lambda (x_2 - x_3); \ \lambda = \frac{y_1 + y_2}{x_1 + x_2} \tag{5.4}$$

- *For $P_1 \neq \mathcal{O}$, the point $P_3 = (x_3, y_3) = P_1 + P_1 = 2P_1$ (doubling) is obtained as:*

$$x_3 = \lambda^2 - 2x_1; \ y_3 = \lambda (x_1 - x_3) - y_1; \ \lambda = \frac{3x_1^2 + a}{2y_1} \tag{5.5}$$

The point $P_3 = (x_3, y_3) = -P_1 = \mathcal{O} - P_1$ (additive inverse) is obtained as:

$$x_3 = x_1; \; y_3 = -y_1 \tag{5.6}$$

- For $P_1 = \mathcal{O}$, the point $P_3 = P_1 + P_2 = P_2$ and $P_3 = 2P_1 = \mathcal{O}$.

5.3.2.4 Elliptic curve point multiplication

As suggested in Figure 5.13, there is an upper layer in the *EC* arithmetic that is constructed based on the operations in Definition 5.3.3 for an *EC* underlaid by a $GF(p)$ finite field. This upper layer consists of the *EC* point multiplication by a scalar, or simply *EC* point multiplication. For a scalar integer s and a point P_1 in $E(a, b, GF(p))$, the *EC* point multiplication P_3 is obtained as:

$$P_3 = sP_1 = \underbrace{P_1 + \ldots + P_1}_{s \; times} \tag{5.7}$$

This operation grants the security of the *EC* cryptography given that knowing P_3 and P_1 it is not possible to compute s in polynomial time, or in other words, it is possible to select reasonable large values of s that allow computing P_3 efficiently but preclude the computation of s from P_3 and P_1 in time short enough to compromise the security of the data being secured within the cryptosystem. This is known as the *elliptic curve discrete logarithm problem (ECDLP)*.

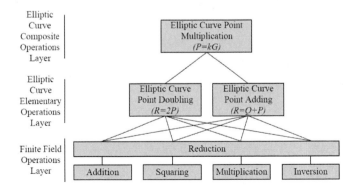

Figure 5.13
EC arithmetic supporting hierarchy

Consider the following example concerning the utilization of the *EC* point multiplication for secure data transfer, usually called El-Gammal protocol for *EC* [406]:

Algorithm 5.7 Double-and-add *EC* point multiplication.

Require: n-bit scalar $s = (s_{n-1} \ldots s_0)$ and P_1 in $E(a, b, GF(p))$;
Ensure: $P_3 = sP_1$;
   $P_3 = P_1$;
   {Assume $s_{n-1} = 1$}
   for $i = n - 2$ down to 0 **do**
      $P_3 = 2P_3$;
      if $s_i = 1$ **then**
         $P_3 = P_3 + P_1$;
      end if
   end forreturn P_3;

- Alice wants to send a message M to Bob, which content is somehow mapped to an *EC* point P_M.

- Alice generates a random secret integer k_A that only she knows. Bob does the same, generating an integer k_B. These two integers are known as Alice's and Bob's private keys.

- There is an *EC* point P_G that is defined in the protocol, thus both Alice and Bob have access to it.

- Alice and Bob compute their public keys, which correspond to *EC* points, by performing the *EC* point multiplications $P_A = k_A P_G$ and $P_B = k_B P_G$, respectively. Thereafter, Alice and Bob publish their public keys so that everyone can access them. Because of the *ECDLP*, Alice and Bob are sure that no one will be able to get their private keys k_A and k_B from their public keys.

- After receiving Bob's public key, Alice is able to compute an *EC* point containing a ciphertext of the message M as $P_C = P_M + k_A P_B$.

- Alice can now send the cryptogram P_C to Bob. Given that no one knows k_A, no one intercepting P_C will be able to compute P_M, except for Bob. Bob can compute $P_C - k_B P_A = (P_M + k_A P_B) - k_B P_A = (P_M + k_A k_B P_G) - k_B k_A P_G = P_M$.

- Knowing P_M, Bob can retrieve the secret message M.

The computation of the *EC* point multiplication can be done with several different methods. One of them, and perhaps the simplest one, is the double-and-add method that is presented in Algorithm 5.7. In this algorithm it is possible to identify that $n - 1$ *EC* point doublings and $h(s) - 1$ *EC* point additions need to be performed in order to compute the final result P_3, where $h(s)$ is the Hamming weight of the binary representation of s.

5.3.2.5 Elliptic curve points: Projective coordinates

Considering Definition 5.3.3, $n + h(s) - 2$ field inversions are required to compute P_3 with Algorithm 5.7. Regarding the high computational demands of the field inversion when compared with the other field operations, a different representation of the *EC* point can be adopted using three coordinates instead of two. These coordinates are called *Projective Coordinates* while the coordinates originally introduced in Definition 5.3.2 are called *Affine Coordinates*.

Definition 5.3.4 (*EC Projective Coordinates*) *The projective coordinates of an EC point $P_i \in E(a, b, \mathbb{F})$, where $\mathbb{F}$ is a finite field, refer to the triple $(\widehat{x}_i, \widehat{y}_i, \widehat{z}_i)$ in $\mathbb{F} \times \mathbb{F} \times \mathbb{F}$ except for the triple $(0, 0, 0)$ in the surface of equivalence classes where $(\widehat{x}_1, \widehat{y}_1, \widehat{z}_1)$ is said to be equivalent to $(\widehat{x}_2, \widehat{y}_2, \widehat{z}_2)$ if there exists a $\lambda \in \mathbb{F}$ such that $\lambda \neq 0$ and $(\widehat{x}_1, \widehat{y}_1, \widehat{z}_1) = (\lambda^{\tau_x}\widehat{x}_2, \lambda^{\tau_y}\widehat{y}_2, \lambda\widehat{z}_2)$, where τ_x and τ_y are positive integers. In this surface, an EC point with affine coordinates (x_i, y_i) can be represented in projective coordinates by setting $\lambda\widehat{z}_i = 1 \Leftrightarrow \lambda = 1/\widehat{z}_i$. Therefore, the correspondence between affine and projective coordinates is given by $x_i = \widehat{x}_i/\widehat{z}_i^{\tau_x}$ and $y_i = \widehat{y}_i/\widehat{z}_i^{\tau_y}$. Different kinds of projective coordinates can be used by setting different values for τ_x and τ_y.*

Different types of projective coordinates for *EC* over $GF(p)$ can be found namely Standard [54], Jacobi [259], and Chudnovsky [77]. This chapter will focus the utilization of standard projective coordinates given that they support the implementation of the Montgomery Ladder [see Algorithm 5.8 ahead] which is an efficient approach to compute the *EC* point multiplication. Using the standard projective coordinates, the *EC* equation in (5.3) becomes:

$$\widehat{y}_i^{\,2}\widehat{z}_i = \widehat{x}_i^{\,3} + a\widehat{x}_i\widehat{z}_i^{\,2} + b\widehat{z}_i^{\,3}, \ a, b \in GF(p) \tag{5.8}$$

Given that the curve's definition with these coordinates is different the projective versions of the *EC* addition equations are also different:

$$\widehat{x_{\text{add}}} = \begin{cases} -4b\widehat{z}_1\widehat{z}_2\,(\widehat{x}_1\widehat{z}_2 + \widehat{x}_2\widehat{z}_1) + (\widehat{x}_1\widehat{x}_2 - a\widehat{z}_1\widehat{z}_2)^2 & \text{if } P_1 \neq P_2 \\ \left(\widehat{x}_1^{\,2} - a\widehat{z}_1^{\,2}\right)^2 - 8b\widehat{x}_1\widehat{z}_1^{\,3} & \text{if } P_1 = P_2 \end{cases} \tag{5.9}$$

$$\widehat{z_{\text{add}}} = \begin{cases} x_{\text{sub}}\,(\widehat{x}_1\widehat{z}_2 - \widehat{x}_2\widehat{z}_1)^2 & \text{if } P_1 \neq P_2, \\ 4\widehat{z}_1\left(\widehat{x}_1^3 + a\widehat{x}_1\widehat{z}_1^{\,2} + b\widehat{z}_1^{\,3}\right) & \text{if } P_1 = P_2 \end{cases}$$

Note that the operations in (5.9) were deliberately written so as to depend on the affine coordinate x_{sub} which is the affine coordinate of the difference between the two *EC* points that are being added. With this, the operations in (5.9) suits the implementation of an *EC* point multiplication algorithm

---

Algorithm 5.8 Montgomery ladder for computing the *EC* point multiplication.

Require: *n*-bit scalar $s = (s_{n-1} \ldots s_0)$ and P_1 in $E(a, b, GF(p))$;
Ensure: $P_3 = sP_1$;
  $(\widehat{x_1}, \widehat{z_1}) = (x_1, 1)$;
  $(\widehat{x_2}, \widehat{z_2}) = \mathtt{Mdouble}(\widehat{x_1}, \widehat{z_1})$;
  for $i = n - 2$ down to 0 **do**
    if $s_i = 1$ **then**
      $(\widehat{x_1}, \widehat{z_1}) = \mathtt{Madd}(\widehat{x_1}, \widehat{z_1}, \widehat{x_2}, \widehat{z_2})$, $(\widehat{x_2}, \widehat{z_2}) = \mathtt{Mdouble}(\widehat{x_2}, \widehat{z_2})$;
    else
      $(\widehat{x_2}, \widehat{z_2}) = \mathtt{Madd}(\widehat{x_2}, \widehat{z_2}, \widehat{x_1}, \widehat{z_1})$, $(\widehat{x_1}, \widehat{z_1}) = \mathtt{Mdouble}(\widehat{x_1}, \widehat{z_1})$;
    end if
  end forreturn $P_3 = \mathtt{Mxy}(\widehat{x_1}, \widehat{z_1}, \widehat{x_2}, \widehat{z_2})$;

---

called Montgomery Ladder (Algorithm 5.8) where Madd and Mdouble stand for the *EC* point addition and doubling, respectively. Algorithm 5.8 has the property that x_{sub} is invariant and corresponds to the affine x coordinate of the *EC* point multiplication input point, i.e., $x_{\text{sub}} = x_1$. Algorithm 5.8 has a final step Mxy that aims at computing the affine coordinates of $P_3 = (x_3, y_3)$. This can be obtained from the projective coordinates as [11]:

$$x_3 = \widehat{x_1}/\widehat{z_1}, \; x_3' = \widehat{x_2}/\widehat{z_2}, \; \frac{-2b + (a + x_1 x_3)(x_1 + x_3) - x_3'(x_1 - x_3)^2}{2y_1} \quad (5.10)$$

If the application underlaid by the *EC* arithmetic does not need the y coordinate, the expression for y_3 does not have to be implemented at all.

5.3.2.6 Modular arithmetic and RNS

This section presents the details of *RNS* arithmetic as well as efficient methods to map the *MA* operations to *RNS*. The fundamentals of the *RNS* dates back to the third-century AD and the proposal of the *CRT* by the mathematician Sun Tzu. This theorem establishes a correspondence between an integer and the remainders of the division of this number by another set of numbers. Today RNS has applications in *HPC* due to the parallelism extraction at data level, which allows one to obtain parallel versions of algorithms even for those crowded with data dependencies, which is the case of the cryptographic Algorithms 5.6 and 5.8 described in Section 5.3.2 (note that each loop's iteration depend on the previous one). This is also the reason for addressing the *RNS* in this chapter, which will be further used in Section 5.3.6 toward the parallel implementation of the *MA* underlying the cryptographic algorithms.

The concept

By defining a basis $B_i = \{m_{1,i}, ..., m_{h_i,i}\}$ of pairwise coprime elements and an associated dynamic range $M_i = \prod_{e=1}^{h_i} m_{e,i}$, an integer $X < M$ has a corresponding *RNS* representation [405]:

$$x_{1,i} = X \bmod m_{1,i}, \tag{5.11}$$

$$\vdots$$

$$x_{h_i,i} = X \bmod m_{h_i,i} \tag{5.12}$$

The main advantage of the *RNS* representation is the possibility to perform in parallel the same operations one would do with integers, $Z = X \odot Y$ (where $\odot$ is either an addition/subtraction or a multiplication and with $Z < M_i$), as:

$$z_{1,i} = x_{1,i} \odot y_{1,i} \bmod m_{1,i}, \tag{5.13}$$

$$\vdots$$

$$z_{h_i,i} = x_{h_i,i} \odot y_{h_i,i} \bmod m_{h_i,i} \tag{5.14}$$

An operation applied over the *RNS* representation mod $m_{e,i}$ is said to be performed on the *RNS* channel defined by $m_{e,i}$. As (5.13) suggests, the *RNS* operations correspond to *MA*. Therefore, when the *RNS* is used to accelerate $GF(p)$ arithmetic, which also corresponds to *MA*, there will be two different layers of *MA* being used, and it is therefore important to clearly distinguish these two layers, e.g., the *EC* computation based on the $GF(p)$ finite field hierarchy presented in Figure 5.13 becomes a four layer hierarchy with the inclusion of the *RNS*-related hierarchy at its bottom, as an extra layer of arithmetic, as presented in Figure 5.14. The *MA* that corresponds to the $GF(p)$ finite field is herein referred to as *SMA* and the *RNS* arithmetic that is used to accelerate the $GF(p)$ arithmetic is referred to as *CMA* because it corresponds to computation on the *RNS* channels. In summary, the *CMA* is used to accelerate the *SMA*.

Figure 5.15 and Example 5.3.5 illustrate the correspondence between an operand X, used in the *SMA*, and the residues $x_{e,i}$ which are used within the *CMA*. The operands used in the description of an *MA* algorithm are converted to the *RNS* representation so that the *SMA* is computed in parallel by several channels, each one computing *CMA*. Thereafter, the results are converted back to the original representation used in the *SMA*. The challenge in the implementation of the *RNS* modular operations is to find efficient ways to compute operations such as the modular reduction described in *SMA* with *CMA*.

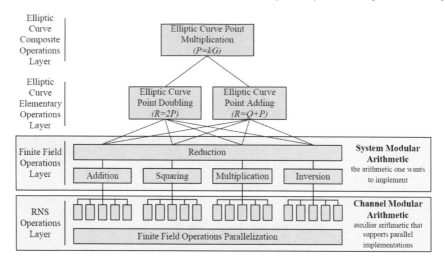

Figure 5.14
EC arithmetic supporting hierarchy including the underlying *RNS* layer

Example 5.3.5 *Consider that one wants to compute the SMA operations $Z_+ = X + Y$ and $Z_\times = X \times Y$ for $X = 14$ and $Y = 23$ using the CMA. The expected results for these operations are $Z_+ = 27$ and $Z_\times = 322$. In order to obtain the CMA equivalent one can define a basis B_1 of $h_1 = 3$ elements $B_i = \{7, 8, 9\}$. One can easily verify that all elements $m_{i,1}$ are pairwise coprime and that the dynamic range of B_i is $M_i = 7 \times 8 \times 9 = 504$. Since $M_i > Z_+$ and $M_i > Z_\times$, a mapping of SMA to CMA can be obtained. One has only to obtain the residues $x_{i,1} = X \bmod m_{i,1}$ and $y_{i,1} = Y \bmod m_{i,1}$: $\{x_{1,1}, x_{2,1}, x_{3,1}\} = \{0, 6, 5\}$ and $\{y_{1,1}, y_{2,1}, y_{3,1}\} = \{2, 7, 5\}$. To compute the operations in CMA one can now compute $z_{+i,1} = x_{i,1} + y_{i,1} \bmod m_{i,1}$ and $z_{\times i,1} = x_{i,1} \times y_{i,1} \bmod m_{i,1}$. Therefore the result is $\{z_{+1,1}, z_{+2,1}, z_{+3,1}\} = \{2, 5, 1\}$ and $\{z_{\times 1,1}, z_{\times 2,1}, z_{\times 3,1}\} = \{0, 2, 7\}$. The correctness of the mapping can be finally confirmed by checking the obtained residues against the values $Z_+ \bmod m_{i,1}$ and $Z_\times \bmod m_{i,1}$. With these operations an SMA operation was accomplished with the CMA in 3 parallel flows.*

Note that in Example 5.3.5, the operations described as *SMA* are not exactly modular operations given that the reduction operation is not applied.

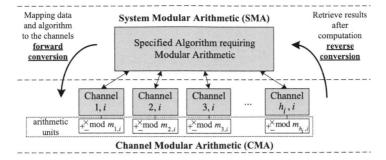

Figure 5.15
Hierarchy for computing the target algorithm based on the *RNS* representation: correspondence between *SMA* and *CMA*

The operations

The main operations to be supported in an MA algorithm are the addition/subtraction, multiplication, and reduction as well as the RNS conversions from/to binary. In particular, these two last operations are far more complex and are thoroughly discussed in this section. The multiplicative inverse operation is also often used, but it is not herein addressed since alternative algorithms based on multiplications and additions can be used with this purpose especially if the modulo of the operation being held is a prime number [1].

5.3.3 RNS Forward/Reverse Conversions

Forward conversion corresponds to the conversion of the input data to the RNS representation so that it can be used in parallel implementations. The conversion of an input X is usually accomplished by computing for every channel e of the basis B_i: $x_{e,i} = X$ mod $m_{e,i}$. Assuming that the moduli $m_{e,i}$ can be rewritten as $2^k - c_{e,i}$, with $c_{e,i} > 0$, the computation of the forward conversion can be partitioned in $\lceil \#_{[X]}/k \rceil$ terms, each obtained with k-bit wide computations:

$$x_{e,i} = X \text{ mod } m_{e,i} \tag{5.15}$$

$$= \left(\sum_{j=1}^{\lceil \#_{[X]}/k \rceil} \langle X \rangle_{k,j} \, 2^{k(j-1)} \right) \text{ mod } m_{e,i}$$

$$= \sum_{j=1}^{\lceil \#_{[X]}/k \rceil} \langle X \rangle_{k,j} \left(2^{k(j-1)} \text{ mod } m_{e,i} \right) \text{ mod } m_{e,i}$$

where $\#_{[X]}$ is the size of X in bits and $\langle X \rangle_{k,j}$ is the j-th word of the binary representation of X split in k-bit words. Considering (5.15), in order to obtain the RNS representation of X, the constants $2^{k(j-1)} \bmod m_{e,i}$ can be precomputed, and the conversion is accomplished with a few RNS channel modular multiply-and-accumulate operations.

Example 5.3.6 (RNS forward conversion with channel arithmetic)
This numerical example concerns the implementation of the RNS forward conversion using exclusively the arithmetic that is available for an RNS channel, i.e., k-bit wide MA for a moduli $m_{e,i} = 2^k - c_{e,i}$. For $k = 8$, the basis of pairwise coprime integers $B_i = \{253, 247, 239\}$ is a valid RNS basis with dynamic range $M_i = 14,935,349$. Consider one wants to convert $X = 1,234,567 = [100101101011010000111]_b$ whose k-bit wide limbs are:

$$\langle X \rangle_{k,1} = 135 \tag{5.16}$$
$$\langle X \rangle_{k,2} = 214$$
$$\langle X \rangle_{k,3} = 18$$

to the RNS representation for the basis B_i. This can be accomplished by precomputing the constants $2^{k(j-1)} \bmod m_{e,i}$:

$$2^k \bmod m_{1,i} = 3 \tag{5.17}$$
$$2^{2k} \bmod m_{1,i} = 9$$
$$2^k \bmod m_{2,i} = 9$$
$$2^{2k} \bmod m_{2,i} = 81$$
$$2^k \bmod m_{3,i} = 17$$
$$2^{2k} \bmod m_{3,i} = 50$$

and use (5.15) for each channel:

$$x_{1,i} = 135 + 214 \times 3 + 18 \times 9 \bmod 253 = 180;$$
$$x_{2,i} = 135 + 214 \times 9 + 18 \times 81 \bmod 247 = 61;$$
$$x_{3,i} = 135 + 214 \times 17 + 18 \times 50 \bmod 239 = 132$$

It can be easily verified that these values for $x_{e,i}$ correspond to the direct computation of $X \bmod m_{e,i}$.

For the reverse conversion, which means converting the computation results back to the binary representation, two main approaches can be used. One of these approaches is the MRC which can be accomplished by computing the following for the basis B_i [291, 405]:

$$X = x'_{1,i} + x'_{2,i}m_{1,i} + x'_{3,i}m_{1,i}m_{2,i} + \ldots + x'_{h_i,i} \prod_{e=1}^{h_i-1} m_{e,i} \qquad (5.18)$$

where the values $x'_{e,i}$ correspond to the MR digits. Each of these digits can be obtained from the RNS residues $x_{e,i}$ using a recursive computation:

$$x'_{1,i} = x_{1,i} \qquad (5.19)$$

$$x'_{2,i} = \left| \left(x_{2,i} - x'_{1,i} \right) \left| m_{1,i}^{-1} \right|_{m_{2,i}} \right|_{m_{2,i}} ;$$

$$x'_{3,i} = \left| \left(\left(x_{3,i} - x'_{1,i} \right) \left| m_{1,i}^{-1} \right|_{m_{3,i}} - x'_{2,i} \right) \left| m_{2,i}^{-1} \right|_{m_{3,i}} \right|_{m_{3,i}} ;$$

$$\vdots$$

Given the recursive nature of the MRC, it is not suitable for efficient parallel implementations as suggested by experimental results using GPU [419]. A different approach, more parallelization-friendly, can be used instead. This approach results directly from the CRT and can be computed as [291, 405]:

$$X = \sum_{e=1}^{h_i} \zeta_{e,i} M_{e,i} \bmod M_i \qquad (5.20)$$

$$= \sum_{e=1}^{h_i} \zeta_{e,i} M_{e,i} - \alpha M_i,$$

$$\alpha < n, \ \zeta_{e,i} = \left| \frac{x_{e,i}}{M_{e,i}} \right|_{m_{e,i}},$$

$$M_{e,i} = \frac{M_i}{m_{e,i}}$$

In (5.20), the subtraction of αM_i allows obtaining the required reduction modulo M_i. Two main methods have been used to compute the constant α: a method proposed by Shenoy and Kumaresan [389] and the other proposed by Kawamura et al. [204].

The method proposed by Shenoy and Kumaresan. [389] consists of the establishment of an extra moduli m_{extra} and requires all the operations to be performed not only on the basis B_i, but also on this extra moduli. Hence, the RNS representation of the integer X is $(x_{1,i}, x_{2,i}, \ldots, x_{h_i,i}, x_{\text{extra}})$. Applying the reduction modulo m_{extra} to (5.20), the following expression is obtained:

$$x_{\text{extra}} = \left| \sum_{e=1}^{h_i} \zeta_{e,i} M_{e,i} - \alpha M_i \right|_{m_{\text{extra}}} \qquad (5.21)$$

Rewriting (5.21), α can be obtained as:

$$\left|\alpha\right|_{m_{\text{extra}}} = \left|\left|\sum_{e=1}^{h_i} \xi_{e,i} M_{e,i} - x_{\text{extra}}\right|_{m_{\text{extra}}} \left|M_i^{-1}\right|_{m_{\text{extra}}}\right|_{m_{\text{extra}}} \tag{5.22}$$

Since $\alpha < n$, choosing $m_{\text{extra}} \geq n$ results in $\alpha = \left|\alpha\right|_{m_{\text{extra}}}$.

The method proposed by Kawamura et al. [204] avoids the utilization of the extra moduli m_{extra} and all computation that need to be performed modulo m_{extra} and has therefore lower complexity [10,147]. This method involves a fixed point successive approximation approach. This method observes that $\frac{X}{M} < 1$ given that $X < M$, which allows (5.20) to be rewritten as:

$$\sum_{e=1}^{h_i} \frac{\xi_{e,i}}{m_{e,i}} = \alpha + \frac{X}{M_i} \Rightarrow \alpha = \left\lfloor \sum_{e=1}^{h_i} \frac{\xi_{e,i}}{m_{e,i}} \right\rfloor \tag{5.23}$$

Since (5.23) requires costly divisions by $m_{e,i}$, an approximation ($\hat{\alpha}$) is suggested to this expression:

$$\hat{\alpha} = \left\lfloor \sum_{e=1}^{h_i} \frac{\text{truncH}_t\left(\xi_{e,i}\right)}{2^k} + \beta \right\rfloor = \left\lfloor \sum_{e=1}^{h_i} \frac{\text{truncH}_t\left(\xi_{e,i}\right)}{2^{k-t}} + \Phi \right\rfloor / 2^t \tag{5.24}$$

where $\text{truncH}_t\left(\xi_{e,i}\right)$ sets the $k - t$ least significant bits of $\xi_{e,i}$ to zero, with $t < k$, and $\Phi = \left\lfloor 2^t \beta \right\rfloor$. The parameter β (and consequently Φ) is a corrective term that should be carefully chosen such that $\alpha = \hat{\alpha}$. A set of inequalities that support the selection of good values for β were stated, supported on the maximum initial approximation errors:

$$\epsilon = \max_e \left(\frac{2^k - m_{e,i}}{2^k} \right), \, \delta = \max_e \left(\frac{\xi_{e,i} - \text{truncH}_t\left(\xi_{e,i}\right)}{m_{e,i}} \right) \tag{5.25}$$

In [204], Theorems 5.3.7 and 5.3.8 are stated concerning the computation of the value $\hat{\alpha}$, its relationship with β, the number of channels h_i, and the errors stated in (5.25):

Theorem 5.3.7 *If* $0 \leq h_i(\epsilon + \delta) \leq \beta < 1$ *and* $0 \leq X < (1 - \beta)M_i$*, then* $\alpha = \hat{\alpha}$.

Theorem 5.3.8 *If* $\beta = 0$, $0 \leq h_i(\epsilon + \delta) \leq 1$ *and* $0 \leq X \leq M_i$*, then* $\hat{\alpha} = \alpha$ *or* $\hat{\alpha} = \alpha - 1$.

Theorem 5.3.8 refers to the computation of a non-exact, although bounded, value of $\hat{\alpha}$ that allows a non-exact RNS to binary conversion, while Theorem 5.3.7 refers to an exact conversion.

Considering that the basis B_i one wants to convert from has m_{min} as the minimum value of $m_{e,i} = 2^k - c_{e,i}$, the value $\Delta = h_i(\epsilon + \delta)$ used in the definition of Theorems 5.3.7 and 5.3.8 can be obtained as:

$$\Delta = h_i(\epsilon + \delta) \tag{5.26}$$

$$= h_i\left(\frac{2^k - m_{min}}{2^k} + \frac{2^{k-t} - 1}{m_{min}}\right) \tag{5.27}$$

$$= \frac{h_i}{2^k m_{min}}\left(2^k\left(m_{min} + 2^{k-t} - 1\right) - m_{min}^2\right)$$

Summarizing, Theorem 5.3.7 allows suitable values of Φ in (5.24) to be obtained such that $\alpha = \hat{\alpha}$, and Theorem 5.3.8 bounds the maximum error of $\hat{\alpha}$ $(+1)$ if Φ is set to zero, thus, an extra term M_i may exist in (5.20).

Example 5.3.9 (RNS reverse conversion using the CRT) *This numerical example concerns the implementation of the RNS reverse conversion using the CRT. Consider the RNS basis with $h_1 = 3$ moduli $B_i = \{253, 247, 239\}$ whose dynamic range is $M_i = 14,935,349$, and the value $X = 1,234,567$ which residues are $\{x_{1,i}, x_{2,i}, x_{3,i}\} = \{180, 61, 132\}$ [see Example 5.3.6]. In order to compute the RNS reverse conversion using (5.20) the following constants should be precomputed:*

$$M_{1,i} = 59,033; \quad \left|M_{1,i}^{-1}\right|_{m_{1,i}} = 250;$$

$$M_{2,i} = 60,467; \quad \left|M_{2,i}^{-1}\right|_{m_{2,i}} = 36;$$

$$M_{3,i} = 62,491; \quad \left|M_{3,i}^{-1}\right|_{m_{3,i}} = 207$$

Given these constants and the RNS residues, it is now possible to compute the values $\xi_{e,i}$:

$$\xi_{1,i} = \left|x_{1,i}\left|M_{1,i}^{-1}\right|_{m_{1,i}}\right|_{m_{1,i}} = 180 \times 250 \bmod 253 = 219;$$

$$\xi_{2,i} = \left|x_{2,i}\left|M_{2,i}^{-1}\right|_{m_{2,i}}\right|_{m_{2,i}} = 61 \times 36 \bmod 247 = 220;$$

$$\xi_{3,i} = \left|x_{3,i}\left|M_{3,i}^{-1}\right|_{m_{3,i}}\right|_{m_{3,i}} = 132 \times 207 \bmod 239 = 78$$

and obtain the resulting converted value $X_{uncorrected}$ without the correction term αM_i:

$$X_{uncorrected} = \sum_{e=1}^{h_i} \zeta_{e,i} M_{e,i} = 219 \times 59,033 + 220 \times 60,467 + 78 \times 62,491$$

$$= 31,105,265$$

It can be observed that for $\alpha = 2$, the resulting value for $X = X_{uncorrected} - \alpha M_i$ will be $1,234,567$ as expected. Therefore, the aforementioned methods for computing α should return the expected value 2.

The method proposed by Shenoy and Kumaresan requires the computation of (5.22). In order to accomplish this computation, an extra moduli m_{extra} is required. In order to simplify the computation of α this moduli is usually set to a power of two given that it results in more efficient arithmetic [11]. This power of two should be the smallest that verifies $m_{extra} > h_i$, which for this example is $m_{extra} = 2^2 = 4 > 3$. This also requires the residue for this extra moduli which is $x_{extra} = 3$. Note that this residue can be obtained like any other RNS residue $x_{e,i}$, and should also be obtained during the forward conversion process. Given the values x_{extra}, $X_{uncorrected}$, and the constant $M_i \bmod m_{extra} = 1$, the value of α can be obtained as in (5.22):

$$\alpha = \left| \left| X_{uncorrected} - x_{extra} \right|_{m_{extra}} \left| M_i^{-1} \right|_{m_{extra}} \right|_{m_{extra}}$$

$$= (((31,105,265) - 3) \bmod 4) \times 1 \bmod 4 = 2$$

which is the expected value.

In order to compute the value of α using the Kawamura et al. method [204], one must first verify if the maximum number that has to be reverse converted to binary is either under the conditions of Theorem 5.3.7 or 5.3.8. Assuming that $X_{max} = 1,234,567$ is the maximum value that will ever be converted, the following holds:

$$X_{max} \leq (1 - \beta) M_i \Leftrightarrow \beta \leq 1 - \frac{X_{max}}{M_i}$$

In order to be under the conditions of Theorem 5.3.7 the following has to be verified:

$$\frac{h_i}{2^k m_{min}} \left(2^k \left(m_{min} + 2^{k-t} - 1 \right) - m_{min}^2 \right) < \beta$$

which is assured if there is a value $t \leq h_i$ that verifies:

$$\frac{h_i}{2^k m_{min}} \left(2^k \left(m_{min} + 2^{k-t} - 1 \right) - m_{min}^2 \right) < 1 - \frac{X_{max}}{M_i} \Leftrightarrow$$

$$h_i M_i \left(2^k \left(m_{min} + 2^{k-t} - 1 \right) - m_{min}^2 \right) < 2^k m_{min} (M_i - X_{max})$$

The lower the value t the less complex the remaining operations for computing α are. Given the values $h_i = 3$, $m_{min} = m_{3,i} = 239$, X_{max}, and M_i it is possible to verify that the value $t = 3$ is the lowest that comply with the aforementioned expression. Knowing the value t, it is now possible to compute alpha using (5.24). The value Φ has to be precomputed as:

$$\Phi = \lfloor 2^t \beta \rfloor = \lfloor \frac{2^t (M_i - X_{max})}{M_i} \rfloor = \lfloor \frac{8 \times 13,700,782}{1,234,567} \rfloor = 7$$

and, with the values:

$$\frac{truncH_t(\xi_{1,i})}{2^{k-t}} = 219 >> (8-3) = 6;$$

$$\frac{truncH_t(\xi_{2,i})}{2^{k-t}} = 220 >> (8-3) = 6;$$

$$\frac{truncH_t(\xi_{3,i})}{2^{k-t}} = 78 >> (8-3) = 2$$

it is possible to obtain α as:

$$\alpha = \frac{\lfloor 6 + 6 + 2 + \Phi \rfloor}{2^3} = \frac{\lfloor 14 + 7 \rfloor}{8} = 2$$

If the conditions of Theorem 5.3.7 are not met but the ones in Theorem 5.3.8 are, the procedure to compute α would be exactly the same but the value $\Phi = 0$ would have to be used and the final conversion result X would need to be compared with M_i in order to detect if the approximated α has an error and the correct conversion obtained using $\alpha + 1$ instead. Note that the selected RNS basis and value X_{max} can impose an approximation error too high so that neither the conditions of Theorem 5.3.7 or 5.3.8 are met. In that situation the computation of α as in (5.24) will not be possible, and an alternative method should be selected.

Another challenge concerning the reverse conversion is the mapping of this conversion to the resources assigned to the channel arithmetic in order to avoid the utilization of a dedicated unit for this purpose. In other words, the challenge is to map the reverse conversion to the k-bit arithmetic used in each one of the RNS channels. Each channel e is able to compute the values $\xi_{e,i} = \left| x_{e,i} \left| M_{e,i}^{-1} \right|_{m_{e,i}} \right|_{m_{e,i}}$ and consequently a value of α can be computed by accumulating the contributions of each channel by using the method in (5.22) or in (5.24). Each channel can also contribute to the final value in (5.20) if the constants $\langle M_{e,i}\rangle_{k,j}$ such that $M_{e,i} = \sum_{j=1}^{\#[M_{e,i}]/k} \langle M_{e,i}\rangle_{k,j} 2^{kj}$ are available to each channel e, where $\#_{[M_{e,i}]}$ is the size in bits of $M_{e,i}$. With these constants, the values $\xi_{e,i}$ $\langle M_{e,i}\rangle_{k,j}$ can be accumulated obtaining k bits of the result at once. The contribution that depends on α has to be computed by one of the channels that should have available the constants $\langle -M_i\rangle_{k,j}$ such that $-M_i = \sum_{j=1}^{\#[M_i]/k} \langle -M_i\rangle_{k,j} 2^{kj}$. This channel can compute α $\langle -M_i\rangle_{k,j}$ as a contribution to the final accumulation. Note that this contribution does not require modular computation; hence, the logic of any channel can be used for that.

5.3.4 RNS Addition/Subtraction and Multiplication

Addition, subtraction, and multiplication in an RNS channel e is straightforward and relies on the corresponding binary operations followed by the final reduction in the channel to assure that the result is in the range $[0, m_{e,i} - 1]$. In spite of the elements of an RNS basis that are not necessarily prime, the RNS addition, subtraction, and multiplication can be implemented with binary arithmetic units for $GF(p)$ by rewriting any modulo $m_{e,i}$ associated to the channel as $2^k - c_{e,i}$.

5.3.5 RNS Reduction

The reduction modulo N as herein discussed is usually introduced as part of a modular multiplication algorithm known as Montgomery modular multiplication [293]. For the purposes of this chapter, the reduction part is considered separately from the multiplication itself. The rationale of this algorithm is to replace the costly reduction mod N, where N is an arbitrary number, by a reduction mod L that is easier to compute. In the original proposal [293], the target was a binary system; hence, the good choices for L were powers of 2 given that reducing modulo a power of 2 is the same as applying a binary mask. However, in RNS it is not possible to apply such a mask, which turns the reduction modulo a power of 2 inefficient. Addressing this problem, an RNS variation of the Montgomery modular multiplication that relies in a BE method [25] has been proposed. The name of this method relates to the requirement to extend the representation of an RNS basis to another one during the computation, i.e., the RNS representation needs to be extended to a second different basis. These two bases are herein referred to as B_1 and B_2. In RNS, reduction modulo the dynamic range M_1 is an easy task since it relies on the modulo $m_{e,1}$ operation in each one of the e RNS channels of the basis B_1. This property is exploited in the BE method by replacing the arithmetic modulo N by arithmetic modulo $L = M_1$. This replacement requires the usage of precomputed constants that require $|N^{-1}|_{M_1}$ and $|N^{-1}|_{M_2}$ to exist, which introduces an extra constraint in the selection of the basis given that $gcd(M_1, N) = 1$ and $gcd(M_2, N) = 1$ must be verified. Moreover, $\left|M_1^{-1}\right|_{M_2}$ must also exist which requires $gcd(M_1, M_2) = 1$ to be verified as well, which is equivalent to force all the moduli of both basis B_1 and B_2 to be pairwise coprime.

In the computation of the modular reduction using the BE, the input data should be represented in an alternative MD [Section 5.3.5.2 provides more details about these domains]. The correspondence between the MD and the data's OD is bijective; hence, any operation performed in the MD has a correspondence in the OD. In order to define the MD, the modulo N arithmetic and the RNS dynamic range M_1 should be defined. For an element A in the OD its correspondence in the MD is given by $_MA = A(M_1 \bmod N) \bmod N$. For the addition and subtraction of two elements in the MD it is easy to see a direct correspondence with the OD since $_MA + {_M}B = (A + B)(M_1 \bmod N)$. For the multiplication of two elements in the MD, the result is given by $_MR = {_M}A{_M}B = (AB)(M_1^2 \bmod N)$, which includes an extra factor $M_1 \bmod N$ in the result. This factor is corrected by assuring that the multiplication is followed by a reduction operation that computes $_MR = {_M}RM_1^{-1} \bmod N$. Also, in order to keep a unified reduction method in RNS, each time the reduction is performed after an addition or subtraction (these do not create the extra factor $M_1 \bmod N$ in the result), the input value to be reduced should be multiplied by $M_1 \bmod N$. Hence, for a value R to be reduced in the OD, the input of the reduction in the MD has the form $R(M_1^2 \bmod N)$. The reduction result $_MU$ is obtained by computing:

$$_MU = (_MR + QN)/M_1 \tag{5.28}$$

Note that since $QN \equiv 0 \bmod N$, (5.28) corresponds to $_MU \equiv RM_1 \bmod N$ as desired. The purpose of the value QN is to assure that $(_MR + QN)$ is a multiple of M_1, which allows the division by M_1 to introduce no approximation error in the result. A condition for $(_MR + QN)$ to be a multiple of M_1 is the verification of $(_MR + QN) \equiv 0 \bmod M_1$, which is the same as computing $_Mr_{e,1} + q_{e,1}n_{e,1} \equiv 0 \bmod m_{e,1}$ for all the RNS channels e of B_1. With the aforementioned property, Q is efficiently computed in the RNS channels as $q_{e,1} = -_Mr_{e,1}n_{e,1}^{-1} \bmod m_{e,1}$.

It is important to note that, although the resulting value $_MU$ is supposed to be reduced modulo N, for $_MR = {_M}A{_M}B$, with $_MA, {_M}B < N$, the following holds:

$$_MU = (_MA{_M}B + QN)/M_1 < (M_1N + M_1N)/M_1 = 2N \tag{5.29}$$

which means that $_MU$ is actually upper bounded by $2N$ instead of N, and the identity $_MU \bmod N = {_M}U - N$ can occur in some cases. Nevertheless, the utilization of the values $_MU < 2N$ does not consist of a significant drawback given that it allows the computation of bounded and correct results modulo

---

Algorithm 5.9 RNS reduction.

Require: ${}_M r_{e,1}, {}_M r_{e,2}$.
Ensure: ${}_M u_{e,1} = (RM_1 \bmod N)_{e,1}$ and ${}_M u_{e,2} = (RM_1 \bmod N)_{e,2}$.

1: In ch. $e, 1$: $q_{e,1} = \left| {}_M r_{e,1} n_{e,1}^{-1} \right|_{m_{e,1}}$

2: In ch. $e, 1$: $\xi_{e,1} = \left| q_{e,1} \left| M_{e,1}^{-1} \right|_{m_{e,1}} \right|_{m_{e,1}}$

3: In ch. $e, 2$: $q_{e,2} = \left| \sum_{j=1}^{h_1} \xi_{j,1} \left| M_{j,1} \right|_{m_{j,1}} - \alpha_1 M_1 {}_{e,2} \right|_{m_{e,2}}$

4: In ch. $e, 2$: ${}_M u_{e,2} = \left| (_M r_{e,2} + q_{e,2} n_{e,2}) M_1^{-1} \right|_{m_{e,2}}$

5: In ch. $e, 2$: $\xi_{e,2} = \left| {}_M u_{e,2} \left| M_{e,2}^{-1} \right|_{m_{e,2}} \right|_{m_{e,2}}$

6: In ch. $e, 1$: ${}_M u_{e,1} = \left| \sum_{j=1}^{h_2} \xi_{j,2} \left| M_{j,2} \right|_{m_{j,2}} - \alpha_2 M_2 \right|_{m_{e,1}}$ return ${}_M u_{e,1}, {}_M u_{e,2}$.

N as long as $M_1 > 4N$, which can be verified by considering ${}_M A, {}_M B < 2N$ and the inequality:

$$
\begin{aligned}
{}_M U &= ({}_M A {}_M B + QN)/M_1 & (5.30) \\
&< ((2N)^2 + M_1 N)/M_1 \\
&< (M_1 N + M_1 N)/M_1 = 2N
\end{aligned}
$$

One of the drawbacks of this reduction algorithm is that M_1 cannot be represented in the moduli set with dynamic range M_1. This is the reason for setting an extra basis B_2 with dynamic range M_2 higher than M_1, with $\gcd(M_2, M_1) = 1$ and $\gcd(M_2, N) = 1$, to compute (5.28). The use of the extra basis B_2 requires the data to be reduced $({}_M R)$ to be also represented in B_2 and, the value of Q to be converted from B_1 to B_2. This conversion can be performed with any of the methods described in Section 5.3.3 for the RNS reverse conversion, with the main difference that for each channel e the operations are computed modulo $m_{e,2}$. After computing the result ${}_M U$ in B_2, it has to be converted to B_1 and the reduction algorithm is complete. Algorithm 5.9 presents in detail the steps of the described reduction algorithm for each RNS channel e.

In Algorithm 5.9 two conversions are performed using (5.20): from B_1 to B_2 using α_1 as the correction term and from B_2 to B_1 with α_2. The computation of the values α_i can be accomplished by any of the methods described in Section 5.3.3. The method by Shenoy and Kumaresan in (5.22) requires that the operations in Algorithm 5.9 to be computed modulo the modulus m_{extra} as well so that the value of the residue corresponding to this modulus is known. The method by Kawamura et al. in (5.24) can also be used to avoid the computation modulo m_{extra}, but special attention must be paid to the approximation errors for both the conversions from B_1 to B_2, and vice-versa, in order to verify whether or not the conditions in Theorems 5.3.7 and 5.3.8 are verified. Concerning the conversion from B_1 to B_2, the value to be converted is Q, which is in the range $[0, M_1 - 1]$, whereas for the conversion from B_2 to B_1 the value converted is $_M U$ in (5.28). Given the range of Q, the conditions of Theorem 5.3.7 cannot be verified because the value $\beta = 0$, thus α_1 should be obtained from (5.24) in the conditions of Theorem 5.3.8 instead, with the associated error that can impose an extra term M_1 in the final conversion result. Therefore, for the computation of α_1 the value Φ in (5.24) should be set to zero and the minimum value of t such that $0 \leq \Delta \leq 1$ should be used [see Theorem 5.3.8]. Note that the minimum possible value of t results in the operands' size in the accumulation in (5.24) to be minimum as well. For $_M U$ it is possible to obtain α_2 under the conditions of Theorem 5.3.7, i.e., a value β such that $0 \leq _M U < (1 - \beta)M_2$ can exist as long as the difference between the dynamic range M_2 and the value to be converted is large enough. Therefore, the minimum value of t such that $\max(_M U) < (1 - \Delta)M_2$ should be computed and α_2 obtained from (5.24) using $\Phi = \lfloor 2^t \Delta \rfloor$. Note that the computation of α_i can also be accomplished by using different methods for $i = 1$ and $i = 2$ (e.g., the Shenoy–Kumaresan method (5.22) for α_1 and the Kawamura et al. method for α_2).

Algorithm 5.9 can be further optimized by merging constants so as to reduce the required precomputed constants and temporary results. Algorithm 5.10 is an optimized version of Algorithm 5.9 that comprises the merging of the constants as:

$$\lambda_{1e} = \left| -n_{e,1}^{-1} M_{e,1}^{-1} \right|_{m_{e,1}}, \tag{5.31}$$

$$\lambda_{2ej} = \left| -M_{j,1} M_1^{-1} \right|_{m_{e,2}},$$

$$\lambda_{3e} = \left| -M_1 n_{e,2} \right|_{m_{e,2}},$$

$$\lambda_{4e} = \left| M_1^{-1} \right|_{m_{e,2}},$$

$$\lambda_{5e} = \left| M_{e,2}^{-1} \right|_{m_{e,2}},$$

$$\lambda_{6ej} = \left| -M_{j,2} M_2^{-1} \right|_{m_{e,1}},$$

$$\lambda_{7e} = \left| -M_2 \right|_{m_{e,1}}$$

Note that the methodology for computing the values α_i in Algorithm 5.10 is exactly the same as the process for Algorithm 5.9 with the difference that the values q in steps 1 and 4 are used instead of the values ξ in steps 2 and 5 of Algorithm 5.9. Also note that some of the constants λ depend on the residues of N and are therefore dependent of the MD they are going to be used in. Hence, there should be a precomputed version of these constants for each target MD.

Algorithm 5.10 Optimized RNS reduction.

Require: $_Mr_{e,1}$, $_Mr_{e,2}$.
Ensure: $_Mu_{e,1} = (RM_1 \bmod N)_{e,1}$ and $_Mu_{e,2} = (RM_1 \bmod N)_{e,2}$.
  1: In ch. e, 1: $q_{e,1} = \left| {}_Mr_{e,1}\lambda_{e,1} \right|_{m_{e,1}}$
  2: In ch. e, 2: $q_{e,2} = \left| \left(\sum_{j=1}^{h_1} q_{j,1}\lambda_{2ej} + \alpha_1 \right) \lambda_{3e} \right|_{m_{e,2}}$
  3: In ch. e, 2: $_Mu_{e,2} = \left| ({}_Mr_{e,2} + q_{e,2})\lambda_{4e} \right|_{m_{e,2}}$
  4: In ch. e, 2: $q_{e,1} = \left| {}_Mu_{e,2}\lambda_{5e} \right|_{m_{e,2}}$
  5: In ch. e, 1: $_Mu_{e,1} = \left| \left(\sum_{j=1}^{h_2} q_{j,1}\lambda_{6ej} + \alpha_2 \right) \lambda_{7e} \right|_{m_{e,1}}$ return $_Mu_{e,1}$, $_Mu_{e,2}$.

Example 5.3.10 (Basis extension computation) *This example addresses the computation of a modular operation using the BE method discussed above to perform the reduction. In this example the conversion of an operand $A = 123,456$ to the MD associated with the moduli $N = 199,999$ is addressed. This conversion is accomplished by computing $_MU = A(M_1 \bmod N) \bmod N$, i.e., this operation involves a multiplication and a reduction modulo N. In order to use the BE method the following operation will be computed instead:*

$$_MU = (AB + QN)\,/M_1,\ B = (M_1^2 \bmod N)$$

The BE method requires two RNS basis, B_1 and B_2, whose dynamic ranges are coprime with $M_2 > M_1$. In this example the basis $B_1 = \{253, 247, 239\}$ and $B_2 = \{255, 251, 241\}$ are used whose dynamic ranges are $M_1 = 14,935,349$ and $M_2 = 15,425,205$. These two dynamic ranges and the modulo N must also be coprime, which is assured given that the value N selected in this example is a prime number. The moduli in the two bases herein used are of the form $2^k - c_{e,i}$ with $k = 8$. With the bases set it is now possible to obtain the residues of the operands of the multiplication:

$$a_{1,1} = 245;\ a_{2,1} = 203;\ a_{3,1} = 132;$$
$$a_{1,2} = 36;\ a_{2,2} = 215;\ a_{3,2} = 64;$$
$$b_{1,1} = 172;\ b_{2,1} = 104;\ b_{3,1} = 83;$$
$$b_{1,2} = 46;\ b_{2,2} = 55;\ b_{3,2} = 132$$

and obtain the residues of the value $_MR = AB$ to be reduced:

$$_Mr_{1,1} = 142;\ _Mr_{2,1} = 117;\ _Mr_{3,1} = 201;$$
$$_Mr_{1,2} = 126;\ _Mr_{2,2} = 28;\ _Mr_{3,2} = 13$$

Before proceeding to the BE computation with Algorithm 5.10 the precomputation of the constants λ should be accomplished:

channel e	1	2	3
λ_{1e}	153	84	21
λ_{2e1}	128	125	20
λ_{2e2}	32	63	40
λ_{2e3}	16	21	121
λ_{3e}	79	90	115
λ_{4e}	254	34	82
λ_{5e}	41	69	136
λ_{6e1}	126	216	224
λ_{6e2}	127	185	219
λ_{6e3}	232	206	119
λ_{7e}	205	192	94

The first step in the computation of Algorithm 5.10 is the calculus of $q_{e,1} = \left| M r_{e,1} \lambda_{e,1} \right|_{m_{e,1}}$:

$$q_{1,1} = 142 \times 153 \bmod 253 = 221;$$
$$q_{2,1} = 117 \times 84 \bmod 247 = 195;$$
$$q_{3,1} = 201 \times 21 \bmod 239 = 158$$

Using the values $q_{e,1}$ it is possible to compute the value α_1 using a method equivalent to the ones in Example 5.3.9. If the method by Shenoy and Kumaresan [389] is used, an extra modulus needs to be used and the operations computed to this modulus as they are computed for any of the moduli in the basis B_1 and B_2. In this example the method by Kawamura et al. [204] is used. Given that the maximum of the value that is converted from B_1 to B_2 in step 2 of Algorithm 5.10 is $M_1 - 1$ it is not possible to use the result in Theorem 5.3.7. Theorem 5.3.8 may be used instead to precompute a value $t_1 < h_1$ that comply with [see (5.26)]:

$$\frac{h_1}{2^k m_{3,1}} \left(2^k \left(m_{3,1} + 2^{k-t_1} - 1 \right) - m_{3,1}^2 \right) \leq 1 \Leftrightarrow$$
$$h_1 \left(2^k \left(m_{3,1} + 2^{k-t_1} - 1 \right) - m_{3,1}^2 \right) \leq 2^k m_{3,1}$$

The value $t_1 = 3$ is the minimum value that complies with this expression for this example. The value α_1 can be obtained in runtime by summing up the contributions from all channels:

$$\frac{truncH_{t_1}(q_{1,1})}{2^{k-t_1}} = 219 >> (8-2) = 3;$$
$$\frac{truncH_{t_1}(q_{2,1})}{2^{k-t_1}} = 220 >> (8-2) = 3;$$
$$\frac{truncH_{t_1}(q_{3,1})}{2^{k-t_1}} = 78 >> (8-2) = 2$$

it is possible to obtain α_1 as:

$$\alpha_1 = \frac{\lfloor 3+3+2+\Phi_1 \rfloor}{2^2} = \frac{\lfloor 8+0 \rfloor}{4} = 2$$

Given that Theorem 5.3.8 is used, the value $\Phi_1 = 0$ and the real value α_1 can be $\alpha_1 + 1$. This is not the case of this example, but the possibility of an error in the value in α_1 should always be considered. Anyway, in terms of the BE final result this error will not have any impact as long as M_1 is sufficiently large (the upper bound obtained in (5.30) would be the same independently of α_1). The value α_1 can now be used to compute step 2 of Algorithm 5.10:

$$q_{1,2} = (221 \times 128 + 195 \times 32 + 158 \times 16 + 2) \times 79 \bmod 255 = 182;$$
$$q_{2,2} = (221 \times 125 + 195 \times 63 + 158 \times 21 + 2) \times 90 \bmod 251 = 200;$$
$$q_{3,2} = (221 \times 20 + 195 \times 40 + 158 \times 121 + 2) \times 115 \bmod 241 = 186$$

Follows the computation of step 3 of Algorithm 5.10, which can be accomplished as:

$$_{M}u_{1,2} = (126 + 182) \times 254 \bmod 255 = 202;$$
$$_{M}u_{2,2} = (28 + 200) \times 34 \bmod 251 = 222;$$
$$_{M}u_{3,2} = (13 + 186) \times 82 \bmod 241 = 171$$

It can be verified that the obtained residues $_{M}u_{e,2}$ correspond to the expected result $_{M}U = A(M \bmod N) \bmod N = 65,482$. Note that the value $_{M}U$ as obtained with the BE method can contain an offset equal to N, i.e., the real value of $_{M}U$ can actually be $_{M}U - N$, [see (5.30) for details]. Notwithstanding, this offset does not occur in this example, this offset (a multiple of N) does not change the value $_{M}U \bmod N$.

*   The value $_{M}U$ can now be converted to the basis B_1 by computing the remaining steps of Algorithm 5.10. In step 4 one can obtain $q_{e,1}$ as:*

$$q_{1,1} = 202 \times 41 \bmod 255 = 122;$$
$$q_{2,1} = 222 \times 69 \bmod 251 = 7;$$
$$q_{3,1} = 171 \times 136 \bmod 241 = 120$$

enabling the computation of α_2. Given that the maximum value to be converted from B_2 to B_1 is less than 2N, the conditions in Theorem 5.3.7 can be verified if there is a value $t_2 < h_2$ that complies with.

$$h_2 M_2 \left(2^k \left(m_{3,2} + 2^{k-t_2} - 1 \right) - m_{3,2}^2 \right) < 2^k m_{3,2}(M_2 - 2N)$$

The minimum value of t_2 that complies with this expression is $t_2 = 2$ which *allows the computation of the contributions of each channel:*

$$\frac{truncH_{t_2}(q_{1,1})}{2^{k-t_2}} = 122 >> (8-2) = 1;$$

$$\frac{truncH_{t_2}(q_{2,1})}{2^{k-t_2}} = 7 >> (8-2) = 0;$$

$$\frac{truncH_{t_2}(q_{3,1})}{2^{k-t_2}} = 120 >> (8-2) = 1$$

and the precomputation of the value Φ_2

$$\Phi_2 = \lfloor \frac{2^{t_2}(M_2 - 2N)}{M_2} \rfloor = 3$$

The value α_2 *can be finally obtained, error free, as:*

$$\alpha_2 = \frac{\lfloor 1 + 0 + 1 + \Phi_2 \rfloor}{2^2} = \frac{\lfloor 2 + 3 \rfloor}{4} = 1$$

The residues of the value $_MU$ *in basis* B_1 *can now be obtained as:*

$$_Mu_{1,2} = (122 \times 126 + 7 \times 127 + 120 \times 232 + 1) \times 205 \bmod 255 = 208;$$
$$_Mu_{2,2} = (122 \times 216 + 7 \times 185 + 120 \times 206 + 1) \times 90 \bmod 192 = 27;$$
$$_Mu_{3,2} = (122 \times 224 + 7 \times 219 + 120 \times 119 + 1) \times 115 \bmod 94 = 235$$

It can be verified that these residues correspond to the value $_MU = 65,482$ *as obtained for the basis* B_2.

5.3.5.1 Basis extension with offset

Among all the RNS operations described above, the BE is certainly the most complex. Therefore, the optimization of this operation has a remarkable impact in the overall performance of an RNS implementation. An optimized implementation of the BE method is presented in Algorithm 5.10. All the steps in this algorithm correspond to operations that are accomplished in parallel for each RNS basis B_1 and B_2. Only the computation of the values α_i requires the serialization of the computation given that each of the channels computing steps 2 and 4 of Algorithm 5.10 require the values α_1 and α_2, respectively, to proceed with the computation. Addressing this problem, Bajard et al. [25], proposed to avoid the computation of α_1 by setting $\alpha_1 = 0$ at the expense of a higher dynamic range M_1 and a larger multiple of N offset in the resulting values $_MU$. Without the computation of α_1 and knowing that $\alpha_1 < h_1$, an extra offset $(h_1 - 1)M_1$ needs to be accommodated during the computation after the conversion from B_1 to B_2 of the value Q. Hence, the upper bounds obtained in (5.29) and (5.30) change. Considering $_MA, {}_MB < N$ and $\alpha_1 = 0$, the following holds:

$$
\begin{aligned}
_MU &= (_MA_MB + (Q + \alpha_1 M_1)N)/M_1 \\
&= (_MA_MB + QN)/M_1 + \alpha_1 N \\
&< (M_1 N + M_1 N)/M_1 + h_1 N \\
&= (h_1 + 2)N
\end{aligned}
\tag{5.32}
$$

In order to keep the value $_MU$ always bounded including when $_MA, {}_MB < (h_1 + 2)N$ the inequality $M_1 > (h_1 + 2)^2 N$ needs to be verified so that the following holds:

$$
\begin{aligned}
_MU &= (_MA_MB + (Q + \alpha_1 M_1)N)/M_1 \\
&< ((h_1 + 2N)^2 + M_1 N)/M_1 + \alpha_1 N \\
&< (M_1 N + M_1 N)/M_1 + h_1 N \\
&= (h_1 + 2)N
\end{aligned}
\tag{5.33}
$$

5.3.5.2   Using several Montgomery domains

The MD are defined as domains where the operands involved in MA modulo N_i can be alternatively represented to enhance the efficiency of these operations in RNS, namely the reduction. The input variables of an algorithm supported on MA are said to be in their OD; thus, conversions between MD and the OD are required for efficiency purposes. Considering a variable X in its OD represented in an RNS basis B_1, the corresponding variable $_MX$ in the MD associated to the modulus N_i is $_MX = X(M_1 \bmod N_i) \bmod N_i$. The inverse conversion of domain can be performed as $X = {}_MX(M_1^{-1} \bmod N_i) \bmod N_i$. Note that the MD depend exclusively on N_i and M_1. Given that the RNS basis is not changed during algorithms' computations, i.e., M_1 is invariant, one can consider that there is an MD for each N_i used in a given algorithm.

The isomorphism between an MD and the OD is exploited in operations such as the modular reduction as discussed above. The following summarizes the properties of the operations that can be performed within an MD:

- For efficiency reasons, the reduction $U = R \bmod N$ is usually computed recurring to a method that reduces the results but at the same time inserts a term $M_1^{-1} \bmod N_i$ in the result [see (5.28)]. Therefore, in order to assure that the output of this operation is in the required MD, the $_MR$ input of the reduction is replaced by $_MR(M_1 \bmod N_i)$ so that the output is $_MR(M_1 \bmod N_i)(M_1^{-1} \bmod N_i) \bmod N_i = {}_MR \bmod N_i$.

- For the addition $R = A + B \bmod N_i$ the correspondence is direct, i.e, $_MA + {}_MB \bmod N_i = (A + B)(M_1 \bmod N_i) \bmod N_i = {}_MR$. For the subtraction the rationale is the same.

- The multiplication $R = AB \bmod N_i$ in the MD results in $AB(M_1^2 \bmod N_i) \bmod N_i = {}_M R(M_1 \bmod N_i)$. Therefore, every time a multiplication is performed, the factor $M_1 \bmod N_i$ should be corrected. This correction can performed using the aforementioned reduction operation in the MD without using the factor $M_1^{-1} \bmod N_i$. The utilization of the reduction to correct this factor is also useful to keep the required dynamic ranges as low as possible. The correction can be avoided if one of the operands ${}_M A$ or ${}_M B$ is used in the OD, which would result in $AB(M_1 \bmod N_i) = {}_M R$.

As suggested by the properties above, the MD representation of the operands within the same operation should be the same, and the utilization of the MD require the precomputation of constants that depend on N_i.

Concerning the consistency of MD between operands, conversions between MD may be required, which can be accomplished with a conversion from the current MD to OD followed by a conversion from the OD to the required MD. The conversion from OD to MD is accomplished with a multiplication of the operand to be converted X by $M_1^2 \bmod N_i$ followed by a reduction:

$$\begin{aligned}
{}_M X &= X(M_1^2 \bmod N_i)(M_1^{-1} \bmod N_i) \bmod N_i \quad\quad (5.34) \\
&= X(M_1 \bmod N_i) \bmod N_i
\end{aligned}$$

while the conversion from a MD to the OD is obtained with a single reduction:

$$\begin{aligned}
X &= {}_M X(M_1^{-1} \bmod N_i) \bmod N_i \quad\quad (5.35) \\
&= X(M_1 \bmod N_i)(M_1^{-1} \bmod N_i) \bmod N_i
\end{aligned}$$

Concerning the different constants for the different MD, versions of the constants that depend on N_i should be obtained for each value i. In particular to the BE method in Algorithm 5.10, there should be a version of the constants λ_{1e} and λ_{3e} for the different N_i that are to be supported.

5.3.6 Mapping Algorithms to the RNS

Even though RNS has been efficiently used to prospect high performance in the implementation of algorithms [12,147], its utilization may not be straightforward to most designers due to several reasons. The designer may not be aware of all the RNS details, namely the ones about the operations [see Section 5.3.2.6] that support an implementation based on this system. Also, the identification of the best implementation guidelines and the best RNS parameters can be difficult to achieve as they are heavily dependent on the algorithm or target implementation platform. Another reason concerns the verification effort of an RNS implementation which is typically complex. Given that RNS introduces another arithmetic layer in the implementation, it may not be obvious why using the RNS will be efficient because a designer looking for high performance would like to be as close as possible to the computing platform instead of inserting arithmetic layers between the main algorithm and that platform.

In order to overcome some of the aforementioned constraints that can prevent designers from benefiting from RNS, a methodology to map algorithms to the RNS is introduced in this section. Notwithstanding, this chapter mainly considers cryptographic applications, the description of the methodology presented herein can be applied to generic algorithms that rely on MA. Nine different steps can be identified in the herein proposed methodology: the first six steps introduce optimizations in the algorithm so that it can support the RNS efficiently whereas the last three steps correspond to an analysis of the algorithm to extract all relevant RNS parameters. In order to support a complete understanding of these steps, in particular the six optimization steps that insert modifications in the algorithm, an illustrative example is provided in Figure 5.16 where (a) corresponds to the graph of the input algorithm (without optimizations), Figure 5.16(h) corresponds to the final algorithm optimized toward an RNS implementation, and Figures 5.16(b) to 5.16 (Cont.)(g) correspond to each one of these six steps. In the following sections each of the mapping process steps are described and mapped to the example of Figure 5.16.

5.3.6.1   Non-RNS standard optimization

In the optimization of an algorithm one can benefit from some standard optimizations that are independent of the implementation. Examples of these optimizations are the *constant folding*, *operand re-association*, and *operations combining*. The goal of these optimizations is to reduce the amount of operations in the final algorithm, by, for example, precomputing some operation that involves constants. In Figure 5.16 some nodes are identified that are a possible target for these optimizations: the two additions can be combined into a single one and constants 14 and 23 folded.

5.3.6.2   Obtain the RNS domains

The modulo operations supported on the RNS are computed by a method that is often referred to as BE. This method requires the data to be converted from its OD to a different domain called MD [see Section 5.3.5 for details]. For each value N_d used in operations modulo N_d, there is an MD. Given that operations in the different domains require different constants to be precomputed, all the domains must be identified before the implementation of the algorithm. Therefore, the list of values N_d has to be extracted from the algorithm in order to precompute and initialize the data to be used during the computation. In Example 5.3.6 this identification process is accomplished by browsing all the nodes and identifying all the different moduli used in the computation. In this particular example only two different moduli can be

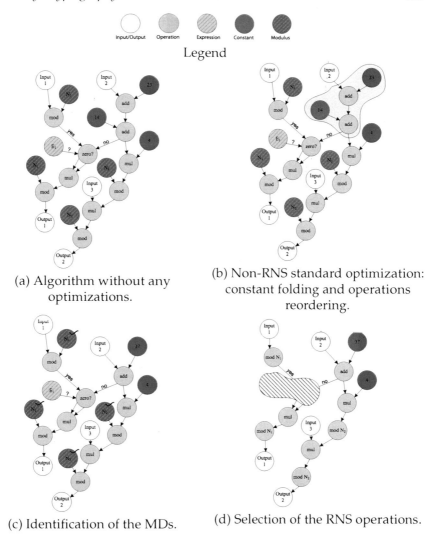

(a) Algorithm without any optimizations.

(b) Non-RNS standard optimization: constant folding and operations reordering.

(c) Identification of the MDs.

(d) Selection of the RNS operations.

Figure 5.16
Optimization steps in the mapping of an algorithm to the RNS

identified: N_1 and N_2. This moduli are then embedded in the modular reduction node.

5.3.6.3 Identify RNS operations

During the evaluation of the algorithm, not all the instructions should be considered. Only the instructions that are dependencies for the outputs need to be considered. For example, all the instructions that are used to define conditions are ignored. This pass creates a worklist with all operations that need

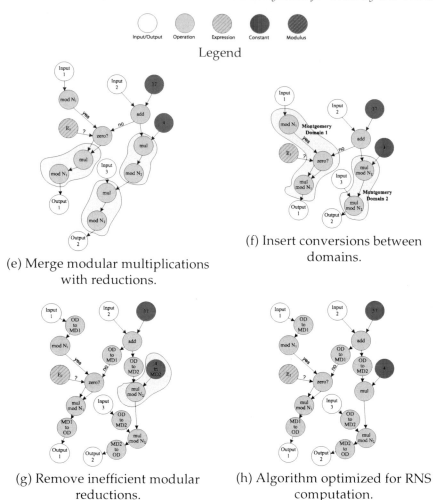

Legend

(e) Merge modular multiplications with reductions.

(f) Insert conversions between domains.

(g) Remove inefficient modular reductions.

(h) Algorithm optimized for RNS computation.

Figure 5.16 (Cont.)
Optimization steps in the mapping of an algorithm to the RNS

to be considered for RNS computation, and disregard all the others. In Example 5.3.6 there is a node that computes a conditional that is controlled by the expression E_1. Usually the expressions used in the conditionals employ bitwise operations that cannot be efficiently computed with the RNS. The computation of a conditional itself can be difficult to accomplish given that it usually requires procedures like magnitude comparison which are hard to compute with the RNS. Due to the aforementioned burdens, this optimization stage removes expressions and conditionals from consideration while optimizing the computational part of the algorithm for the RNS, and expects that all the other operations are evaluated statically prior the computation.

5.3.6.4 Merge multiplications with base extensions

The establishment of a dynamic range which accommodates the operands is mandatory to obtain an efficient RNS implementation. Maintaining this dynamic range as low as possible results in several advantages both in terms of resources and performance. Therefore, this optimization step aims at the reduction of the program's dynamic range by detecting the operations immediately followed by a reduction (employing a BE), since the required dynamic range of a combination of any operation and a reduction is potentially lower than the sequence of the two separated operations. In particular, for the modular multiplication $R = AB \bmod N$ of two numbers $A, B < N$, the multiplication itself results in a maximum value of $N^2 - 1$, while the multiplication followed by the reduction is bounded by a small multiple of N [see (5.30)]. For the particular case of the multiplication, the number of bits required to represent the dynamic range of the multiplication is almost twice the number of bits required when the multiplication is followed by the reduction. Therefore, this pass merges the two operations into a single one. In Example 5.3.6 any consecutive nodes corresponding to a multiplication and a reduction are merged into a single node and interpreted as a single operation with lower dynamic range requirements.

5.3.6.5 Montgomery domains and conversions

In the target algorithm, the designer may be interested in using MA modulo $N_1, N_2, ..., N_d$. For the RNS implementation to support the reduction operation for these different moduli, it must also support the conversion of the data to/from the d different MD, given that there is one MD for each modulus N_i in the SMA [see Section 5.3.5.2 for details]. Each time an operand is involved in a reduction, it is previously converted to the corresponding MD. Another case that requires conversion from/to MD occurs when data in the two different domains (MD and OD) is used in the same operation. To assure the operations' correctness, the inputs' domains must be the same. Therefore, the data is converted from the MD to the OD, and the operation is computed using then OD's representation each time the operands' domains do not match. This optimization step introduces additional operations that perform the conversions in order to assure that all the MD can be used while keeping data consistency. Another task of this pass is to identify the constants that are involved in the operations of any of the MD so as to annotate the constant and assure that there is an instance of this constant stored in memory in the appropriate domains. In Example 5.3.6 two domains are identified domains, each referring to N_1 or N_2. In the links that cross the boundaries of the domains, operations for the conversions are inserted. Also, the constant "4" is annotated so as to be precomputed in the respective domain, avoiding the operations for the conversion.

5.3.6.6   Remove inefficient basis extensions

The BE used in the reductions and domains' conversions are the most computationally demanding operation supported by the framework. Therefore, the BE should be avoided unless strictly required or if it does not have a remarkable impact in the performance. This pass detects the BE that can be avoided, namely the ones that result from the reduction of multiplications by constants. In this case, the constant can be a small number and the width of the result of the multiplication be only slightly higher than the non-constant operand itself, i.e., the dynamic range required by this multiplication with or without the BE following it will be almost the same. The criterion for removing the BE in these cases may vary, but a valid approach is to remove the BE if the constant's number of bits is smaller than the channel width k. With this criterion, the result of the multiplication with this constant would be unlikely to significantly increase the required dynamic range and, therefore, it would not potentially require an increase in the number of channels. Example 5.3.6 identifies a modular multiplication with reduction where the constant 4 is involved. Assuming that 4 is small regarding the channel width, the multiplication with reduction is replaced by an ordinary integer multiplication and the assigned domain for the constant is modified, in this example, to the OD.

5.3.6.7   Evaluate dynamic ranges

After inserting all of the algorithm's modifications, it is possible to evaluate the required dynamic ranges. One of the tasks of this stage is to find the RNS bases that support the required dynamic ranges M_1 and M_2: *i)* the dynamic range of B_1 (M_1) must be greater than the maximum number to be represented during the algorithm's execution; and *ii)* B_2 is obtained such that $M_2 > M_1$ and $gcd(M_2, M_1) = 1$. Several methods can be used to pick the bases. One possible method to obtain the basis is the identification of the values $m_{e,i} = 2^k - c_{e,i}$ for increasingly higher values of $c_{e,i}$ that are pairwise coprime with all the other moduli that are already part of the bases. By considering A and B with S_A and S_B bits, respectively, the dynamic range for the: *i)* addition is $\max(S_A, S_B) + 1$; *ii)* multiplication is $S_A + S_B$; *iii)* reduction of A is $S_{\lceil R \rceil}$ where $R = (2^{S_A} + (Q + M_1)N)M_1^{-1}$. The reason for using the value $Q + M_1$ instead of Q in the dynamic ranges' evaluation relates with the methods that can be adopted to compute the offsets α_1 and α_2 in the BE method in Algorithm 5.10, namely the method proposed by Kawamura et al. that can be used to compute both α_1 and α_2. This method requires the lowest dynamic range and for some implementations, namely hardware ones [147], it does not introduce any computational overhead. Given that the maximum value for Q is $M_1 - 1$ the conversion of Q from B_1 to B_2 has to be accomplished under the conditions of Theorem 5.3.8, which can insert a unitary error in the offset α_1. This error makes the result of the conversion of Q to have an extra term M_1; therefore, the maximum value after the conversion is

$Q + M_1$ instead of Q. If a method without error is used, the term M_1 can be ignored while computing the dynamic range.

In other applications such as fixed point digital signal processing, operand bit-width estimation can be accomplished using semi-analytical approaches and error models [241]. A method that relies on the computation of the algorithm to be implemented using the operations' worst-case dynamic ranges can be used. By taking into account the aforementioned operation's dynamic ranges and assuming the inputs have its maximum bit-width size defined, it is possible to accurately evaluate the required dynamic range before the implementation. Note that the dynamic range also depends on other properties associated with the BE, and the dynamic range of prior operations estimated so far.

5.3.6.8   Extract conditions

In an early step of the optimization process, the evaluation of conditions was disregarded in terms of the RNS optimizations. Now that the algorithm is optimized, these conditions need to be statically evaluated so that this information can be used to run the algorithm in the target device. Therefore, since the implementation of the RNS algorithms does not foresee the possibility of data evaluation, conditionals must not depend on the input data, which means that all conditions must be evaluated prior to the implementation. Also, given that the dynamic ranges may depend on these conditions, evaluating the conditions statically enables the accurate computation of the required dynamic range without the need of overestimating its value. From the conditions extraction, it is possible to obtain a string of zeros and ones that can be used by the target implementation to control the program flow according to the content of this string.

5.3.6.9   Extract register and constant map

The evaluation of the number of registers and constants required by the algorithm is required. The variables and constants can be assigned to memory addresses: two addresses need to be assigned to each variable to store the value modulo $m_{e,1}$ and $m_{e,2}$, respectively. The identification of addresses for temporary variables must be also accomplished. These temporary addresses should be reused in order to keep the required number of memory locations as small as possible, turning the upcoming implementation more efficient.

5.3.7   Implementing an RNS-based Algorithm

In the previous section, a methodology to obtain an algorithm optimized to an RNS implementation was described. The final step is to obtain an efficient parallel implementation starting from the optimized version of the algorithm. The implementation is different for each target platform given that the way the parallelization capabilities are exploited differ, but the main concepts are similar. In this section two target devices are considered as examples to obtain an efficient parallel implementation of an algorithm based on the RNS: GPU and FPGA.

5.3.7.1   GPU implementation

The utilization of RNS arithmetic in GPU has been successfully exploited in the literature [11, 419]. These implementations suggest not only that an RNS approach can enhance the throughput of the operations being held, but also reduce the latency of such operations. However, the proposals in the related state of the art consist of static implementations (using CUDA or OpenCL) targeting a given algorithm and did not result from a generic methodology as the one herein presented. Furthermore, none of these proposals implements the RNS forward and reverse conversions. All these operations are herein addressed and obtained using the PTX language which is similar to an assembly language for NVIDIA GPU, enabling the fine tuning of the implementation for these devices. In order to use the PTX kernel efficiently a C program was developed for the GPU's host using the CUDA API which allows one to load the PTX kernel effortlessly.

The support for general purpose computation with NVIDIA GPU started with the Tesla architecture [254] which has evolved to slightly different architectures but all sharing the same basic attributes. Concerning the implementation herein presented, the relevant information about the architecture is the hierarchy that supports data parallelism: a GPU kernel is computed by several threads grouped in blocks. The threads within the same block run in the same multiprocessor and can therefore share the same resources, namely a fast shared memory and a local cache that can efficiently operate as constant memory. In order to extract full advantage of the microprocessors, the threads are grouped in 32 simultaneous running threads (warp). The full power of a multiprocessor is not completely exploited if less than 32 threads are simultaneously available to run. Data dependencies or divergent control flows between the different threads in the same block can limit the number of threads ready to run and should therefore be avoided to maximize performance. While implementing the RNS algorithm the desired level of parallelism has to be decided: *i*) by setting the RNS channel width k so that the number of threads required to support the algorithm's dynamic range can be derived; *ii*) by specifying the number of operations $N_{ops.\ per\ block}$ to be run within a block (same multiprocessor); and *iii*) by specifying the number of blocks N_{blocks}. Hence, $N_{ops.\ per\ block} \times N_{blocks}$ instances of the same algorithm in the GPU can be run for different input data sets. Note that some combinations of k, $N_{ops.\ per\ block}$, and N_{blocks} can exhaust the available resources in the GPU or in a multiprocessor (e.g., shared memory).

The utilization of optimized PTX to compute the RNS-related arithmetic translates to several advantages in terms of performance when compared to implementations obtained with high-level languages. An advantage of the

PTX description is that registers can be more efficiently assigned and reused. The value for the channel width k can be any value but the target NVIDIA GPU have a 32-bit datapath. Therefore, for any value $k > 32$, multiprecision arithmetic has to be implemented which requires the application of masks and operations with carry flags. On the contrary to the C description supported by CUDA, the PTX directly supports the manipulation of carry flags. On the other hand, the utilization of PTX with operations requiring the carry flags results in several data dependencies, e.g., by propagating the carry in a multiprecision addition. This can limit the efficiency of the program by exposing the deep pipeline in each of the several scalar processors inside a GPU's multiprocessor. Nevertheless, assigning a large number of threads (more channels or more operations per block) allows the GPU scheduler to interleave the execution of threads each time a data dependency is identified among the running threads, mitigating the impact of the data dependencies in the performance.

Each thread can easily handle the carry values and use the built-in instructions for 32-bit addition, subtraction, multiplication (high and low part), and shift (left and right) to implement the RNS operations as discussed in Section 5.3.2.6. Also, the RNS forward conversion can be easily implemented since it corresponds to a sequence of modular multiplications and additions for each thread (channel). For the BE and reverse conversion, the implementation is more complex given that the threads need to cooperate. Figure 5.17 presents the scheduling of the operations on each thread for the computation of the BE taking into account the steps of Algorithm 5.10. In order to assure correctness, several synchronizations need to be inserted each time a thread needs data that was previously computed by any other thread. The values $\hat{\alpha}_j$ are computed by a single thread using (5.24), as a simple accumulation and shift of all the truncated contributions of each thread. The data sharing between all the running threads is assured by the device's shared memory. In the related state of the art the method in Section 5.3.5.1 has been used to compute the BE, which avoids the computation of $\hat{\alpha}_1$, reducing the size of the divergent section in the BE computation [11]. Nevertheless, as discussed in Section 5.3.6.7, the implementations herein presented adopt the method by Kawamura et al. [204] discussed in Section 5.3.5 [see Example 5.3.10] because it assures a lower dynamic range.

Figure 5.18 presents the operations' schedule for the RNS reverse conversion resulting directly from the aforementioned reverse conversion method in (5.20) split up by several threads as introduced in Section 5.3.3: each thread computes its contribution to each position of the final result, and then a single thread computes the correction offset $\hat{\alpha}_2 M_2$ and accumulates all the contributions. Note that Figure 5.18 presents the thread's contribution to final result as k-bit words. In order to enhance the efficiency, these contributions are computed as 32-bit words instead, which is the width of the datapath of the target GPU.

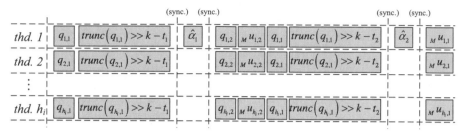

Figure 5.17
Schedule of the operations to compute a BE according to Algorithm 5.10

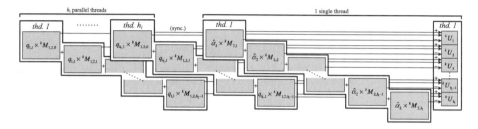

Figure 5.18
Schedule of the operations to compute the RNS reverse conversion with several threads

5.3.7.2   FPGA implementation

Two different approaches have been proposed to accelerate MA based on RNS with dedicated hardware [317, 374]. In [374] an accelerator for EC cryptography over a finite field $GF(p)$ with p prime was proposed. The architecture of this accelerator consists of three main components: *i*) an RNS forward converter; *ii*) a register file connected with buses to the channel's arithmetic units; and *iii*) an RNS reverse converter followed by a projective to affine converter (EC coordinates conversion). The multiplication is performed by a Horner scheme [175] and no optimizations are introduced regarding the reductions. This accelerator performs an initial conversion of the input data to RNS, uses the dedicated arithmetic units to implement the modular operations with the operands stored in the registers and, finally, converts the obtained results back to the binary weighted representation.

An accelerator for the modular exponentiation, a main building block of the RSA cryptographic algorithm, was proposed in [317]. The architecture of this accelerator is composed by as many multiply-accumulate units as the number of RNS channels, connected to each other in a ring shape. Each one of these units is fed by a RAM that stores the input dependent data and a ROM that stores the required constants. This architecture implements the RNS version of the modular reduction by using a BE scheme such as the one presented in Algorithm 5.10. In order to handle the required conversions between bases, the conversion offset values $\hat{\alpha}_j$, similar to the ones in (5.24), are computed by a unit present in the logic that implements each channel. This architecture was upgraded to handle the EC point multiplication, a key operation of EC-based cryptography [147]. Although these architectures allow the utilization of the RNS properties, they consist of dedicated implementations that can only support a specific application.

The implementation herein presented consists of a fully functional and programmable accelerator specified with an HDL language able to compute the desired application, which can be synthesized for an FPGA device. The architecture of the accelerator is depicted in Figure 5.19. It consists of several REs each one responsible for computing all the arithmetic for one of the RNS channels in basis B_1 and B_2. The architecture also contains one input buffer and one output buffer, implemented as FIFO queues, which are the only components the designer has to interface with. There is a dedicated unit ('α accumulator') that computes the RNS conversion offset ($\hat{\alpha}_j$) required for the BE method as in (5.24). The control of the operation is assured by the broadcast of microinstructions to all the RE. This architecture is similar to the architectures in [317] and [147] but with the following main improvements: *i*) the RE contain a simple pipeline with several data forwarding paths instead of a regular pipeline designed and balanced to suit a specific application; *ii*) the RE are microcontroled with a sequence of instructions generated by the framework for the target application; *iii*) the RE not only support the standard modular operations but also the forward and reverse conversions; and *iv*) an accumulator and FIFO buffers are introduced to implement the required conversions, providing a simple interface. The overview of the framework's supported architecture is presented in Figure 5.19. Figure 5.20 depicts the structure of the RE used in the architecture in Figure 5.19. The RE include the required resources for computing all the arithmetic. These resources include a binary multiplier and adder, as well as two bitwise operation blocks which are responsible for multiplexing, complementing, and padding the operands. A pipeline is used, along with some data forwarding paths, to reduce the effective number of clock cycles per operation and increase the throughput. Furthermore, the RE includes a dual port RAM, which is directly addressed by the broadcasted instructions and is responsible for storing input/output and temporary data, as well as the RNS constants. The control of the RE is directly included in the instructions that feed the hardware structure. Note that no multiprecision computation is directly instantiated within the RE given that the physical implementation of the arithmetic units is to be decided by the synthesis tool.

The reverse conversion is computed as specified in Figure 5.18 with the difference that: *i*) the computation is performed by the RE instead of threads; *ii*) the data is shared not recurring to a memory but traveling in the ring; and *iii*) the final data accumulation is accomplished by a hardwired accumulator,

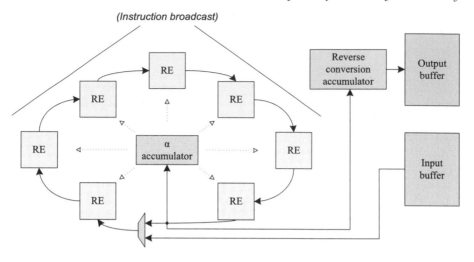

Figure 5.19
Architecture supported by the proposed HDL specification

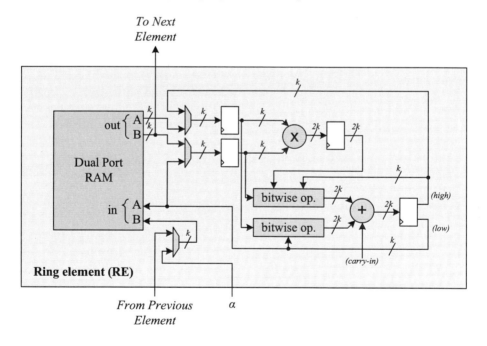

Figure 5.20
The RE: the basic processing element of the architecture employed in the HDL specification

that feeds the output buffer with the data already in the binary weighted format.

Figure 5.21 presents a schematic of the computation of the BE in Algorithm 5.10 with the architecture in Figure 5.19. The shape of the architecture is mainly motivated by the need to improve the performance of the BE, which is the most demanding operation in the RNS-based arithmetic. This operation has $O(h_i^2)$ complexity ($O(h_i)$ in a single channel), with h_i the number of channels, due to the summation in the steps 2 and 5 of Algorithm 5.10. In this algorithm, each channel e needs to gather data from all the other channels j. The ring shape of the architecture optimizes the computation of $\sum_{j=1}^{h_1} q_{j,1} \lambda_{ej}$ enabling the forwarding of the values $q_{j,1}$ to the next RE after it is used by the current one. When a value $q_{j,1}$ completes a trip around the ring, it is assured that all the RE have gathered all the information they need to proceed with the computation.

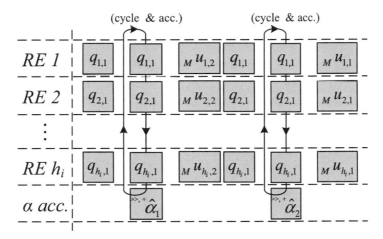

Figure 5.21
Schematic distribution of the BE workload presented in Algorithm 5.10 by the RE available in the architecture implemented with the HDL specification

Another, yet not less important, characteristic of the architecture is the scalability. It is straightforward to tradeoff between the number of RE and the datapath's size inside the RE, which introduces extra flexibility in the design and allows the exploration of the best configuration given not only the timing and resource constraints but also the RNS algorithm's dynamic range.

5.3.8 Experimental Results

Even though the methodology proposed in Section 5.3.6 and implementation guidelines in Section 5.3.7 are generic and can fit the demands of any algorithm based on MA, the public key cryptography is probably the most prominent application that can benefit from this approach [14]. Therefore, the experimental assessment of the proposed methodology is herein accomplished by addressing two typical public key cryptographic operations: the 1024-bit wide modular exponentiation used in the RSA and the EC point multiplication over an underlying 224-bit wide finite field. Both target algorithms are thoroughly described in Section 5.3.2. The implementation of the modular exponentiation and EC point multiplication results directly from the steps in Algorithms 5.6 and 5.8, respectively. The methodology discussed in Section 5.3.6 was applied to both algorithms and an optimized RNS version was obtained. The implementation of the EC point multiplication is more complex given that additions, subtractions, multiplications, and conditions are required. As discussed in the previous sections, implementation for two different devices are herein addressed: *i*) GPU (NVIDIA), through the generation of an optimized PTX description of the algorithm, and *ii*) reconfigurable hardware, through the generation of a technology independent HDL specification of a processor and the microcode required to program it. In the following subsections, results obtained for each of these devices are assessed and discussed. Table 5.3 summarizes the experimental setup used.

5.3.8.1 GPU implementation

The following parameters are required to obtain the final GPU implementation: *i*) the channel width k; *ii*) the number of thread blocks B to be used during the computation; and *iii*) the number of Operations Per Block (OPB). With these three variables it is possible to define all the levels of parallelism to be exploited in the GPU: *i*) with larger k, less threads are required to operate a single instance of the target algorithm but higher precision arithmetic is required for each thread; *ii*) B specifies how many multiprocessors of the GPU are to be used; and *iii*) OPB specifies the number of parallel instances of the algorithm that are run in a single multiprocessor (block). The analysis herein presented focuses on the performance of the implementation of a single block, given that there is no cooperation between blocks. Therefore, in particular for the target algorithms which are not data intensive and therefore not likely to be constrained by memory-bandwidth issues, the execution time for one block will be similar to the execution time of two blocks and the maximum throughput will be close to the throughput of a single block multiplied by the number of multiprocessors available in the GPU. Note that, notwithstanding several thread blocks and operations per block are addressed in the following results, the RNS parallelism can only be explored in a single operation, i.e., it does not enhance the operation level parallelism.

Figure 5.22 presents the experimental latency and throughput results for the two target algorithms for a single block and the experimental setup 580GTX (see Table 5.3). The modular exponentiation is referred to as RSA. Note that for $OPB = 20$ and $OPB = 12$ there are some channel width val-

Table 5.3
Experimental setup used in the evaluation of the RNS algorithms

(a) GPU Back-end

Ref.	GPU	Host CPU	CUDA API
580GTX	NVIDIA GeForce 580 GTX (Fermi arch 2.0)	Intel Core 2 Quad Q9550	version 3.2
285GTX	NVIDIA GeForce 285 GTX (Tesla arch 1.3)	Intel Core i7 3930K	version 4.2
8800GTX	NVIDIA GeForce 8800 GTX (Tesla arch 1.0)	AMD Opteron 170 Dual Core	version 3.2

(b) HDL Back-end

Ref.	Xilinx FPGA part	Synthesis and place and route tools
V4	Virtex 4 - xc4vsx55ff1148-12	Synopsis Simplify - version E-2010.09-SP2
V5	Virtex 5 - xc5vlx220ff1760-2	Xilinx ISE tools - version 12.4

ues that do not present results. This is due to the lack of enough resources, namely shared memory, for those configurations. All configurations' results comprise the host to/from GPU data transfers through the PCI bus. Due to the interleaved execution of the threads, given the data dependencies caused by the multiprecision arithmetic, the latency figures for different OPB are almost overlapped which causes the throughput to be approximately proportional to the OPB value.

From Figure 5.22(a) it can be concluded that the latency is almost invariant for different values of OPB. This observation has a direct connection with the PTX implementation of the multiprecision arithmetic, which results in several data dependencies and, consequently, pipeline exposure. Given that the GPU has a fast scheduler and thread switcher, each time there is a data dependency which prevents the running thread from continuing, another thread is issued during the time the pipeline would be stalling. Therefore, for configurations with larger values of OPB, there will be more threads available to run, but this will have almost no impact in the latency, given that all operations in the block will be running interleaved, occupying the time slots the other threads would be stalling. The same effect can be observed in Figure 5.22(b). This effect allows one to conclude that the GPU implementation may not be as competitive in terms of latency as it is in terms of throughput. Due to the latency behavior properties, the values of throughput in Figures 5.22(c) and 5.22(d) are approximately proportional to OPB.

Concerning the behavior of the implementation performance with the channel width k, depicted in Figure 5.22, there are values of k that result in local latency minimums (throughput maximums). These values are usually the values of k that are multiples of 32, the GPU's datapath size, e.g., in Figures 5.22(c) and 5.22(d) the most competitive configurations are obtained for $k = 32$ and $k = 64$, respectively. Nevertheless there are also other values with similar behavior. These values usually result in a more efficient exploitation of the computation power available by distributing the threads by the GPU warps (the 32 threads that can run simultaneously) more efficiently.

5.3.8.2 FPGA implementation

In order to obtain the HDL specification of the accelerator for a given algorithm, only the channel width k must be defined. After obtaining the resulting files the HDL can be synthesized to a given technology [see Table 5.3(b) for the tools used in the experiments]. Once the implementation is obtained, the accelerator can be easily integrated in any design by interfacing with the FIFO buffers, and by streaming the required microinstructions into the accelerator to execute the algorithm. Figure 5.23 presents experimental results for the latency and resource occupation for each of the two addressed applications. In this case, the accelerator only performs a single operation at once; therefore, the throughput T can be obtained from the latency L as $T = 1/L$. The results in Figure 5.23 are normalized to the most competitive configura-

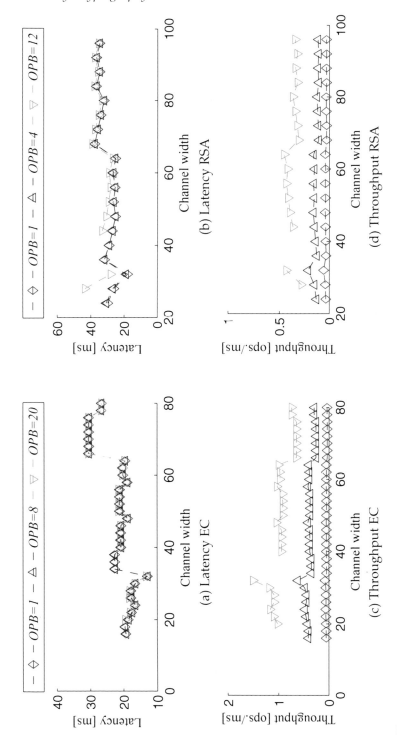

Figure 5.22
Latency and throughput results of the RNS implementations using a GPU device (580GTX) for a single block and different numbers of operations per block

tion, i.e., for each metric the configuration with less latency or resources is set to one and for the other configurations the metric is provided as a proportion of the best configuration (an X result means that the configuration is X times slower or requires X times more resources than the best configuration). The obtained minimum latencies are 5.8 and 5.3 ms for EC and RSA, respectively.

As Figure 5.23 suggests, by varying the channel width it is possible explore the latency/resources tradeoff. Concerning latency, low-width channels correspond to a higher number of channels, which results in more clock cycles to implement the BE (the ring size increases while the channel width decreases), and also more clock cycles to reduce inside the RE (maximum size of $c_{e,i}$ in $m_{e,i} = 2^k - c_{e,i}$ also increases with the number of channels). For channels with larger width the number of cycles to compute the algorithm will decrease but, since the complexity of the RE increases, the operating frequency decreases. Hence, as depicted in Figure 5.23(a) shows, the minimum latency is achieved at 96- and 64-bit channel width mark for the RSA and EC, respectively, and then the latency starts increasing. Concerning the RAM resources, the FPGA has fixed sized Block RAM that are inferred by the synthesis tool and can be configured with up to 32 bits per position. Hence, some widths, multiples of 32, will better suit the Block RAM configuration and the behavior in Figure 5.23(c) is not monotonic. The DSP48 slices in the target FPGA are only used to multiply since they possess a dedicated multiplier and, recalling that the HDL specification is technology independent, no extra effort in fully utilizing the FPGA DSP48 slices capabilities was taken. Since the DSP48 slices are the RE multipliers, their amount will increase with the complexity of the RE. By increasing the channel width linearly, it results in a linearly decreasing number of REs, but the number of required DSP48 slices increase due to the multiplication complexity. The number of required DSP48 slices increases when the channel width is increased, as Figure 5.23(d) shows. The number of slices is a tradeoff between the number of RE and the cost of a single RE, since they are used for the additions and bitwise operations, as well as in DSP48 slices' interconnections. The slices are expected to increase for large values of the channel width as an effort of the synthesis tool to enhance timing. The number of resources of the RSA application (the operations' modulus is 1024-bit wide) is larger than the EC's (the operations' modulus is 224-bit wide), but since the resources are strongly related with the channel implementation, the normalized figures for the resources are similar for both applications.

5.3.8.3　Discussion

In Sections 5.3.8.1 and 5.3.8.2, results for two typical operations that can be implemented with the presented methodology are introduced. In order to contextualize the obtained results, they are compared with related state-of-the-art implementations that share common characteristics with the presented implementations, i.e., are programmable solutions and/or employ

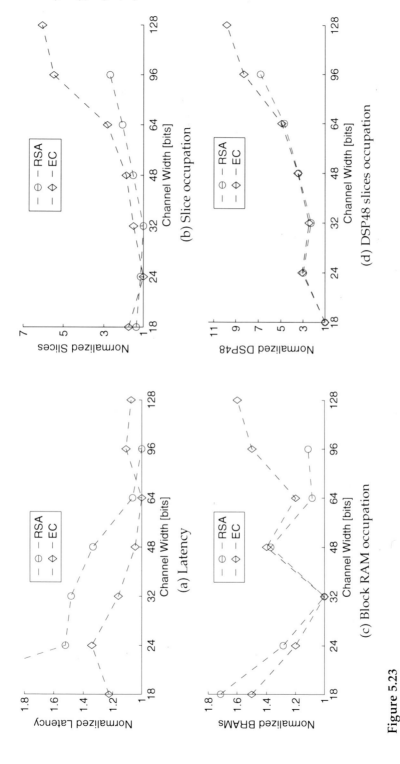

Figure 5.23
Normalized experimental results of the proposed framework targeting a Xilinx Virtex 4 FPGA for a modular exponentiation (RSA) and EC point multiplication. The value 1 stands for the most competitive configuration for each one of the metrics.

RNS-based computation. For the experimental setup herein employed, Table 5.4 summarizes the results for the most competitive configurations in terms of channel width k and, for the GPU back-end, B and OPB, along with the aforementioned related state-of-the-art results. Note that none of the related state-of-the-art implementations compute the RNS forward and reverse conversions. In [11] a GPU implementation of the EC point multiplication using RNS and the BE method to compute the modular reduction as described in Section 5.3.5 is presented. To the best of one's knowledge, this implementation provides the lowest latency and highest throughput known to date for GPU platforms. It was possible to obtain results for the same experimental setup [see Table 5.3(a)] employed in this chapter. In [419] the evaluation of several BE methods is accomplished and results for both an RNS and non-RNS implementation of the modular exponentiation are stated. The lowest latency of 144 ms is obtained with the RNS method, however, the highest throughput per multiprocessor (67.75 ops./s) is obtained with the non-RNS method. These results are stated for an NVIDIA 8800 GTS GPU that corresponds to the first GPU supporting the Tesla architecture. The results for the 8800GTX experimental setup suggest an improvement of 21% in latency but lower throughput. Regarding [11], the comparison suggest a different conclusion with similar latency values but 76% improvement for the throughput. Given that the implementation herein proposed requires storing the same amount of constants in the shared memory independently of the number of instances of the algorithm running in each multiprocessor, for devices with less shared memory resources the ratio of memory assigned to the constants is higher, and less memory is available to run more parallel instances of the same algorithm and increase throughput. Therefore, the proposed implementation is more competitive for the more recent GPU architectures [see results for the implementations 285GTX and 580GTX in Table 5.4].

The FPGA-based implementations more that halve the latency of the GPU implementations but cannot supersede them in terms of throughput. Given that the dedicated hardware implementations in [317] and [147] refer to different technologies than the ones herein addressed [see Section 5.3.7.2 for a detailed description of these implementations], a fair direct comparison cannot be supported both in terms of latency and resources. Furthermore, the proposed implementation is technology independent and is not fine tuned as the reference implementations. Therefore, the results in Table 5.4 only serve to contextualize the obtained results. In order to complement this contextualization, the following presents a summary of the resource utilization of the FPGA implementation regarding [147], which also refers to an FPGA implementation: the presented implementation requires 12 Block RAM (18 KB each), 64 DSP and 3435 Slices, 3%, 12%, and 13%, of the total available resources in the device, respectively; and the implementation in [147] requires 96 DSP and 9177 ALM (the slice equivalent for Altera devices).

Table 5.4

Summary of the best results for the experimental setup in Table 5.3 between all the configurations and comparison with the reference designs. The improvement figures refer to the experimental setup with closer characteristics to the reference implementations.

		GPU Back-end			HDL Back-end		
	Ref.	Latency [ms]	T.put [ops./s][1]	Ref.	Latency [ms]	T.put [ops./s][2]
EC point mult.	[11] NVIDIA 580 GTX	12.0	1,252	[147]	0.68[3]	1470[3]
	580GTX	12.9	2,208	V4	5.77	173.4
	(improvement)	(-1.08)	(+1.76)		(-8.48)	(-8.48)
	285GTX	63.43	151.6	V5	6.41	156.0
	8800GTX	65.2	146.9			
Modular exp.	[419] NVIDIA 8800 GTS	144	67.75	[317]	4.2[4]	238[4]
	8800GTX	119	23.68	V4	5.31	188.4
	(improvement)	(+1.21)	(-2.86)		(-1.26)	(-1.26)
	285GTX	116	24.41	V5	3.54	282.3
	580GTX	17.9	703.2			

[1] The throughput is evaluated for a single multiprocessor.
[2] The HDL back-end only runs a single operation at once, thus the latency is the inverse of the throughput.
[3] The results were obtained for an Altera Stratix II technology and a 256-bit wide prime field.
[4] The results were obtained for a $0.25\mu m$ CMOS technology.

5.3.9 Summary

This section thoroughly discusses a methodology for implementing public key cryptographic algorithms based on *MA* with the *RNS*. This methodology supports the prospection of data-level parallelism in the algorithms suiting the demands of an efficient implementation for devices that typically tend to provide increasing parallel computation capabilities. Implementations of the modular exponentiation (main operation in the *RSA* algorithm) and the *EC* point multiplication (main operation of *EC*-based cryptography) were obtained based on this methodology targeting *GPU* and *FPGA*, either by the development of an optimized *PTX* kernel and an *HDL* specification, respectively. The resulting implementations are easily scalable for different algorithms and performance/resource demands enabling the tuning of the implementation by deciding the number of parallel flows that take part in the computation. On top of that, the obtained implementations support an extra level of flexibility given that they are programmable and can benefit from the reconfiguration capabilities of the devices to scale the implementation to different algorithms. Besides its flexibility, the implementations obtained with this methodology suggest to have competitive performance figures. The throughput marks of the related state-of-the-art implementations for an *EC* point multiplication were superseded by a factor of 1.76 with the implementations obtained with this methodology for an NVIDIA Fermi architecture.

6

Physically Unclonable Functions: A Window into CMOS Process Variations

Raghavan Kumar

University of Massachusetts, Amherst

Xiaolin Xu

University of Massachusetts, Amherst

Wayne Burleson

University of Massachusetts, Amherst

Sami Rosenblatt

IBM Systems and Technology Group

Toshiaki Kirihata

IBM Systems and Technology Group

CONTENTS

6.1 Introduction

The role of embedded systems in everyday life has increased significantly in the last decade and will likely continue to in the forthcoming days. Some major embedded systems include smartphones, tablets, payment systems, smart cards, and medical devices. The trend of *ubiquitous computing* has therefore seen a great scope of improvement and has led to an era of *Internet of Things (IoT)*, where the devices have communication ability in addition to computation power. However, because of their ubiquitous nature, they also bring out security and privacy issues as they pose an ideal target for attackers. In particular, protection of sensitive data stored on the devices has brought forth an alarming issue and demands cryptographic protection.

Most of the embedded systems pose tight constraints on area and energy because of their low cost. So, the trend towards lightweight cryptographic primitives is becoming extremely popular in the design market. Most of the primitives based on classical cryptography are based on the concept of a *secret binary key* embedded on the device. However, they pose some serious security vulnerabilities especially against physical attacks (invasive, non-invasive, and side-channels) and software attacks. The fact that the key has to be stored in a non-volatile memory further aggravates the problem. All these issues have served as one of the major motivations towards the development of Physically Unclonable Functions (PUFs) in the literature.

A PUF is a partially disordered system that maps a set of external inputs C to a response R. In silicon PUFs, the mapping function is decided by process variations arising during the manufacturing process. The manufacturing process is extremely complicated especially in sub-45 nm design space and is hard to control even by the manufacturer. This ensures that a manufacturer cannot produce an identical tuple of an IC with the same layouts. PUFs can be employed in several security-related applications, and the scope is often limited by the number of responses that can be generated. PUFs can be broadly classified into "strong" and "weak" PUFs based on the number of independent responses that it can generate. Please note that the terms "strong" and "weak" do not have any reference to the security level of a PUF. A strong PUF can produce a large number of responses (R_i) for different inputs C_i and can be used in security protocols, key establishment, and device authentication. A classical example for a strong PUF is an arbiter PUF [416]. A weak PUF, on the other hand, has a reduced response space and produces only a single response in the worst case scenario. So, the response(s) must be kept secure from the external world and must never be shared with a third party. The weak PUFs can be used in classical cryptosystems for deriving the secret key. One of the typical examples for a weak PUF is an SRAM PUF [172].

PUFs, although primarily developed for hardware security purposes, also provide insights into process and timing variability in CMOS circuits. This chapter looks at PUFs from both the perspective of a variation-aware CMOS circuit designer and a hardware security primitive designer. Sources of variation include manufacturing process variations such as doping and geometry, as well as environmental variations such as temperature and supply voltage. The former are generally considered static but vary across the die, wafer, and lot, while the latter are typically common-mode for a given PUF and will vary over time as different challenges are applied. These characteristics should be exploited in the design and use of PUF circuits to provide the best metrics of uniqueness, reliability, and unpredictability.

This chapter focuses on delay-based PUFs and the lessons learned go beyond PUF design and apply more generally to the timing issues in CMOS circuits. Techniques for simulating process variations based on Monte Carlo techniques are presented. Measurements on a 45nm test chip are used to validate these techniques, thus allowing future research to proceed with simulation-based techniques that overcome the costs, inconvenience, and limitations of test chips. Other recent PUF test chips [176] and FPGA-based PUFs [416] confirm these results.

This chapter is structured as follows. We begin with an introduction to common PUF terminologies and performance metrics used throughout this chapter in Section 6.2. Section 6.3 briefly describes the sources of process variations in CMOS circuits. Section 6.4 provides a brief overview of some of the most popular silicon-based PUF circuits along with their performance metrics. Next, we discuss the design flow adopted for any silicon-based PUF implementation in Section 6.5. Based on the design flow, we discuss the sub-threshold mode of operation of arbiter PUFs and compare its performance to super-threshold PUFs in Section 6.6. Section 6.7 describes the impact of technology scaling on performance metrics of PUF circuits. Sections 6.6 and 6.7 help to understand the techniques for simulating variation-aware CMOS circuits. We provide outlook and conclusions in Sections 6.8 and 6.10, respectively.

6.2 PUF Terminologies and Performance Metrics

The terminologies and performance metrics used in the field to evaluate PUF devices are briefly summarized in the following sections.

6.2.1 Challenge-Response Pairs (CRP)

As the name suggests, PUF circuits can be envisioned as a function mapping a set of inputs to outputs. However, as identified in [265], PUF circuits do not implement a true function as they can produce different outputs for an input under different operating conditions. The inputs to a PUF circuit are known as *challenges* and outputs are referred to as *responses*. A challenge associated with its corresponding response is known as a *Challenge-Response pair (CRP)*. In an application scenario, responses of a PUF circuit are collected and stored in a database. This process is generally known as *enrollment*. Under *verification* or *authentication* process, the PUF circuit is queried with a challenge from the database. The response is then compared against the one stored in the database. If the responses match, the device is authenticated.

6.2.2 Performance Metrics

There are three important metrics used to analyze a PUF circuit, namely uniqueness, reliability, and unpredictability. Uniqueness is addressed in

Section 6.2.2.1, reliability in Section 6.2.2.2, and unpredictability in Section 6.2.2.3.

6.2.2.1   Uniqueness

PUF devices are primarily used to generate unique signatures for device authentication. In this application, it is desirable to have a large difference between responses from any two PUF instances. Here, the two PUF instances may be from the same wafer or different wafers. A typical measure used to analyze uniqueness is known as *inter-distance* and is given by [214]:

$$d_{inter}(C) = \frac{2}{k(k-1)} \sum_{i=1}^{i=k-1} \sum_{j=i+1}^{j=k} \frac{HD(R_i, R_j)}{m} \times 100\% \qquad (6.1)$$

In equation (6.1), $HD(R_i, R_j)$ is the Hamming distance between two responses R_i and R_j of m bits long for a particular challenge C, and k is the number of PUF instances under consideration. The desired inter-distance is 50%. By carefully looking at equation (6.1), one can correspond the inter-distance $d_{inter}(C)$ to the mean of the Hamming distance distribution obtained over k chips for a challenge C. It is also useful to obtain the standard deviation of Hamming distance distribution given by $\sigma_{inter}(C)$, which measures the extent of deviation in Hamming distance from the desired inter-distance. Lower $\sigma_{inter}(C)$ is preferable for PUF design.

While designing a PUF circuit, inter-distance is often measured through circuit simulations. A common practice is to perform Monte Carlo simulations over a large population of PUF instances. Though there is no single concrete number for the number of PUF instances to be considered for simulation purposes, it is safe to assume around 1000 samples to obtain a good estimate of uniqueness. In simulations, care must be taken to efficiently model various sources of manufacturing variations in CMOS circuits, as they directly translate into uniqueness. During the simulations, manufacturing variations are modeled using a gaussian distribution. In such cases, mean and standard deviation of the gaussian distribution under consideration must correspond to either inter-die or inter-wafer variations' statistics. variations are more correlated than inter-die variations [213]. Through circuit simulations as per the framework described in this chapter, we found that inter-distance obtained using inter-wafer variations is lower than the inter-distance obtained using inter-die variations.

6.2.2.2   Reliability

A challenge applied to a PUF operating on an integrated circuit will not necessarily produce the same response under different operating conditions as the circuit is subject to environmental variations. The robustness of PUF's responses is measured in terms of reliability. Reliability of a PUF refers to its ability to produce the same responses under varying operating conditions.

Reliability can be measured by looking at the average number of flipped bits in responses for the same challenge under different operating conditions. A common measure of reliability is *intra-distance* given by [129,214]:

$$d_{intra}(C) = \frac{1}{s} \sum_{j=1}^{s} \frac{HD(R_i, R'_{i,j})}{m} \times 100\% \qquad (6.2)$$

In equation (6.2), R_i is the response of a PUF to challenge C under nominal conditions, s is the number of samples of response R_i obtained at different operating conditions, $R'_{i,j}$ corresponds to j-th sample of response R_i for challenge C, and m is the number of bits in the response. Intra-distance is expected to be 0% for ideal PUFs, which corresponds to 100% reliability. The terms intra-distance (d_{intra}) and reliability have been used interchangeably further in this chapter. Given d_{intra}, reliability can always be computed ($100 - d_{intra}(\%)$).

6.2.2.3   Unpredictability

Responses from a PUF circuit must be unpredictable in order to ensure that the signatures/keys are safe from adversaries possessing information about the responses to different challenges from the same device. One of the measures of unpredictability is the amount of randomness in responses from the same PUF device. This can be evaluated using NIST tests [129, 265]. Silicon PUFs produce unique responses based on intrinsic process variations, that are very difficult to clone or duplicate by the manufacturer. However, by measuring responses from a PUF device for a subset of challenges, it is possible to create a model that can mimic the PUF under consideration. Several modeling attacks on PUF circuits have been proposed in the literature [176,249,252,253,267,359]. The type of modeling attack depends on the PUF circuit. A successful modeling attack on a PUF implementation may not be effective for other PUF implementations. Modeling attacks can be made harder by employing some control logic surrounding the PUF block, that prevents direct read-out of its responses. One such technique is to use a secure one-way hash over PUF responses. However, if PUF responses are noisy and have significant intra-distance, this technique will require some sort of error-correction on PUF responses prior to hashing [461].

6.3   Sources of CMOS Process Variations

The sources of process variations in ICs are summarized in this section. The reader is referred to the book chapter by Inyoung Kim et al. [214] for a detailed description on sources of CMOS variability. From the perspective of PUF circuits, the sources of variations can be either *desirable* or *undesirable*.

The desirable source of variations refers to process manufacturing variations (PMV) as identified in [214]. The environmental variations and aging are undesirable for PUF circuits.

6.3.1   Fabrication Variations

Due to the complex nature of manufacturing process, the circuit parameters often deviate from the intended value. The various sources of variability include proximity effects, Chemical-Mechanical Polishing (CMP), lithography system imperfections, and so on. The process manufacturing variations consist of two components as identified in the literature, namely systematic and random variations [46].

6.3.1.1   Systematic variations

Systematic component of process variations includes variations in lithography system, nature of layout, and CMP [46]. By performing a detailed analysis of the layout, the systematic sources of variations can be predicted in advance and accounted for in the design step. If the layout is not available for analysis, the variations can be assigned statistically [46].

6.3.1.2   Random variations

Random variations refer to non-deterministic sources of variations. Some of the random variations include random dopant fluctuations (RDF), line edge roughness (LER), and oxide thickness variations. The random variations are often modeled using random variables for design and analysis purposes.

6.3.2   Environmental Variations and Aging

Environmental variations are detrimental to PUF circuits. Some of the common environmental sources of variations include power supply noise, temperature fluctuations, and external noise. These variations must be minimized to improve the reliability of PUF circuits.

Aging is a slow process, and it reduces the frequency of operation of circuits by slowing them down. Circuits are also subjected to increased power consumption and functional errors due to aging [448].

6.4   CMOS PUFs

In this section, we provide an overview of some of the silicon-based PUF instantiations proposed in the literature. Though there are several non-electronic PUFs in the literature, e.g., optical PUFs [135, 322], CD PUFs [155], paper PUFs [33,59], Magnetic PUFs [182], they are not taken into account in this section. We refer to the chapter by Roel Maes and Ingrid Verbauwhede in [265] for a complete discussion on non-electronic PUFs.

We present an overview of two popular constructions of strong PUFs, namely arbiter PUFs and ring oscillator (RO) PUFs in Sections 6.4.1 and 6.4.2, respectively and one construction of weak PUFs known as SRAM PUFs in Section 6.4.3.

6.4.1   Arbiter PUFs

The concept of arbiter PUFs was introduced in [249,250,416]. An arbiter PUF architecture is shown in Figure 6.1. Arbiter PUFs are based on delay variations in logic gates arising from process manufacturing variations in integrated circuits. The general idea is to trigger a race condition in two identically laid out paths and decide the winner among the paths using an *arbiter*. From an implementation point of view, the identical paths are built using switches or multiplexers, that accept an external challenge bit (C_i). The challenge bit decides whether the switch *passes* the input to output or *switches* it, as shown in Figure 6.1(a). The switches/multiplexers are connected in series to form a digital delay path. A switch/MUX implementation in terms of logic gates is shown in Figure 6.1(b). When a signal (say a rising pulse) is applied to the delay path, the signal undergoes different delays through every switch. At the end of the delay paths, the arbiter decides which of the paths is faster based on the instantaneous arrival times of the racing signals. If we denote the delay difference between the arrival time of racing signals at outputs (A & B) and the input trigger time as T_A and T_B respectively, then the response computation is given by equation (6.3). A common practice is to use a set-reset (SR) latch as an arbiter by connecting one of the racing paths to *set* and the other path to *reset*. It is important to note that the number of CRP pairs is exponential to the number of challenge bits. So, arbiter PUFs fall under the category of strong PUFs.

$$r = \begin{cases} 0 & \text{if } T_A > T_B \\ 1 & else \end{cases} \tag{6.3}$$

To increase the sensitivity of arbiter PUFs to process variations, it is better to use minimum sized transistors in the delay stage circuits. However, the arbiter should be designed using up-sized transistors to tolerate process variations. The arbiter should fairly evaluate the response based on the delay difference between the racing signals (which can be positive or negative) and must not introduce any bias. To evaluate the fairness in response computation, arbiters built using D-type flip-flop and SR NAND latch were compared. D-type flip-flop arbiter in 45nm technology node has a setup time of around 20–35 ps, that introduces a bias in response computation. However, SR NAND latch arbiter has a bias of less than 3 ps because of its cross-coupled

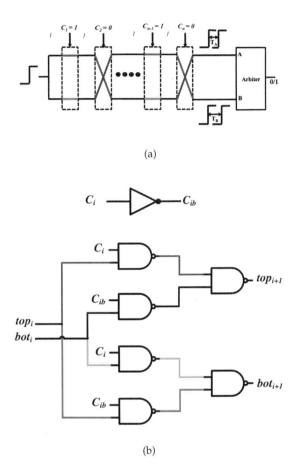

(a)

(b)

Figure 6.1
Arbiter PUF architecture: (a) Arbiter PUF architecture with n stages; (b) NAND gate-based implementation of a single MUX/switch stage. The path of propagation of two signals top_i and bot_i is determined by the challenge bit C_i. If $C_i = 1$, then $top_{i+1} = top_i$ and $bot_{i+1} = bot_i$. Else, $top_{i+1} = bot_i$ and $bot_{i+1} = top_i$.

structure, and it enables fair arbitration. The comparison results are shown in Figure 6.2.

Several results on arbiter PUF's functionality have been published based on test chip measurements. Experimental validation of an arbiter PUF fabricated using 45nm Silicon-On Insulator (SOI) technology in our research group yielded $d_{inter} \approx 39\%$ [253]. The low d_{inter} is partially due to on-chip substrate noise, that is very hard to control. A similar arbiter PUF design in $0.18\mu m$ technology node showed $d_{inter} \approx 23\%$ [250]. The low d_{inter} in [250]

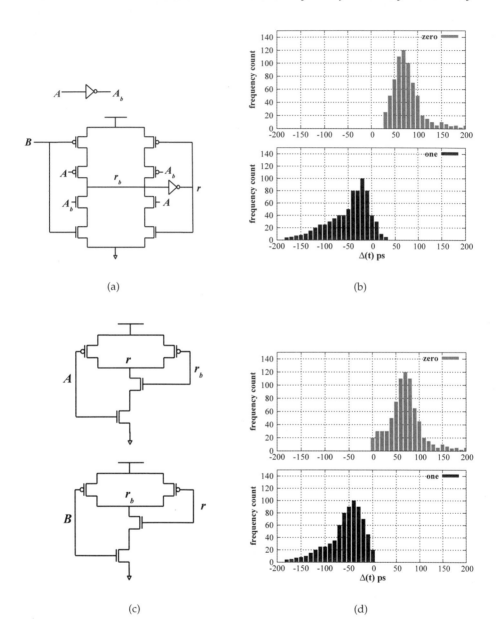

(a) (b)

(c) (d)

Figure 6.2
Fairness evaluation of arbiters based on bias point for choosing 0/1 output.
The bias point is given by $\Delta(t) = T_A - T_B$ (a) A simple D-Type flip-flop
arbiter; (b) Plot showing the bias point of d-type flip-flop arbiter for choosing
0/1 around 30 ps; (c) A simple SR NAND latch arbiter; (d) Plot showing the
bias point of SR NAND arbiter for choosing 0/1 approximately at 0ps

when compared to 45nm test chip [253] can be attributed to different technology nodes being used. In general, the extent of process variations is high in sub-90nm technology nodes. The intra-distance of arbiter PUFs in 45nm test chip [253] was found to be less than 7%, when both supply voltage and temperature fluctuations were considered. Similar results were shown for arbiter PUF implemented in 0.18μm node as well.

Because of the additive nature of delay, several successful attempts have been made to attack arbiter PUFs using *modeling attacks* [176, 249, 267, 359]. The basic idea is to observe a set of CRP pairs through physical measurements and use them to derive runtime delays using various machine learning algorithms. In [359], it was shown that arbiter PUFs composed of 64 and 128 stages can be attacked successfully using several machine learning techniques, achieving a prediction accuracy of 99.9% by observing around 18,000 and 39,200 CRPs, respectively. These attacks are possible only if the attacker can measure the response, i.e., the output of a PUF circuit is physically available through an I/O pin in an integrated circuit. If the response of a PUF circuit is used internally for some security operations and is not available to the external world, modeling attacks cannot be carried out unless there is a mechanism to internally probe the output of the PUF circuit. However, probing itself can cause delay variations, thereby affecting the accuracy of response measurement.

Several non-linear versions have been proposed in the literature to improve the modeling attack resistance of arbiter PUFs. One of them is feed-forward arbiter PUFs, in which some of the challenge bits are generated internally using an arbiter as a result of racing conditions at intermediate stages. However, such a construction has reliability issues because of the presence of more than one arbiter in the construction. This is evident from the test chip data provided in [250]. It was reported that feed-forward PUFs have $d_{intra} \approx 10\%$ and $d_{inter} \approx 38\%$. Modeling attacks have been attempted on feed-forward arbiter PUFs, and the attacks used to model simple arbiter PUFs were found to be ineffective when applied to feed-forward arbiter PUFs. However, by using evolution strategies (ES), feed-forward arbiter PUFs have been shown to be vulnerable to modeling attacks [359]. Non-linearities in arbiter PUFs can also be introduced using several simple arbiter PUFs and using an XOR operation across the responses of simple arbiter PUFs to obtain the final response. This type of construction is referred to as XOR arbiter PUF. Though XOR arbiter PUFs are tolerant to simple modeling attacks, they are vulnerable to advanced machine learning techniques. For example, a 64-stage XOR arbiter PUF with 6 XORs has been attacked using 200,000 CRPs to achieve a prediction accuracy of 99% [359]. All these modeling attacks impose a strong pressing need on design of a modeling attack resistant arbiter PUF.

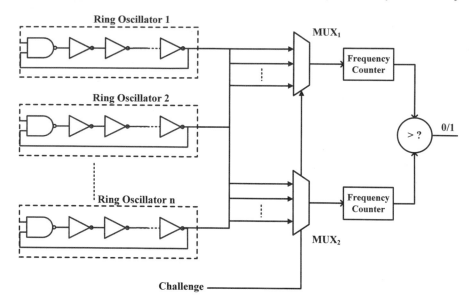

Figure 6.3
Ring oscillator PUF architecture. The external challenge selects two ring oscillators for frequency comparison to generate a single bit response

6.4.2   Ring Oscillator PUFs

Ring oscillator PUF (RO PUF) was introduced in [135, 322]. They also use delay variations in standard logic gates, but the delay variations are mapped to frequency variations using a ring oscillator. The oscillation frequency (f_{osc}) of a ring oscillator with n stages is given by the equation

$$f_{osc} = \frac{1}{2n\tau} \tag{6.4}$$

where τ is the delay of an inverter and is highly sensitive to process variations in inverters. So, the oscillation frequencies of two identical ring oscillators will not be the same. This concept is used in ring oscillator PUF construction. By comparing frequencies of two identical ring oscillators, a digital response is obtained. The ring oscillators for frequency comparison are chosen using an external challenge. A ring oscillator PUF architecture is shown in Figure 6.3.

It is easy to infer that ring oscillator PUFs do not offer the same advantages as arbiter PUFs, especially in the number of response bits generated from the circuit. If there are N ring oscillators, then $N(N-1)/2$ comparisons can be made. However, some of these comparisons will yield redundant response bits and hence the entropy of a ring oscillator PUF circuit is less than $N(N-1)/2$ [416]. For example, if comparison of ring oscillators A and B produces '1' ($f_{osc}(A) > f_{osc}(B)$) and comparison of ring oscillators B and C yields '0' ($f_{osc}(B) < f_{osc}(C)$), then it is not hard to deduce that comparison of oscillators A and C will yield '1' ($f_{osc}(A) > f_{osc}(C)$). The number of useful bits that can be obtained from ring oscillator PUFs is $log_2(N!)$ [416].

Similar to arbiter PUFs, several real time data on ring oscillator PUFs have been published. Since it is easy to instantiate a ring oscillator in FPGA and duplicate it as a hard-macro, majority of the published results are from FPGA prototyping. Note that arbiter PUFs are generally hard to instantiate on FP-GAs because of symmetric routing requirement. This is not a serious concern in ring oscillator PUFs, as small differences in routing can be masked by counting many oscillation cycles. Experiments conducted using 1024 ring oscillators on 15 FPGAs yielded $d_{inter} \approx 46\%$ and $d_{intra} \approx 0.5\%$ [416].

Ring oscillator PUFs can be attacked using sorting techniques, in which an attacker use standard sorting algorithms to sort the oscillation frequencies in ascending or descending order. Reference [359] shows that a RO PUF using 1024 ring oscillators can be attacked to achieve 99.9% prediction accuracy using around 345,000 CRPs. Quick Sort (QS) algorithm was employed for this purpose.

6.4.3 SRAM PUFs

The concept of SRAM PUFs was first introduced in [145, 172, 173]. Static random-access memory (SRAM) is a type of semiconductor memory, which is composed of CMOS transistors. SRAM is capable of storing a fixed written value "0" or "1", when the circuit is powered up. An SRAM cell capable of storing a single bit is shown in Figure 6.4. Each SRAM bit cell has two cross coupled inverters and two access transistors. The inverters drive the nodes q and q_b as shown in Figure 6.4. When the circuit is not powered up, the nodes Q and $\bar{Q}$ are at logic low (00). When the power is applied, the nodes enter a period of metastability and settle down to either one of the states ($Q = 0$, $\bar{Q} = 1$ or $Q = 1$, $\bar{Q} = 0$). The final settling state is determined by the extent of device mismatch in driving inverters and thermal noise. For PUF operation, it is desirable to have device mismatch dominant over thermal noise, as thermal noise is random in nature. The power up states of SRAM cells were used for generating unique fingerprints in [172,173]. Experiments were conducted over 5,120 64-bit SRAM cells from commercial chips and d_{inter} was found to be around 43%. Similarly, d_{intra} was found to be around 3.7%. Similar results were obtained in [145]. An extensive large-scale analysis over 96 ASICs, with each ASIC having 4 instances of 8kB SRAM memory was performed in [202]. The experimental results from [202] show that SRAM PUFs typically have high entropy and responses from different PUF instances were found to be highly uncorrelated.

Since SRAM PUFs based on power-up states have a single challenge, the response must be kept secret from the external world. Modeling attacks are

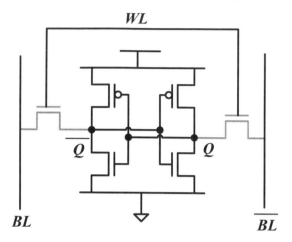

Figure 6.4
An SRAM cell using standard CMOS transistors. BL is the bit line, WL refers to word line, Q and $\bar{Q}$ are the state nodes for storing a single bit value. The transistors that make up inverters are shown in black and access transistors are shown in grey color.

not relevant for SRAM PUFs. However, other attacks such as side-channel and virus attacks can be employed, but they are not covered in this chapter.

6.5   PUF Design Flow

This section focuses on a generalized design flow adopted for any type of PUF instantiation using standard CMOS gates. PUF design flow includes two broad steps, namely

- Pre-fabrication phase and

- Post-fabrication phase

We describe pre-fabrication phase in Section 6.5.1 and post-fabrication phase in Section 6.5.2.

6.5.1   Pre-fabrication Phase

Any PUF design flow is built upon traditional ASIC design flow, with some specific steps for variation modeling and statistical verification. A complete PUF design flow is shown in Figure 6.5. As shown in Figure 6.5, PUF design

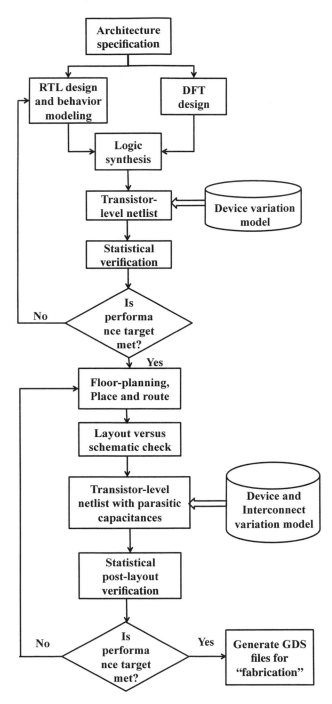

Figure 6.5
PUF design flow

starts from the architecture specification by defining general design goals and performance requirements. This is followed by a register-transfer level (RTL) design for the PUF circuit under consideration using Verilog and/or VHDL. Together with design for testing (DFT) techniques, a logic synthesis procedure is performed upon the developed RTL code to generate a gate-level netlist along with sub-circuit modules to describe the transistor-level netlist. A PUF-specific design process of introducing device variation models is important to enable statistical verification of PUF metrics such as uniqueness and reliability. If this verification fails, i.e., the performance metrics target of the PUF under consideration is not met, then the process to this step should be repeated with modifications to improve the PUF metrics. Upon statistical verification failure, a common placein the design process to inspect is the standard cell library. Often, the standard cell libraries are optimized to tolerate process variations. In such cases, standard cell libraries specific for PUF design process must be built. If the verification process succeeds, the design process continues further to physical design phase including floorplanning, placement, and routing and layout-versus-schematic verification. Once the physical model of PUF circuit is available, the more realistic device and variation models with extracted parasitics can be used for post-layout statistical verification. If this final verification fails, either the layout or the original design has to be modified. A successful verification at this phase should result in the delivery of GDSII files for fabrication.

To introduce variations to a specific device parameter during the verification phase, Monte Carlo (MC) based approach can be used to instantiate random values from a normal distribution or from a post-silicon parameter variation profile available through chip measurements. Each MC iteration produces one PUF instance. Usually, a large number of PUF instances (MC iterations) are required for uniqueness validation process. Supply voltage and temperature fluctuations are assigned over the netlist with extracted parasitic capacitances and simulations are performed for reliability and security/unpredictability analysis. Note that the focus of verification should be the post-layout stage and sufficient number of MC iterations should be performed, while the verification of pre-layout netlist can have reduced number of MC iterations.

6.5.2   Post-fabrication Phase

Once the pre-fabrication phase is complete, the generated GDS files are sent for fabrication. The fabricated PUF circuit must be tested for exact real time performance analysis. The challenges for the PUF circuit under consideration can be generated using an external equipment or generated inside the chip using a pseudo-random number generator like Linear Feedback Shift Register (LFSR), as shown in Figure 6.6. LFSR will start generating pseudo-random challenges, when *set* signal is pulled high. Usually post-silicon validation of PUF is done by storing the waveform of responses along with challenges. The

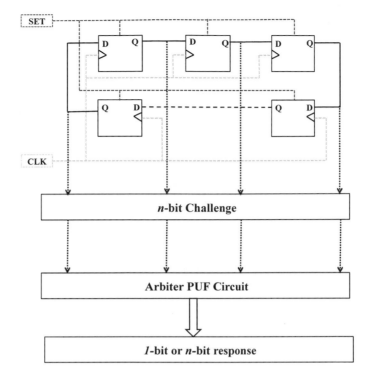

Figure 6.6
Arbiter PUF circuit with on-chip LFSR for challenge generation

CRPs are then extracted from the waveform files. The CRP extraction methods for pre-silicon and post-silicon validation are almost similar in nature. The only minor concern is that pre-silicon waveforms have accurate time reference and post-silicon measurements may have some time uncertainties due to clock jitter and external noise. Hence, automatic data synchronization must be employed to use the same CRP extraction method for pre- and post-silicon validations. A typical response measurement waveform from 45-nm test chip [253] is shown in Figure 6.7.

6.6 Sub-threshold PUFs (sub-V_{th} PUFs)

Since the notion of ubiquitous computing is becoming a reality, technologies like Radio Frequency Identification (RFID) and Wireless Sensor Networks (WSN) are massively deployed for many applications including healthcare, military, environmental, and wildlife monitoring. For a majority of these applications, providing security is of utmost importance and quite challenging

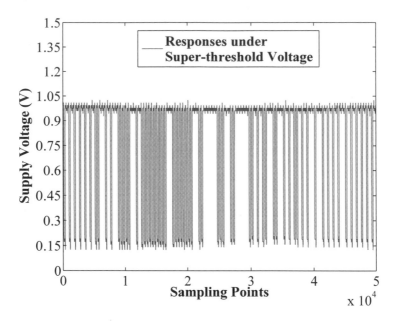

Figure 6.7
Response waveform from 45nm test chip [253]

too. For example, the power budget of digital part of passive RFID tags is limited to within $10\mu W$ [191]. Also, a nominal specification of an Electronic Product Codes (EPC) tag states that a maximum of 2000 gates can be used for security purposes [191]. Although such a tight budget can be met using lightweight cryptography algorithms [112], their low implementation complexities make them vulnerable to power side-channel analysis attacks [332]. Also, side-channel attack resistance of lightweight cryptographic algorithms is degraded by process variations [251]. In such cases, PUFs can be used as a security equivalent to lightweight cryptographic algorithms with minimal design overheads. PUF responses can also be used for key generation in low power embedded devices using post-processing algorithms [264].

In this section, we analyze the usage of PUFs in low power embedded devices by exploring sub-threshold mode of operation of PUFs. It is well known that operating a circuit at sub-threshold supply voltage results in low power consumption. Also, devices in sub-threshold conditions are more sensitive to process variations, thereby favoring PUF designs. However, it may also pose reliability issues. To that extent, the design of sub-threshold arbiter PUFs and its performance metrics are analyzed. We also provide data from sub-threshold arbiter PUF implemented in 45nm test chip to validate our results [253]. First, the impact of sub-threshold operation on stage delays is shown in Section 6.6.1, followed by sub-threshold arbiter PUF's performance metrics in Section 6.6.2.

6.6.1   Impact of Sub-threshold Supply Voltage

In this section, we present the impact of sub-threshold supply voltage on stage delays of an arbiter PUF circuit. A 64-stage arbiter PUF architecture shown in Figure 6.1 is considered for illustration purposes. To begin with, the propagation delay of a CMOS gate is given by

$$t_{pd} = \frac{V_{DD}C_L}{I_{on}} \tag{6.5}$$

In equation (6.5), t_{pd} refers to the propagation delay, V_{DD} is the supply voltage, C_L is the load capacitance, and I_{on} is the on-current through a transistor. The current through a transistor depends on the supply voltage and region of operation and is given by equation (6.6) [253].

$$I_{on} \approx \begin{cases} \mu C_{ox} \frac{W}{L} \left((V_{gs} - V_{th}) V_{Dsat} - \frac{V_{Dsat}^2}{2} \right), & V_{DD} >> V_{th} \\ \mu C_{ox} \frac{W}{L} I_0 \exp \left(\frac{V_{gs} - V_{th}}{n V_T} \right), & V_{DD} < V_{th} \end{cases} \tag{6.6}$$

In equation (6.6), μ is the carrier mobility, V_{gs} is the gate-to-source voltage, V_{th} is the threshold voltage, V_{DD} is the supply voltage, W and L are the transistor width and length respectively, C_{ox} is the oxide capacitance, n is the sub-threshold slope factor, V_T is the thermal voltage, and I_0 is the on current when $V_{gs} = V_{th}$.

The impact of process variations on stage delays of an arbiter PUF at sub-threshold operating voltage (0.4V) is shown in Figure 6.8. The reason for operating the circuit at 0.4V is demonstrated in Figure 6.9. It can be seen from Figure 6.8 that increasing the number of stages improves the deviation in stage delays. Approximately 100 PUF instances were generated through Monte Carlo simulations for this analysis. The simulation framework described in Section 6.5.1 was adopted for this purpose. The same experiment was performed using a similar setup to analyze the impact of process variations on the stage delays in super-threshold arbiter PUFs operating at 1V (nominal supply voltage at 45nm technology node). The percentage improvement in stage delay deviation (1σ) of 64 stage sub-threshold arbiter PUF over super-threshold PUF is shown in Table 6.1. Note that, the low deviation of stage delays in super-threshold PUFs do not necessarily imply reduced uniqueness. The deviation in stage delays of super-threshold arbiter PUFs is sufficient enough to produce unique responses.

The above experiments show that sub-threshold operation favors PUF design. However, the optimum supply voltage for sub-V_{th} PUFs has to be determined. Figure 6.9 shows the optimum supply voltage for an arbiter PUF architecture shown in Figure 6.1. It can be inferred from Figure 6.9 that at a supply voltage of around 0.4V, the power-delay product (PDP) is minimum. Note that even lower power consumption can be achieved by reducing supply voltage further. However, this will degrade PDP and is not an efficient design strategy. This optimum supply voltage fluctuates around 0.4V under

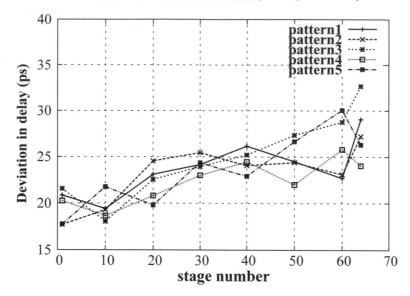

Figure 6.8
Stage delay variation in a 64-stage sub-threshold arbiter PUF operating at
$V_{DD} = 0.4$V

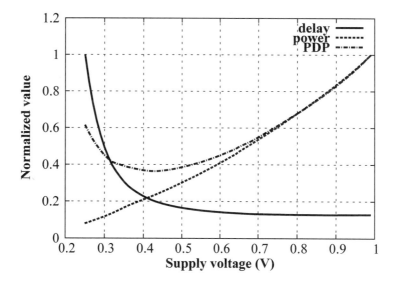

Figure 6.9
Impact of supply voltage on delay, power, and power-delay product. Supply
voltage at which PDP is minimum (0.4V) is chosen as the optimum supply
voltage for sub-threshold arbiter PUFs.

Table 6.1

Percentage improvement in stage delay deviation of sub-threshold arbiter PUF when compared to super-threshold arbiter PUF

Stage number	% improvement in stage delay deviation of sub-V_{th} PUF				
	pattern1	pattern2	pattern3	pattern4	pattern5
1	17.8	26.2	10.18	20.2	21.1
10	2.1	8.41	14.8	7.31	13.17
20	31.1	13.66	-1.82	11.71	20.04
30	22.92	18.9	10.13	6.31	29.43
40	8.42	8.58	29.2	17.7	13.3
50	3.21	-3.16	14.01	13.09	21.1
60	19.7	18.6	20.9	-3.6	-8.2
64	12.04	16.3	20.89	12.69	28.78

process variations. The impact of such fluctuations on the performance of sub-V_{th} PUFs will be discussed in Section 6.6.2.

6.6.2   Performance Metrics of Sub-threshold Arbiter PUFs

The uniqueness of sub-threshold arbiter PUF in terms of d_{inter} and σ_{inter} is shown in Table 6.2. Since the difference in d_{inter} for both sub-threshold and super-threshold arbiter PUFs is small, it is worthwhile to take a look at σ_{inter}. Lower σ_{inter} for sub-threshold PUF makes it slightly better than super-threshold PUFs in terms of uniqueness. Table 6.3 shows d_{inter} for both sub- and super-threshold PUFs as measured from the 45nm test chip [253]. Note that d_{inter} for sub-threshold PUF is significantly higher than super-threshold PUFs. So, σ_{inter} is not shown.

Table 6.2

Uniqueness of sub- and super-threshold arbiter PUFs from simulations

Type of PUF	d_{inter}	σ_{inter}
super-threshold	0.46	8.8
sub-threshold	0.47	3.7

Table 6.3

Uniqueness of sub- and super-threshold arbiter PUFs from 45nm test chip

Type of PUF	d_{inter}
super-threshold	0.39
sub-threshold	0.43

Table 6.4
Worst case reliability data of both sub- and super-threshold arbiter PUFs from 45nm test chip

Type of PUF	Reliability (%) (d_{intra})
super-threshold	91 (0.09)
sub-threshold	82 (0.18)

Figure 6.10(a) shows the impact of supply voltage fluctuations on reliability of both super- and sub-threshold PUFs. The impact of supply voltage fluctuations on reliability of super-threshold PUF is slightly lower than sub-threshold PUF. Similarly, Figure 6.10(b) presents the effect of temperature fluctuations on sub-threshold PUF's reliability. It can be observed that sub-threshold PUF is more susceptible to temperature fluctuations than supply voltage fluctuations. The reliability drops to as low as 90% at extreme temperatures. Worst case reliability data from test chip measurements are shown in Table 6.4. High correlation between simulation results and test chip data was observed. Notice that reliability can also be expressed in terms of d_{intra}, which is given by

$$d_{intra} = 1 - \frac{\text{reliability}(\%)}{100} \tag{6.7}$$

But, the data is shown in terms of reliability (%) instead of d_{intra} to be consistent with [253]. The worst case d_{intra} of PUFs from the test chip computed directly from reliability (%) is also shown in Table 6.4.

As is the case with super-threshold arbiter PUFs, the sub-threshold counterpart is also vulnerable to modeling attacks. We base our attack as per the framework described by Daihyun Lim in [249]. An arbiter PUF's challenge-response relationship can be expressed through an additive delay model. The top and bottom paths of an arbiter PUF shown in Figure 6.1 can be expressed as the sum of delays of individual stages. A direct measurement of stage delays is extremely difficult. However, they can be estimated with the help of a machine learning algorithm by observing a subset of CRPs. Each stage of an arbiter PUF can be expressed through two delay difference parameters as shown in Figure 6.11, that encode the four individual delay paths of a stage shown in Figure 6.12. The notations of individual delay components are consistent with [249]. Now, the delay parameter for C_i=1 in terms of individual delay components is shown in the following equations:

$$\delta_{top}(i) = p_i + \delta_{top}(i-1), \delta_{bot}(i) = q_i + \delta_{bot}(i-1) \tag{6.8}$$

$$\begin{aligned}
\Delta t_i &= \delta_{bot}(i) - \delta_{top}(i) \\
&= (q_i - p_i) + (\delta_{bot}(i-1) - \delta_{top}(i-1)) \\
&= \delta t_i^1 + \Delta t_{i-1}
\end{aligned} \tag{6.9}$$

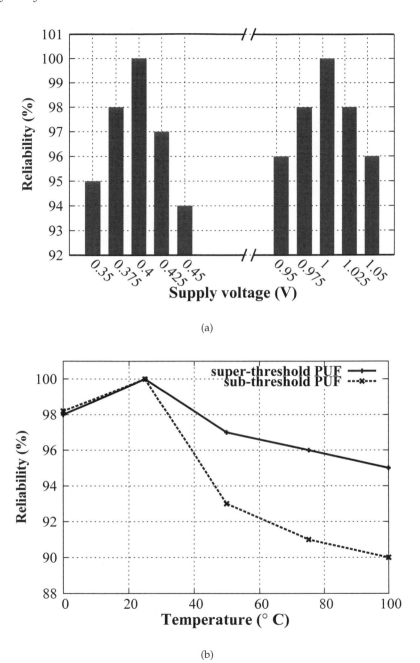

Figure 6.10
Reliability of super- and sub-threshold arbiter PUFs under environmental variations (a) Impact of supply voltage fluctuations on reliability (b) Impact of temperature fluctuations on reliability

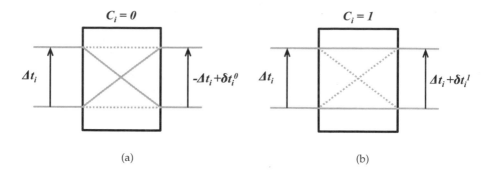

Figure 6.11
Delay difference parameters for (a) $C_i = 0$ and (b) $C_i = 1$

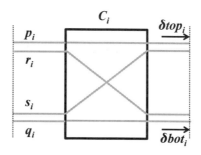

Figure 6.12
Individual delay components of a single stage of an arbiter PUF

Similarly, the delay difference parameter for $C_i=0$ can be expressed.

$$\delta_{top}(i) = s_i + \delta_{bot}(i-1), \delta_{bot}(i) = r_i + \delta_{top}(i-1) \qquad (6.10)$$

$$\begin{aligned}
\Delta t_i &= \delta_{bot}(i) - \delta_{top}(i) \\
&= (r_i - s_i) + (\delta_{top}(i-1) - \delta_{bot}(i-1)) \\
&= \delta t_i^0 - \Delta t_{i-1}
\end{aligned} \qquad (6.11)$$

In equations (6.9) and (6.11), δt_i refers to the delay difference introduced by the unit stage. The delay difference parameter indicates that the delay differences (δt_i) are all that are needed to model the PUF rather than the absolute delay values (p_i, q_i, r_i, s_i). The total delay difference at the input of the arbiter denoted by Δt_n is given by equation (6.12).

$$\Delta t_n = p_0 \Delta t_0 + \sum_{i=1}^{n} p_i \delta t_i \qquad (6.12)$$

In equation (6.12), $(p_0, p_1 ... p_n)$ represents the parity vector computed using equation (6.13) with $p_n=1$.

$$p_i = \prod_{j=i+1}^{n} c_j \tag{6.13}$$

For computing the parity vector, the challenge bits are mapped from $(0,1)$ to $(-1,1)$. The CRP relationship can be expressed as follows:

$$\Delta t_n \lessgtr 0 \tag{6.14}$$

Figure 6.13 shows the prediction accuracy of support vector machine (SVM) based attack on sub-threshold arbiter PUFs. The parity vector shown in equation (6.13) is used as the feature vector along with the observed outputs. From Figure 6.13, it can be observed that even with a smaller subset of CRP pairs (<2500 CRPs), a sub-threshold arbiter PUF can be modeled using SVM to achieve a prediction accuracy of 95%. However, if the PUF is used directly as a seed generator for some sort of cryptographic algorithm, the modeling attack can be made very difficult as the outputs are not directly accessible. This assumption holds good only if the PUF is tamperproof.

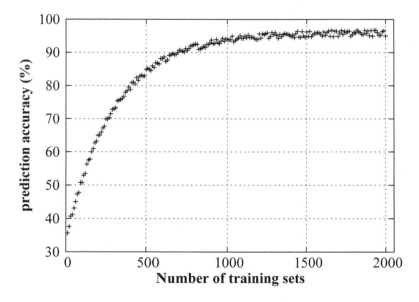

Figure 6.13
SVM-based attack on sub-threshold arbiter PUF. With less than 2500 CRPs, sub-V_{th} arbiter PUF can be attacked to achieve a prediction accuracy $\approx 95\%$

6.7   Impact of Technology Trends

As the transistor dimensions keep shrinking, it is worthwhile to analyze the impact of scaling on performance of PUF circuits. It has been predicted that process variations will increase with shrinking [61]. The 3-σ value of threshold voltage variations is projected to be around 160mV and 210mV for 32 and 22nm technology nodes, respectively. These variations can be mitigated to certain extent using various techniques, including design time [43] and lithographic techniques such as optical proximity correction. However, these techniques can be effectively applied to the logic section of the IC excluding the portion containing the PUF as reported in [129]. This will allow the designers to focus on designing better PUFs in sub-45nm technology nodes that can meet or exceed PUF designs in 45nm node.

In general, technology scaling does not impact uniqueness much. Though the delay variations are slightly higher in sub-45nm nodes, they do not necessarily contribute much towards uniqueness. So, we analyze the impact of technology scaling in the presence of a bias in an arbiter PUF circuit. A common way to model any bias in arbiter PUF circuits is to introduce a delay-difference in the racing paths. For an arbiter PUF circuit with a bias of 20 ps introduced between the delay paths, the uniqueness in terms of d_{inter} for all technology nodes is shown in Table 6.6. We can infer that technology scaling helps to mask the bias in arbiter PUF circuits because of improved delay variations. In case of ring oscillator PUFs, the bias can be masked by counting large number of oscillation cycles.

Table 6.5

Impact of technology scaling on uniqueness of super- and sub-threshold arbiter PUFs

Type of Arbiter PUF	Technology node	d_{inter}
Super-threshold	45nm	0.46
	32nm	0.47
	22nm	0.46
Sub-threshold	45nm	0.47
	32nm	0.48
	22nm	0.47

Table 6.7 presents the impact of technology scaling on reliability of sub- and super-threshold arbiter PUF circuits. Reliability with respect to operating conditions follow the same trend as in Figure 6.10. So, we present only the worst case reliability data in Table 6.7. Note that the worst case data may correspond to extreme operating conditions. In general, technology scaling degrades reliability by as much as 4% in super-threshold arbiter PUFs in the presence of supply voltage variations. Due to increased delay variations in

Table 6.6
Impact of technology scaling on uniqueness of super- and sub-threshold arbiter PUFs in presence of 20 ps delay difference bias

Type of Arbiter PUF	Technology node	d_{inter}
	45nm	0.38
Super-threshold	32nm	0.41
	22nm	0.45
	45nm	0.39
Sub-threshold	32nm	0.42
	22nm	0.46

Table 6.7
Impact of technology scaling on reliability of super- and sub-threshold arbiter PUFs. Only, the worst case data is shown here.

Type of Arbiter PUF	Technology node	Reliability (%)
	45nm	95
Super-threshold	32nm	94
	22nm	93
	45nm	91
Sub-threshold	32nm	90
	22nm	90

sub-threshold arbiter PUFs, the impact of technology scaling on reliability of sub-threshold arbiter PUFs is not much in the presence of supply voltage fluctuations. However, in the presence of temperature fluctuations, the reliability of sub-threshold PUFs degrade and is also impacted by technology scaling. The reliability of sub-threshold PUF in 22nm node is almost 3–4% lower than sub-threshold PUF designed in 45nm node under some operating conditions. This argument need not be necessarily true for worst case reliability. This is reflected by the same worst case reliability observed in both 32- and 22nm sub-threshold arbiter PUFs as shown in Table 6.7.

Note that we again represent reliability in (%) and d_{intra} for the corresponding reliability data can be computed using equation (6.7).

6.8 Open Research Challenges

The field of PUFs has seen several interesting innovations and findings over the past decade. However, there exist several issues to be investigated. The major challenges for PUFs can be briefly addressed with the help of their performance metrics. The perspective of process variations is gaining significant attention with sub-45 nm technologies. Although the extent of process variations in increasing, the presence of correlations can severely degrade a PUF's performance [235]. This can be addressed by effectively bridging the gap between Design for Manufacturability (DFM) concepts and PUFs for better designs. Reliability of PUFs is also a serious concern for PUF-based key generation. However, unreliable CRP data can be effectively used as a side-channel information for breaking PUF circuits [98, 99]. Although there exist several contributions in the literature to improve the reliability of PUFs from circuit and system perspectives [60, 234, 236, 442, 461], the emergence of unreliable CRP data as a tool for modeling attacks has brought an important aspect to be addressed. Also, the vulnerabilities posed by strong PUFs toward modeling attacks will pave the way for an immense competition between codemakers and codebreakers, with a hope that the process will converge to a PUF design that is highly resilient to known attacks.

6.9 Hardware Security Using Embedded DRAM Intrinsic Bitmap IDs

The recent increase in number of counterfeit ICs in the market has escalated the demand for innovation in supply chain security. Extrinsic markers such as bar codes or nonvolatile memory, once considered hard to copy or break in, can be easily circumvented by a determined hacker. Process variations in a VLSI chip may originate unique electrical fingerprints, and these constitute an alternate approach to chip security known as Physically Unclonable Functions (PUFs). As reviewed in this chapter, PUF applications to hardware security may employ intrinsic bitmaps of embedded Dynamic Random Access Memory (DRAM), the same product IP as IBM's microprocessors. These bitmap-based intrinsic chip IDs are discussed by various authentication methods, in particular dynamic key challenges and field-tolerant authentication, along with analytical modeling and simulation. To conclude the chapter, a Self-Authenticating Intrinsic ID (SAIID) architecture and the hardware implementation using embedded DRAM are proposed to constitute a realistic application of memory PUFs that prevents the IDs themselves from leaving the chip and becoming exposed.

6.9.1 Introduction

Rapid progress in science and technology has brought great convenience for business and personal life alike. In particular, evolutionary advancements in nano-scale semiconductor technology [303], while improving performance, are key contributors to the miniaturization of electronic products. These improvements, in turn, have introduced consumers to personal computing and the information era. Lately, consumer products for mobile computing and communication have been the fastest growing segments to drive the semiconductor industry. These new products further incorporate wireless, internet, and revolutionary social network applications, allowing inter-personal communication without restrictions of time and place. However, these new products and services have created various privacy and security concerns. Although communicating with people who one has never met is already common in a person's virtual life, it may result in a breach of privacy as well as in virus attacks to the personal system. Therefore, developing high security products and services is an urgent task not only for business, but also for the semiconductor industry.

Security and privacy in the digital network employ cryptographic protocols [102] with open and private keys, which are now commonplace in the industry. However, this high level of security cannot be realized without establishing a root of trust, which includes the use of highly reliable hardware. Thus, hardware security [333] is an equally, if not more important concern for high security systems. Hardware security requires identification, authentication, encryption, and decryption engines [106]. Unfortunately, a recent surge in counterfeit hardware has resulted in the occasional distribution of cheap, fake chips in the industry [215]. Distribution of these counterfeit chips not only leads to reduced product reliability, but further increases privacy and security risks. These counterfeits are typically the result of either discarded chips reintroduced into the supply chain, or of fabrication of cheap copies that pass as authentic, reducing cost without significant security. Systems using unauthentic chips are consequently exposed to a high risk of failure in the field. More importantly, using fake chips for banking or defense systems results in a potentially unacceptable risk for national security. Thus, the reliability of the task of identifying hardware has become a necessity in contemporary security, which in turn requires a method for highly reliable chip identification and authentication using semiconductor components.

Figure 6.14 summarizes the hardware identification method. The simplest approach uses an identifier (ID) such as text which is uniquely assigned to each hardware unit. However, since the ID can be read by anyone, a counterfeit part can be created by simply writing the ID on the fake hardware. This approach is, therefore, not secure at all. Recent hardware, in particular a VLSI chip, uses embedded fuses for IDs which are uniquely assigned and programmed in each of the VLSI chips. This fuse approach is somewhat secure, because the ID cannot be read without breaking the module, while a special tool is necessary for writing the ID to the fuses. Older fuse technology uses laser [404] for blowing the fuses, which is a large and expensive process. The evolution of fuse technology has resulted in the electrically blowable fuse approach (eFUSE) [349]. This makes it possible to create a high-density One-Time Programmable Read Only Memory (OTPROM) [363, 430] that can be

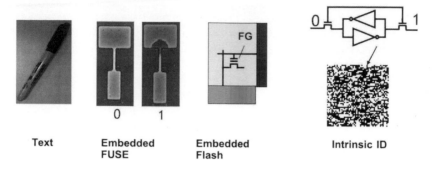

| Text | Embedded FUSE | Embedded Flash | Intrinsic ID |

Figure 6.14
Evolution of chip identification (ID)

used for chip ID, as well as for other programming elements such as redundancy replacement. Regardless of laser fuse or OTPROM using eFUSE, the fuses are visible by optical or electronic microscopy, and therefore the programmed ID information can still be obtained after de-layering. Using nonvolatile memory such as embedded flash memory [100,169] makes it difficult to obtain the ID by the de-layering method, however direct readout from the chip is still possible if the read method is known.

There are two fundamental problems with the existing ID approaches. First, all of them employ an ID which is extrinsic to the hardware. These extrinsic IDs are not integral to the hardware, since the information needs to be written or programmed, and they can be subsequently read by some method. Second, the ID can be cloned as long as it is known. These weaknesses make it possible to create counterfeits by emulating the ID in multiple fake chips. Therefore, the challenges to creating secure IDs are to make it difficult for unauthorized people to detect the ID, and to make it impossible, if not impractical, to create a clone of the ID even if it can be made known with reasonable effort. An intrinsic chip ID employs intrinsic VLSI features, constituting an ideal solution for high security applications and making it prohibitive to make fake copies.

In this chapter, we discuss an intrinsic chip ID using bitmaps of an embedded DRAM macro. Section 6.9.2 introduces the concept of intrinsic ID, followed by a review of the existing approaches. In Section 6.9.3, principles of embedded DRAM are discussed. Section 6.9.4 is concerned with retention-based bitmap fingerprints of a one-transistor and one-capacitor DRAM cell. Section 6.9.6 introduces the dynamic intrinsic ID approach using retention signatures. In Section 6.9.8, a field-tolerant method is explored which improves success of intrinsic ID authentication in a field. Section 6.9.9 introduces the Self-Authenticating Intrinsic ID (SAIID) chip architecture and the hardware implementation method. Section 6.9.10 summarizes the chapter and briefly discusses the future challenges for intrinsic ID research and development.

6.9.2 Intrinsic Chip ID

Fingerprints are widely used for secure identification of individuals. A human fingerprint is a unique and unclonable feature that each person possesses. In a similar fashion, a secure intrinsic ID exploits intrinsic features of a VLSI chip. Such features arise from random process variations, and can be used to generate an ID which cannot be reverse-engineered or easily emulated, also called a "Physically Unclonable Function (PUFs)" [73,85,86,119,131,158,159,173,190,203,248,256,258,263,322,323,352,353,377, 380,396,414,416,432,433,462]. This thus greatly improves chip security over the existing extrinsic ID approach. In this section, we will discuss the intrinsic ID generation and authentication concept using random process variations in manufacturing, and their challenges.

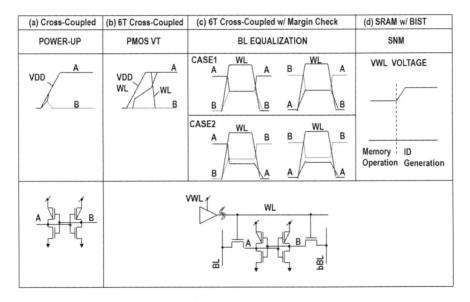

Figure 6.15
Physically unclonable fuse using cross-coupled inverter

An intrinsic chip ID converts process variations in manufacturing into a digital binary vector for chip identification. Figure 6.15 shows intrinsic ID generation methods using CMOS cross-coupled inverters. In the simplest approach (a), each of the inverters drives the other's output, nodes A and B respectively. This structure is commonly used as a temporary storage element or as a bus keeper [396] for VLSI design, and constitutes an ideal element for intrinsic ID. When used in ID generation, the cross-coupled inverter is powered up without node initialization. This power-up method, accomplished by raising VDD as shown, naturally determines the states of A and B as a consequence of the PMOS and NMOS threshold mismatch of the cross-coupled inverters. The A and B states after power-up can be used as a bit of an intrinsic ID. These states, however, may not be stable, as they may be sensitive to noise during power-up. Therefore, a preferred approach (b) includes two additional access transistors which force nodes A and B to GND level as soon as power is turned on. The states can therefore be determined by the PMOS threshold mismatch of the cross-coupled inverters, as nodes A and B are forced to GND level until one of the PMOS is strongly turned on, eliminating instability resulting from power-up noise. This approach (b) has two advantages. First, the VT mismatch of the cross-coupled inverters depends on the local CityplacePMOS StateVT mismatch, and can potentially minimize the systematic impacts of lithography, temperature, or noise. Second, this structure is identical to that of the 6 transistor Static Random Access Memory (SRAM) cell, and as such enables the use of SRAM arrays and their bitmap [73, 85, 131, 158, 159, 173, 377, 380, 414] for intrinsic ID generation. The end result is a significant increase of density with the added benefit of allowing a product memory array to be used for intrinsic ID generation.

Approach (b) may be improved by employing an ID margin detection method to eliminate unstable bits. Unlike the power-up intrinsic ID generation, approach (c) [73] generates the ID after completing a power-up sequence similar to that of conventional memory operation. Prior to ID generation, the A and B nodes are initially set to predetermined states ("1" and "0", or "0" and "1"). This initialization is realized by means of a SRAM write mode. Writing "1" sets nodes A and B to "1" and "0", and writing "0" sets nodes A and B at "0" and "1", respectively. Activation of the wordline (WL) opens the access transistor, while the bitlines (BL) of the pair are held at GND level. This results in a short-circuit of nodes A and B, and the BL and bBL equalization levels are determined by intrinsic features such as threshold voltage mismatch or transistor strength of the cell. Cells containing balanced cross-coupled inverters allow nodes A and B to be equalized, and result in a small voltage difference between BL pairs during equalization, as shown in case 1. On the other hand, cells having largely imbalanced cross-coupled inverters remain unequalized, or flip the states in a preferred direction, as shown in case 2.

Evaluation or generation of the intrinsic ID is enabled by deactivation of the WL, which in turn disables the equalization of nodes A and B. This results in determining the node A and B states naturally. For the well-equalized BL cell (case 1), a generated bit is likely unstable. Otherwise, the generated bit is determined by the initial predetermined voltages of the nodes. On the other hand, a bit generated from the cell which is not well-equalized (case 2) is always stable regardless of the initial predetermined voltages of the nodes. Generation and evaluation must be performed using both initial node states

(A and B at "0" and "1", and "1" and "0") in order to assess the intrinsic stability of the bit.

On-chip intrinsic ID generation requires a Built-In Self-Test (BIST) engine [217]. Fortunately, the on-chip memory in a logic chip is typically supported by BIST, and therefore is readily available for intrinsic ID generation without the introduction of additional circuits or at the cost of a small silicon overhead. The fourth approach (d) employs a BIST engine in the generation of a random binary vector derived from checking the static-noise margin (SNM) of a SRAM array [131]. For memory operation, the wordline voltage (VWL) is adjusted to have a sufficient SNM when the WL goes high. For ID generation, the VWL voltage is increased to reduce the SNM. This results in a fail for a weak cell when the WL goes high. In order to weed out the weakest bits, this approach includes a feedback loop between the number of fails and the VWL voltage supply. This feedback remains active until the fail count reasonably matches a predetermined number. The result is the generation of a stable random bit pattern comprised of the fail bit addresses, which are detected and recorded using the on-chip BIST.

An intrinsic chip ID can be further implemented using a delay-based PUF. Ring oscillator (RO) based PUFs [86, 416] compare the delay between ROs while generating unclonable random bit strings. Similar to the RO PUF, an arbiter-based PUF [248] is composed of delay paths for signals A and B and an arbiter located at the end of the delay path. The arbiter outputs "1" when signal A arrives earlier than B, otherwise it outputs "0". Because the delay path is determined by intrinsic features in manufacturing, the output bit can be random. The challenge of this approach is to lay out the signal path symmetrically in order to minimize the normal delay difference between the two paths. Otherwise, the output will be skewed. Another approach [258] detects an analog voltage determined by the threshold of the MOS transistors, which is subsequently converted to a binary identification sequence using an auto-zeroing comparator. The one-time oxide breakdown PUF [256] leverages the fact that a stress condition such as high voltage can break weak cells with a higher confidence than strong cells, resulting in permanent random intrinsic ID generation. The concern with this permanent approach is that the ID bits can be detected by de-layering the chip in a manner similar to the extrinsic fuse approach. Bit-string generation based on resistance variation in metals and transistors has also been reported [190].

The ultimate intrinsic ID is a fingerprint derived from product intrinsic features without need for allocating additional silicon for chip identification. VLSI Chips, in particular recent multi-core and multi-thread microprocessers [362, 454], include tens or hundreds of on-chip cache memories. Using the on-chip cache as a fingerprint intrinsic ID may therefore offer an ideal solution for highly secure identification, because the memory can be used for the generation of enormous intrinsic IDs without requiring additional expensive silicon from a high performance system. The challenges for this application reside in providing the intrinsic ID function without degrading the product chip features.

In this chapter, we will discuss intrinsic chip ID generation and authentication using embedded Dynamic Random Access Memory (DRAM) product developed for IBM multi-core and multi-thread microprocessor. The next section describes the principle of embedded DRAM product macro operation and the retention signature, which will be used for intrinsic bitmap ID generation and authentication to be discussed in following sections.

6.9.3 High Performance Embedded Dynamic Random Access Memory

For nanometer technology, it is desirable and essential to integrate more functions within a single die. DRAM integration with a high performance logic process (embedded DRAM) [217] not only reduces packaging cost, but it also significantly increases the memory bandwidth while eliminating IO electrical communication that is noisy and power hungry. Because of the smaller memory cell size, the embedded DRAM can be ~3 to 6 times denser than embedded SRAM, and operates with low power dissipation and 1000 times better soft error rate. Embedded DRAM macros [32] that are based on high performance logic are extremely vital microelectronic components, making it possible to integrate 32MB on-chip cache memory on POWER7^{TM} processors [454].

Embedded DRAM employs a one-transistor and one capacitor (1T1C) as a memory cell which stores a data bit for a read and write operation. In order to reduce the cell size, the capacitor is built using either stack [420] or trench capacitor [56] structures. A deep trench capacitor approach is the preferred structure for embedded DRAM because the capacitor is built before device fabrication. This facilitates the implementation of a process fully compatible with logic technology, since transistor performance does not degrade due to capacitor fabrication, and design rules for back-end-of-lines (or metal wiring) remain the same as that of the logic technology. This results in an ideal technology solution for DRAM integration on a high-performance logic chip.

Figure 6.16 shows the overview of the embedded DRAM macro [32] for POWER7^{TM} micro processor [454]. The macro consists of four 292Kb memory arrays stacked in the Y-direction, resulting in a density of 1.168Mb. These arrays are controlled by a peripheral circuit block (IOBLOCK) arranged at the bottom of the macro. The peripheral circuit block, in turn, consists of command and address receivers, decoders, and macro IO circuitries used to control the memory arrays with given input commands.

Each 292Kb memory array consists of 1T1C cells (CELL) arranged in a 2 dimensional matrix. The memory cells in the array including row and column redundancies are accessed by 264 wordlines (WLs) and 1200 bitlines (BLs) for row and column, respectively. The architecture is optimized for a L3 cache application while taking into account the performance, power, and

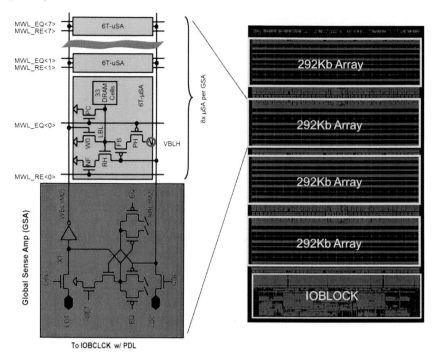

Figure 6.16
High performance embedded dynamic random access memory (embedded DRAM)

IO requirements. Unlike in conventional memory, the wordline drivers are placed in the area adjacent to the global sense amplifiers (GSA), which are in turn controlled by the global wordline drivers located in the peripheral circuit block (IOBLOCK). This orthogonally segmented wordline architecture [217] is the key to realize the wide IO organization.

The 292Kb array is organized using eight 36.5Kb micro-arrays for transistor micro-sense amplifier architecture [32]. 32 cells with an additional redundant cell (total 33 cells) are coupled to the Local Bitline (LBL), and read or written using a local micro-sense amplifier (mSA). Eight groups of LBLs, each one with a mSA arranged in the same column, are coupled to a global sense amplifier (GSA) through the global read and write bitlines (RBL and WBL). This hierarchical approach is important both to improve performance and to reduce power dissipation for a 500MHz random cycle.

While in the stand-by state, the WL is held at a wordline low voltage (VWL), a negative voltage. This is key to turning all access transistors sufficiently off in high performance SOI logic technology. Otherwise, the threshold of the access transistors must be increased, resulting in a higher wordline boost voltage (VPP) requirement to turn the devices on. Signals MWL_EQs and MWL_REs are respectively high and low, to turn the LBL precharge device (PC) on, the NMOS foot device (NF) off, and the PMOS head device (PH) off. This signal configuration precharges all LBLs to the low level (GND), while disabling the NMOS read head device (RH) and the PMOS feedback device (FB). RBL and WBL are held high and low, respectively. When the 36.5Kb micro-array is selected for WL activation, the signal MWL_EQ goes low. This disables the NMOS (PC), floating the LBL in the selected sub-array. The MWL_EQ signal in other unselected arrays remains at the high level, keeping the precharge device of the unselected LBLs at the low level. The low-going MWL_EQ also turns on the PMOS head device (PH), enabling the PMOS feedback device. For writing 1 to the cell, RBL goes low. This allows the LBL to go high, resulting in a high voltage written to the corresponding capacitor. For writing 0 to the cell, WBL goes high. This keeps the LBL at the low level, resulting in a low voltage written to the corresponding capacitor.

When a read command is accepted, read data are transferred from the cell to the BL when the WL goes high. For reading 1 data, LBL goes high, which turns the read head NMOS (RH) on. This results in making RBL low when the signal MWL_RE goes high and the NMOS footer device (NF) turns on. At the same time, low-going RBL turns on the PMOS feedback device (FB). This results in making the LBL go high. The high voltage on the LBL is written back to the corresponding cell. For reading 0 data, LBL stays at the low level, and the NMOS Read Head device (RH) remains disabled. RBL therefore stays at the high level. The RBL data are subsequently multiplexed by a column select signal (CSL) and communicated to the peripheral circuit block (IOBLOCK) through a Local Data bus pair (LDC and LDT) and then the Primary Data Line (PDL) for data transfer. At the same time, the Global Sense Amplifier (GSA) senses the RBL in the high state, making WBL go high. This makes the write 0 device (W0) to turn on, forcing the LBL to the low level. Therefore, the low level voltage on LBL is written back to the corresponding cell.

Precharge operation starts when the WL falls to the VWL voltage (negative voltage). MWL_EQ and MWL_RE go high and low, respectively, precharging the LBL to the low level. This concludes one 500MHz random access memory cycle for <1.5ns latency.

After the deactivation of the WL, the data bits are kept inside the capacitors. However, the charge stored in them leaks as time goes on, and therefore they should be periodically read and written back. This is a unique but important requirement for DRAMs and is known as the refresh operation. As long as the refresh is executed before the voltage of the storage nodes drops below the detection threshold of the sense amplifier, the data bits are maintained. The time interval during which a cell can hold the data bit is called retention time. The key idea behind the embedded DRAM intrinsic chip ID approaches covered in the following sections employ this retention signature to generate an unclonable random bitmap pattern as a chip ID.

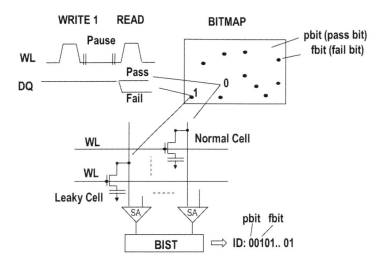

Figure 6.17
Retention-based intrinsic chip ID

6.9.4 Retention-Based Bitmap Intrinsic Chip ID

6.9.4.1 Generation of retention bitmaps

Figure 6.17 shows the concepts of intrinsic chip ID generation and authentication using retention signatures from embedded DRAM. Prior to ID generation, a logic 1 is written to all bit cells in an array, stored in the form of charge in the capacitors. However, the charge in each cell leaks as a function of time. Therefore, after waiting for a predetermined amount of time, the read data bits from the corresponding array may remain as 1 or change to 0. Preferably, the read operation may be managed by a Built-In Self-Test (BIST) engine or, if not implemented, by an external tester. If the charge remains sufficiently high, the output (DQ) from the sense-amplifier (SA) results in the expected value of 1, resulting in a PASS. If the charge leaks beyond the SA detection point of 1, then DQ is 0, resulting in a FAIL. Because each cell has a different retention time, the PASS and FAIL address locations in the array are random and physically unclonable. The array bit pattern (BITMAP) may then be recorded as the intrinsic chip ID in a bit string format such that the length of the string is equal to the array size, and where a passing bit (pbit) is stored as a 0, and a failing bit (fbit) is stored as a 1, ordered from the first to last logical addresses. Notice that the bits of interest for the intrinsic ID binary string are the fbits, and therefore their binary values are conveniently set to 1 as opposed to their logical 0 values after the read operation.

Since generation of fbits can be done by changing the predetermined pause time in a non-destructive manner, this retention-based intrinsic ID approach allows for use of the embedded DRAM array IP as is. However, the retention time depends significantly on various process or design parameters, including sub-threshold leakage, junction leakage, GIDL, bitline capacitance, cell capacitance, and noise. Therefore, chip authentication using retention time becomes highly sensitive to the precise test condition used during ID generation. In addition, creating fbits at low temperatures requires a long pause time, increasing the time for ID generation.

In order to improve bit stability, retention-based ID generation employs a higher wordline low voltage (VWL) than the product target voltage. This is the voltage applied to the wordline in an off-state for data retention. This higher voltage is equivalent to emulating a low-threshold access transistor, forcing the retention signature to be sub-threshold driven. Figure 6.18 employs a wordline low voltage (VWL) controlled by the VWL generator (VWLG) [233] using a feedback loop driven by the array's fail count (FC). Using higher VWL voltage increases the sub-threshold leakage of the access transistor, which in turn generates more fbits in the bitmap. A typical product allows for control of the VWL voltage for performance and retention optimization. In exercising the on-chip VWL adjustment function option, this VWL voltage can therefore be used as a tunable array input parameter controlling the number of the fbits in the BITMAP using a product macro.

Specifically for applications, product use is optimized using a negative VWL resulting in no retention failures within a predetermined retention target. This is labeled "Memory mode" in Figure 6.18. In the "ID mode", significantly higher VWL voltage is applied to the array such that the retention time is determined by sub-threshold leakage. In order to reduce the dependency on temperature and voltage conditions during ID generation, the retention-based intrinsic ID approach includes a counter for the number of fbits in the BITMAP. VWLG adjusts the VWL voltage such that the number of fbits matches the predetermined target number. Consequently, implementing feedback between the BIST-counter pair and VWLG enables the generation of an intrinsic BITMAP (or binary string) for any target fail count (FC) of the retention failures (or 1 in bit string). The target FC may be given by the Original Equipment Manufacturer (OEM) as a challenge to the chip. The generated intrinsic ID vector with the given FC is recorded in the OEM database to identify the chip for subsequent authentication.

6.9.5 Bitmap Authentication

Authentication is realized by extracting the ID string from the intrinsic ID binary vector using the same target FC as previously used during the recording phase. Once the bit string with the target FC is achieved, the chip outputs the generated ID bit string as a response to the OEM, and the database searches for the ID from within its list until it identifies the corresponding chip.

The uniqueness of a set of IDs for a given number of parts can be calculated exactly. The probability of generating n different IDs by randomly

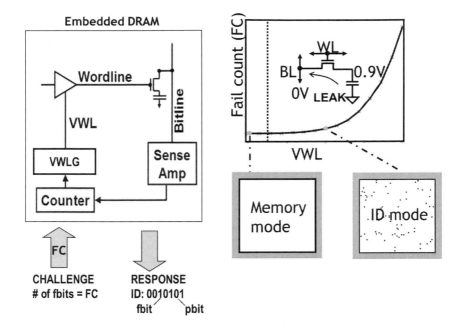

Figure 6.18
Method to generate the bitmap with wordline low voltage (VWL) control. (FC) is used as a challenge. VWL voltage is controlled by the VWL generator (VWLG) or an external tester such that the FC in the bitmap satisfies the target FC. The response is the bit string created by the bitmap, where 0 and 1 are pass and fail address locations in the array, respectively.

choosing j failing bits out of an array of size i is given by the following expression:

$$P_{BASE}(i,j,n) = \prod_{\alpha=0}^{n-1} \frac{\dbinom{i}{j} - \alpha}{\dbinom{i}{j}}$$

For practical implementations, the inherent information entropy contained in an ID set is already very large. For example, a set of 4Kb strings containing 100 fbits each will be unique up to roughly 10^{100} parts.

6.9.6 Dynamic Intrinsic Chip ID

6.9.6.1 Generation and authentication rules

The retention-based intrinsic chip ID allows for control of the number of fails in the array bit pattern (BITMAP) by means of the wordline low voltage (VWL). During generation, the number of the fails (fbits) in the ID matches a

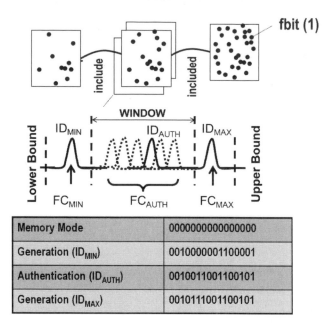

Figure 6.19
Concept of dynamic chip ID generation and authentication method

target number, and therefore that number can be adjusted by the VWL voltage. It is not practical, however, to generate the same exact number of fails as set by the target, and even if it were, it would be too expensive or time consuming. In addition, it may not be safer from a security standpoint to use the exact same ID binary string for each authentication request. The dynamic intrinsic ID approach is introduced to eliminate the necessity to generate the exact same fail count while enhancing ID security.

Figure 6.19 shows the concept of dynamic chip ID generation and authentication. It consists of a window-based authentication method using retention time in which the failed bit (fbit) locations corresponding to an ID with a larger fail count number include the fbit locations corresponding to an ID with a smaller number. Unlike the direct intrinsic ID method of the previous section, the dynamic intrinsic ID method employs a pair of IDs (ID_{MIN} and ID_{MAX}). ID_{MIN} is the intrinsic ID binary string corresponding to a target minimum fail count FC_{MIN} (i.e., 10). ID_{MAX} is the intrinsic ID binary string corresponding to a target maximum fail count FC_{MAX} (i.e., 100). During ID generation, the ID_{MIN} and ID_{MAX} pair is generated by adjusting the VWL voltage while counting the fails until the respective FC_{MIN} and FC_{MAX} targets are achieved. These fail count targets are used to determine the target window (i.e., 10 to 100) for subsequent authentication. Because of the nature of the DRAM cell, the fbits in the ID_{MAX} bit string include those in the ID_{MIN} bit string (Generation Rule).

Authentication uses the fail count target FC_{AUTH} inside the target window, such that $FC_{MIN} \leq FC_{AUTH} \leq FC_{MAX}$. The FC_{AUTH} target is dynamically changed within the window at each authentication request (i.e., 30, then 50, then 70), significantly improving hardware security. By construction, ID_{AUTH}, the regenerated intrinsic ID corresponding to FC_{AUTH}, should have a fail count at least larger than ID_{MIN} and at most smaller than ID_{MAX}. Again because of the nature of the DRAM cell, the fbits in ID_{AUTH} (i.e., 0010011001100101) obtained from the FC_{AUTH} target (i.e., 50) include those in ID_{MIN} (i.e., 0010000001100001) obtained from the FC_{MIN} target (i.e., 10), and are included in the fbits in ID_{MAX} (i.e., 0010111001100101) obtained from the FC_{MAX} target (i.e., 100). Authentication using ID_{AUTH} is therefore possible as long as the generated fail count is within the target window defined by the pair ID_{MIN} and ID_{MAX} (Authentication Rule).

Since ID generation does not need to satisfy the exact FC target provided by the challenge, window-based authentication simplifies ID generation. As shown in Figure 6.19, the FC in typical hardware has a distribution centered about the target FC. It is therefore important to determine a target window suitable for authentication while considering the FC distribution, in order not to violate the authentication rule. One possible violation of the rule can happen if too few fails or too many fails in the ID pair reduces the uniqueness of the chip ID. Hence, it is necessary to establish a lower bound for FC_{MIN} and an upper bound for FC_{MAX} of the generated ID pair in order to avoid false positive authentication (misidentification of the ID with others), which can be regarded as a Window Rule. As an example of what can happen without boundaries, it is possible to authenticate 0 fails or 100% fails for any chip, which is not an acceptable situation. The number of fails in the ID should therefore be within a predetermined window in order to satisfy ID uniqueness for the target application while remaining guaranteed with an analytically chosen degree of confidence.

The uniqueness of an authentication key ID_{AUTH} in recognizing a valid chip is minimized when $ID_{AUTH} = ID_{MIN}$. After all, an authentication key with $FC_{AUTH} = FC_{MIN}$ has the least number of combinations to choose from and is by construction the least unique. It follows that authentication of a pair of ID_{MIN} and ID_{MAX} using the ID_{MIN} bit string has the lowest uniqueness rating for dynamic authentication. For n parts, with a binary string of i bits, $FC_{MIN} = k$ and $FC_{MAX} = j$, an approximate analytical expression which provides a lower bound for the uniqueness of the pair of IDs is given by:

$$P_{MIN}(i,j,k,n) = \left(\prod_{\alpha=0}^{n-2} \frac{\binom{i}{k} - \binom{j}{k} - \alpha}{\binom{i}{k}} \right)^n$$

$$\approx Exp\left(-n^2 \left(\frac{\binom{j}{k} + \frac{n}{2}}{\binom{i}{k}} \right) \right), i \gg j, \binom{i}{k} \gg n$$

$$(6.15)$$

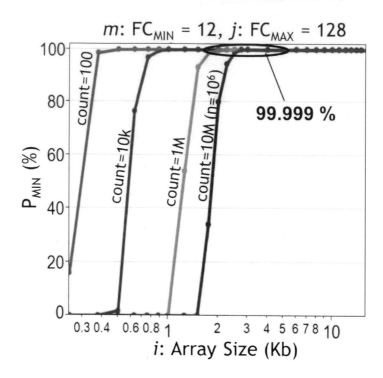

Figure 6.20
Calculated probability of ID uniqueness with equation

As seen in Figure 6.20, the analytical model shows $> 99.999\%$ ID uniqueness using a 4Kb segment ($i = 4096$), lower bound of 12 ($k = 12$, or $FC_{MIN} = 12$), and upper bound of 128 ($j = 128$, or $FC_{MAX} = 128$) for 10^6 parts ($n = 10^6$). The actual uniqueness is characterized by the probability P_{UNIQUE}, which is typically larger than P_{MIN}. They are equal only when all the ID_{MAX} in the set have no fails in common.

6.9.7   Dynamic Chip ID Hardware Results

The concept of dynamic ID generation and authentication has been studied using 32nm SOI eDRAM [142], which was a base design for an embedded DRAM macro IP for an IBM micro-processor [450]. Figure 6.21 shows a chip micro-photograph of an embedded DRAM array (a) and the test site (b) of the IP used for this feasibility study along with its features. The embedded DRAM employs high-performance SOI embedded DRAM design features including 6T micro sense amplifier and orthogonal wordline architecture developed for POWER7^{TM} [32].

The IP used in this feasibility demonstration does not include either BIST engine or a VWL generator, and therefore the VWL voltage is controlled ex-

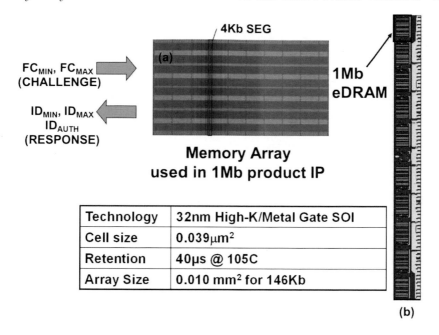

Figure 6.21
Chip micro-photograph of an embedded DRAM array (a) and the test site (b) of the IP used for this feasibility study along with its features

ternally by the tester with a resolution of 5 mV. Prior to generating the ID, confirmation of no fails in the target 4Kb segment is done using a nominal (within product specifications) wordline low voltage (VWL), also known as Memory mode. In this study, challenges of $FC_{MIN} = 25$ and $FC_{MAX} = 95$ were used to generate a pair of intrinsic IDs. The embedded DRAM was tested repeatedly, varying the VWL in the tester in order to search for a binary string with fail bit (fbit) count close to, but not necessarily equal to, the desired target, as long as the Generation Rule was satisfied. If a pair of 4Kb binary strings (ID_{MIN} and ID_{MAX}) was found which:

1. Reasonably matched the corresponding challenges: $FC_{MIN} = 25$ and $FC_{MAX} = 95$, and

2. Satisfied the Generation Rule: the fbit locations of ID_{MAX} included the fbit locations of ID_{MIN}

the ID pair was stored in an ID management system for authentication; else, the generation scheme was applied to successive 4Kb segments (SEG) of the array until a suitable area was found. Beginning with the Memory mode condition of no fails, the typical time to find the target number of fails required

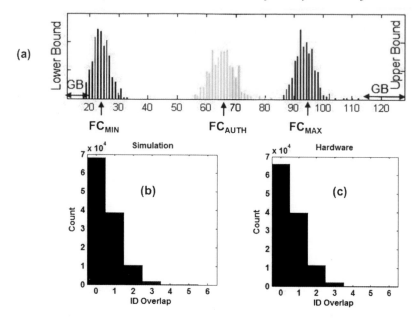

Figure 6.22
Hardware results. (a) measured ID distribution with challenge of $FC_{MIN}=$ 25, $FC_{MAX} = 95$, and $FC_{AUTH} = 65$. ID overlap for different IDs: (b) Monte Carlo simulation, and (c) measured results.

was 30 ms per chip. The search time can be reduced significantly by means of a binary search [3].

For the purposes of authentication security and counterfeit prevention demonstration, chip authentication was realized by using different challenges, selected as a different fail count value at each authentication request, e.g., a FC_{AUTH} of 65 that is between FC_{MIN} of 25 and FC_{MAX} of 95. Similar to ID_{MIN} and ID_{MAX} pair generation, VWL was adjusted across multiple tests of the 4Kb SEG manager circuit originally used for ID_{MIN} and ID_{MAX} pair generation. The voltage was adjusted until an ID_{AUTH} binary string was generated with a number of fbits close to FC_{AUTH} of 65. The ID_{AUTH} string vector was the output of the system, and the string was compared to the list of ID_{MIN} and ID_{MAX} pairs previously stored in the system in order to identify the corresponding chip. A match occurred when the ID_{AUTH} string included all of the fbit locations in ID_{MIN} and was included in the fbit locations in ID_{MAX} for a pair of strings in the database (Authentication Rule from previous section). If the bit string ID_{AUTH} did not follow the rule for any ID_{MIN} and ID_{MAX} pair in the list, then the chip was deemed invalid.

Figure 6.22(a) shows the measured ID distribution generated by challenges ($FC_{MIN}=$ 25, $FC_{MAX} =$95, and $FC_{AUTH} = 65$) for 346 chips at fixed temperature of 25°C and voltage of $V_{DD} = 0.9$ V. It was confirmed that the

346 chips can be uniquely identified by the dynamic retention-based chip ID Generation and Authentication rules. Detailed analysis of the data shows that the mean/standard deviation of overlapping fbit locations for Monte Carlo simulation (b) and hardware (c) are 0.55/0.73 and 0.58/0.76, respectively, demonstrating a good correlation between simulation and hardware results.

6.9.8   Field-Tolerant Intrinsic Chip ID

6.9.8.1   Pattern recognition and model

The dynamic intrinsic ID approach allows for some bit changes as long as the failing bit addresses (fbits) in ID_{AUTH} are included in the ID_{MAX} fbits. It may, however, result in an error if some newly generated fbits in ID_{AUTH} violate the authentication rule. This may happen if all the fbits in ID_{MIN} do not constitute any longer a subset of the fbits in ID_{AUTH}, or if all of the newly generated fbits in ID_{AUTH} cease to be a subset of the ones in ID_{MAX}. This results in false negative authentication. Therefore, it is necessary to exert precise voltage and temperature control, so that ID_{AUTH} generation can emulate the conditions used for ID generation. This is impractical, if not very expensive, for use in the field. Even if the exact same conditions are used in the field, some unstable fbits may still change, resulting in an authentication error. A field-tolerant intrinsic ID is introduced to overcome this problem.

Figure 6.23 illustrates the concept of dynamic ID (a) and field-tolerant intrinsic ID (b) generation and authentication. As discussed in the previous section, the dynamic intrinsic chip ID approach (a) uses an authentication window using ID_{MIN} and ID_{MAX}. This results in successfully authenticating the corresponding chip as long as the fbits in the ID_{AUTH} vector are within the fbit location window bounded by ID_{MIN} and ID_{MAX}. Imperative to enabling the field-tolerant ID approach (b) is the guarantee of unique ID generation and successful chip recognition even if some bits change across various conditions. This can be realized by detecting common fbits between the IDs in a database and the ID for authentication. In order to increase the probability of common fbits, FC_{AUTH} is made larger than the FC_{GEN} used for generation of ID_{GEN} by means of a higher VWL voltage during authentication than during generation. In contrast, if FC_{AUTH} is smaller than FC_{GEN}, the number of fbits in common between ID_{AUTH} and ID_{GEN} is at most FC_{AUTH}, which by construction is a fraction of FC_{GEN}. This makes it harder to authenticate ID_{AUTH} since there is potentially too little common fbit overlap between the two IDs as a percentage of ID_{GEN}. Therefore, as FC_{AUTH} increases relative to FC_{GEN}, a predetermined target percentage of FC_{GEN} common fbits may be used to authenticate ID_{AUTH} with a required level of confidence.

Successful recognition may be achieved even if some unstable fbits change due to variations in field test conditions and/or device shifts over the product's lifetime, e.g., NBTI [459]. As discussed in the dynamic ID approach

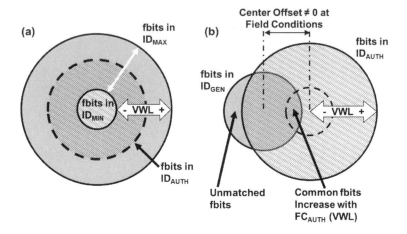

Figure 6.23
Concept of dynamic and field-tolerant ID approaches. Dynamic ID (a) uses
ID_{MIN} and ID_{MAX} to improve ID security. Field-tolerant ID (b) detects common failing bits while accepting some unmatched failing bits which may be
the result of a different environment in the field.

(a), the field-tolerant approach (b) must also keep the number of retention
fails, FC_{AUTH}, less than that of a predetermined maximum value – roughly
half the total array bits – in order to avoid false positive authentication or ID
spoofing. This limit comes from the fact that a binary string with many fbits
is in itself a combination of many different binary strings with fewer fbits.
Therefore, there are more chances of multiple ID_{GEN} combining to form a
larger ID_{AUTH}, or of an attacker using an ID_{AUTH} with an arbitrarily large
number of fbits for spoofing.

The field-tolerant approach requires choosing proper array and ID sizes
in order to avoid false authentication. While false negative authentication is
the main goal of this approach, an increase in retention fail count will also increase the probability of false positive authentication, in which a chip is recognized as a different chip in the database. For randomly-generated IDs, this
probability is modeled by the number of expected fbits in common within
any pair of IDs belonging to the set of stored IDs. Given a string representing 100 random fail locations in 4096 bits, the probability that another similar
random string has k fails in common with the first obeys the well-known and
documented hypergeometric distribution [122]. This can be interpreted as all
the ways of choosing k fail locations in common with the original 100 while
simultaneously choosing the remaining 100-k fail locations from the remain-

ing $4096 - 100$ passing locations, and it is given by:

$$
P(k) = \frac{\left[\begin{array}{c} \#\text{ ways for} \\ k\text{ "successes"} \\ \text{picked} \end{array}\right] \times \left[\begin{array}{c} \#\text{ ways for} \\ (100-k)\text{ "failures"} \\ \text{picked} \end{array}\right]}{[\text{Total }\#\text{ picks}]}
$$

$$
= \frac{\dbinom{100}{k}\dbinom{4096-100}{100-k}}{\dbinom{4096}{100}}
\tag{6.16}
$$

The calculated probability density function $P(k)$ of 100 bit ID pairs as a function of the k overlapping fbits and array size i is shown in Figure 6.24. The function peaks near the theoretical mean of the distribution. The solid lines represent fixed 100-bit IDs, while the dashed lines represent the actual hardware distribution with an average over all ID sizes in a 90 to 110 bit window (accomplished by additional averaging of the function over that range). The level of accuracy in meeting the target fail count is therefore expected to affect the value of the mean (peak) and the tightness of the distribution of overlapping fails (standard deviation).

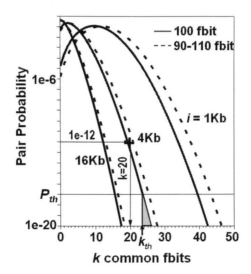

Figure 6.24
Calculated probability to have the corresponding ID overlap with respect to ID bit string length (i = 1Kb, 4Kb, and 16Kb)

Statistical extrapolation can be used to establish a proper guard band of common bits k_{th} for false positive prevention. For a set of 10^6 parts, the total number of pairs is on the order of $10^6 \times 10^6 = 10^{12}$. Referring to the 4Kb distribution of Figure 6.24, the odds of 1 occurrence in 10^{12} pairs correspond to a probability of 10^{-12}, which in turn occurs when the number of common fbits is $k = 20$. By construction, the probability from the model of finding a pair within the set with more than 20 common fbits decreases exponentially. Because this model is also derived from a discrete probability curve, it is reasonable to assume that the cumulative probability above this point never adds to the probability of the point itself. To illustrate this, $k_{th} = 23$ fbits in common represents a probability $P(k_{th}) = P_{th}$ that is roughly 5 orders of magnitude lower than that of 20 fbits in common. This probability also corresponds to the shaded area to the right of k_{th}. Therefore, if no ID is ever authenticated with less than 23 fbits in common, this represents a guard band providing over 10^5 lower odds ($P_{th} = 10^{-17}$) than finding 1 pair with 20 common fbits in a set of 10^6 parts. In other words, the final guard-banded threshold therefore satisfies unique chip IDs 99.999% ($1.0 - 10^{-5}$) of the time for 10^6 parts.

In order to ensure 100% successful authentication, hardware analysis using voltage and temperature fluctuations must be used to take into account unstable retention fails and prevent a false negative result.

6.9.8.2 Field-tolerant hardware results

Field-tolerant intrinsic ID generation and authentication have been demonstrated using 32nm SOI embedded DRAM product IP. The 4Kb ID_{GEN} bit strings with target of 100 retention fails were extracted using a 4Kb segment from a 292Kb array in each of 266 embedded DRAM chips. A nominal voltage condition of $V_{DD} = 0.9$ V and 85°C was used. These ID_{GEN} bit strings identifying each one of the 266 chips were recorded in the local database.

Field-tolerant authentication was then emulated with ID_{AUTH} string generation for a target of 200 retention fails using the same 4Kb segments. Two temperatures, 25°C and 85°C, and three voltage conditions of $V_{DD} = 0.9$V $\pm10\%$, representative of voltage tolerance in the field, were used. The 6 generated ID_{AUTH} from each of the 266 chips were used to search the corresponding chips with previously stored ID_{GEN} in the local database .

In searching for false positives in hardware, every ID_{AUTH} (total 6 $\times$ 266 = 1596) was compared to the stored 266 ID_{GEN} of a *different* chip as shown in Figure 6.25 for a total of ~420k (= 266 $\times$ 1596) pairs. Figure 6.25 therefore shows the probability that a pair ID_{AUTH}-ID_{GEN} has a given ID overlap in number of common failing bits (fbits). The normalized fbit count is overlaid with the hypergeometric model of (6.16) with no fitting parameters. The model assumes that the set of stored IDs have 100 random target fails ($FC_{GEN}=100$) and the set of authentication keys have 200 random target fails ($FC_{AUTH}= 200$), with a $\pm10\%$ fail count tolerance window for each set. The excellent agreement with the random fail model is a measure of the uniqueness and randomness of the set of IDs.

False negative authentication occurs when a previously stored chip ID cannot be recognized due to physical instability of the bit cells in the field.

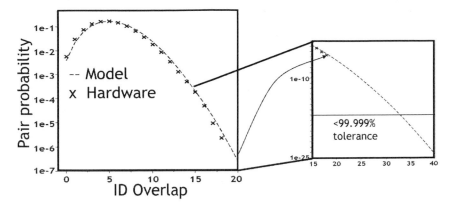

Figure 6.25
Probability to have the corresponding ID overlap for any of ~420K pairs. A
40% overlap between a 100-fail stored ID and a 200-fail authentication key is
sufficient to guarantee 99.999% ID uniqueness for 10^6 parts. Fail count windows of $\pm10\%$ are ensured with a ±5 mV VWL window.

False negatives were characterized by comparing each ID_{AUTH} to the stored
ID_{GEN} of the *same* chip. The results for all 266 chips are summarized in Figure 6.26, which shows the number of authenticated fbits versus the stored ID
size (= fbits in ID_{GEN}). No more than 10% of fbits in the authentication set
fail to match the original ID. Most fbit loss occurs at 25°C, and as recorded
ID size approaches authentication ID size. Tighter fail count control can be
used to limit this problem.

    Whereas false negative characterization depends on the physical proper-
ties of the circuit and the test methodology, false positives rely on the randomness of the ID set and can be characterized with use of the model of
(6.16). Conservatively, it is sufficient to use as few fbits as possible for recognition, as long as it ensures that all parts can be successfully authenticated
without misrecognition. For 10^6 parts, an approach requiring that a minimum of ~40% of the bits from the stored ID_{GEN} (40 fbits for 100-fail FC_{GEN}
target) are common with the authentication key ID_{AUTH} of a different chip
(200-fail FC_{AUTH} target) guarantees authentication with at least 99.999% success rate. On the other hand, security is enhanced when a larger number of
fbits is used for recognition, since it makes counterfeiting harder. Unfortunately, false negative results are made more likely by increasing the number
of fbits that must be regenerated. Since no more than 90% of the stored ID_{GEN}
fbits can be ensured to match the ID_{AUTH} fbits in the simulated field authentication of Figure 6.26 while avoiding a false negative, the resulting 40–90%
window of overlapping fbits contains all the redundant authentication fbits
for this hardware. A compromise between false authentication boundaries

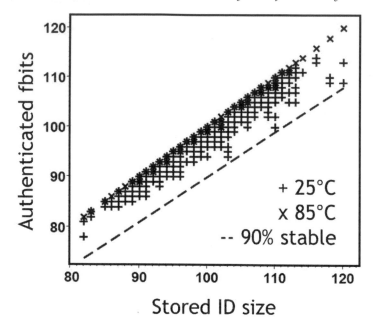

Figure 6.26
Measured ID overlap for each of the same 266 chips during authentication using V_{DD} = 0.9 V and V_{DD} = 0.9 V ±10%, at 25°C and 85°C (total 6 conditions). The results show at least 90% of the ID string 1 bits are stable even if the test condition is changed.

can be therefore reached with a fixed 65% fbit target to be used for field-tolerant authentication.

Use of larger keys for authentication can further reduce the number of unstable fbits that are not matched by forcing those weak cells to fail eventually. Since all cells will fail for sufficiently high VWL or retention time, the target value must be bounded in order to ensure uniqueness of the set and can be calculated using the hypergeometric model. A composite field-tolerant / dynamic key approach follows: a 200 fbit ID needs only to match roughly 65% of fbits in a 100 fbit ID, while having only 65% of its fbits in common with a 400 fbit maximum ID that is also stored in order to improve security.

6.9.9   The Self-Authenticating Chip Architecture

6.9.9.1   The search and storage problem

All the intrinsic chip ID approaches discussed in the previous sections rely on the same fundamentals: that a fingerprint obtained from a memory chip is searchable in a database. However, the resources necessary to allocate,

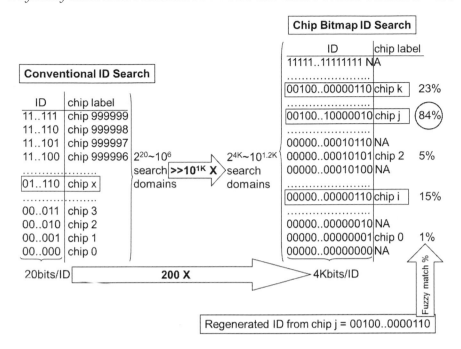

Figure 6.27
Conventional ID search versus chip bitmap ID search. The bitmap search introduces a large overhead in number of bits/ID and an exponentially larger overhead for total virtual storage allocated for search domains. To identify chip j, the regenerated ID 00100..0000110 should be compared to the entire valid ID list in the database to satisfy a bit match criterion (>50% of "1" bits in common). Identification of a memory PUF requires fuzzy bitmap recognition and is unfeasible.

search, and recognize the corresponding bitmap IDs can be prohibitively expensive, especially if the bitmap IDs are as large as the ones used for dynamic chip ID in Section 6.9.6. As seen in Figure 6.27, in order to address one million parts, memory address allocation in the OEM database is accomplished directly with 20 bits, but using 4K binary bits it takes 200 times more memory. And while each conventional ID within one million IDs is uniquely identified by an address (i.e., address 01...110 = chip x), bitmap IDs cannot be identified uniquely by an index because they may include irreproducible bits. The bitmap IDs are randomly distributed within the 2^{4K} ($\sim 10^{1.2K}$) search domain. This requires $> >10^{1K}$ larger virtual storage allocation than a conventional ID approach for unique identification of all the bitmaps. Therefore, the address search method is impossible with the field-tolerant approach of Section 6.9.8 because the regenerated ID contains irreproducible bits and does not match any of the IDs in the OEM database, requiring identification through fuzzy match detection. In this section, we will discuss a Self-Authenticating Intrinsic ID (SAIID) architecture and its hardware implementation, which uses a combination of on- and off-chip encoding and encryption of random bit strings acquired from the embedded memory bitmap-based PUF [352]. In order to understand the architecture, the various concepts are introduced in terms of its fundamental functions: indexed search, on-chip authentication, question/answer pairs, and lie detector.

6.9.9.2   Indexed search

The fuzzy authentication requirement of a bitmap ID in the manner discussed so far is a key disadvantage for practical PUF implementation in hardware security. Since the search operation can be made optimal when there is an index associated with each ID, an improved secure chip architecture should also possess an indexed search using a public ID (eFUSE). As exemplified in Figure 6.28, the index must exist both in the chip itself as (a) a public ID (i.e., 42) and in (b) the OEM database. Therefore, a simple implementation of the fuzzy authentication requires the comparison of the (c) bitmap ID indexed in the OEM database and the (f) bitmap ID regenerated by the chip as (e) a response to (d) the challenge provided by the OEM.

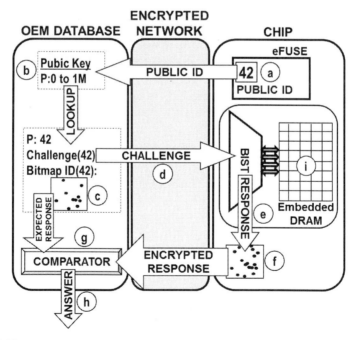

Figure 6.28
Indexed search intrinsic ID architecture using embedded DRAM. Circles indicate essential items for this architecture as listed in the text.

The use of a public ID solves the search problem but retains other inconveniences of using PUFs for authentication: 1) the OEM overhead remains large because a (c) bitmap ID must be stored for every public ID, and 2) the (f) regenerated bitmap ID must be communicated from the chip to the OEM through an encrypted network in order to be (g) compared and (h) authenticated. Security of the architecture shown in Figure 6.28 therefore depends on the strength of the network's encryption, and may expose the bitmap ID to side-channel attacks. A counterfeiter in possession of both the (a) public ID and the (f) regenerated bitmap ID can create a fake chip that will generate the same response to the OEM database as that of the original chip. To protect the bitmap ID, the (g) comparison and (h) fuzzy authentication must not occur in the OEM and therefore must happen in the chip itself.

6.9.9.3   On-chip authentication

Since the (a) public ID in Figure 6.28 does not offer any security, a feedback mechanism must exist between the chip and the OEM. For this reason, the secret information contained in the (i) embedded DRAM PUF must be hidden. One way of accomplishing this is by splitting the secret itself, represented by a (d) challenge and (e) response pair. Therefore, the secret contained in the PUF could have two parts: an indexed challenge in the OEM database, and a fuzzy authentication of the bitmap response on the chip.

Moving the authentication step to the chip requires a reference bitmap for the fuzzy comparison. This reference bitmap (Sections 6.9.6 and 6.9.5) can be generated and subsequently stored in the non-volatile, local OTPROM eFUSE [349,363,430]. This procedure has two immediate problems: 1) a counterfeit chip can always provide a fake answer (or respond "authenticated" to the database without performing the comparison between the reference bitmap (c) and the regenerated bitmap (f)) to the authentication request, and 2) the OEM cannot verify the authenticity of the chip any further. Additionally, the purpose of the PUF is circumvented by storing a physical copy which can be extracted by de-layering of the chip. A solution is provided next.

6.9.9.4   Question/answer pairs

Referring to Figure 6.29, a (X) single fixed challenge, for simplicity, may apply to the (Y) entire embedded DRAM PUF at every authentication request, while the (e) response may be handled by means of encoding and encryption. Exploiting the use of an indexed search, a (a) public ID can be used for lookup of domain locations (D) or bit offsets (O) within the (b) OEM's database. Together, these functions form an (c) individual question (i.e., D: 7, 12, O: 5, 4). Questions (d) passed on to the chip operate on the chip's (e) response inside the (f) SAIID generator. Each resulting encrypted string can then be (g) stored in the OTPROM eFUSE as a local SAIID copy for future retrieval, or (h) compared to the local SAIID copy. The result of the comparison is an (i) answer of the type true/false (T/F) which may constitute a final authentication step.

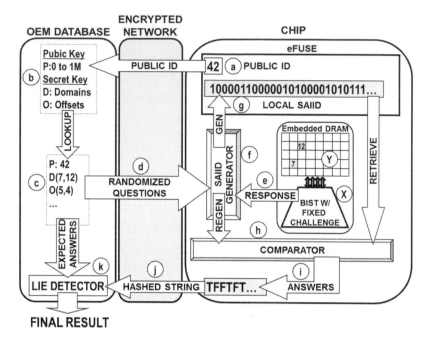

Figure 6.29
Self-authenticating intrinsic ID (SAIID) architecture using embedded DRAM.
Circles indicate essential items for the self-authenticating architecture as
listed in the text.

But this procedure so far does not prevent its use by counterfeit chips, because a fake chip has 50% odds of providing the correct (i) answer to a (c)
single question. To fix this, the answer must also be protected.

The odds of a successful counterfeit chip can be reduced by increasing
the number of question/answer pairs. A true answer must correspond to a
valid single challenge/response pair from the PUF. But each individual answer may be true or false, intentionally, depending on the group of (d) questions chosen. To improve security even further against eavesdropping, the
network may be encrypted, while the (d) group of questions may be randomized. This last step ensures that no two groups of (i) answers will ever
be the same. A method for generating multiple questions is discussed next.

6.9.9.5   Lie detector

The (i) group of answers of Figure 6.29 is then verified by the OEM in a (k)
lie detector. The OEM possesses knowledge of the (i) answers to each of the
(d) randomized questions because the only true answers result from the (c)
individual questions which were used to generate the data stored in the (g)

local SAIID in first place. These true questions are associated with the same (a) public ID, which is known to the OEM.

Due to the properties of PUFs, the collision rates between bitmap (e) responses from different chips or from different areas of the same chip (such as asking a different (c) question) are extremely low. Therefore, there are plenty of choices of questions to choose from resulting in false answers. On the other hand, only a few selected questions may result in true answers, as long as the encrypted bitmap responses were previously stored in the (g) local SAIID. If the concatenated string of (i) answers (i.e., TFFTFT...) provided by the chip matches exactly the string of expected answers in the OEM, *the chip has aided in authenticating itself*. In that case, the chip passes the (k) lie detector test and the authentication process is complete.

Security of the string of answers may be further improved by means of a (j) hashing function [322], [68, 151]. Without it, a side-channel attack may be able to link each (c) question with each (i) answer and build a model from which to counterfeit the chip. Additionally, the (X) challenge may have a dynamic fail count target to enable dynamic and/or field-tolerant authentication (Sections 6.9.6 and 6.9.8).

6.9.9.6 SAIID algorithm and schematic

These elements help explain how the SAIID architecture works. In order to detail the SAIID generator, on the other hand, a discussion of the SAIID algorithm and schematic representation follow:

SAIID Algorithm

Figures 6.30(a) and 6.30(b) show how subdividing the response, also called a Fingerprint String (FPS), of a memory into domains can be used to distribute fails uniformly by logical and physical address scrambling. 1K domains of 4Kb each in a 4Mb array can be selected with 10 binary address bits A0–A9, which are swapped with binary address bits A21'–A12' in this order. This avoids systematic fails from a bitmap which may later degrade the rate of successful recognition.

Figure 6.30(c) shows the application of a question to two domain selections A and B as an example. The question specifies memory domain locations and offsets to the starting position of the corresponding FPSs for superimposition. The resulting offset-superimposed bitmap is the Self-Authenticating Intrinsic ID (SAIID). If the number of subsets and bit offsets available are large, the number of combinations permitted makes the SAIID more unique and harder to counterfeit.

SAIID Schematic

Figure 6.31 shows a timing diagram (a) and a block diagram (b) of a possible VLSI implementation of a SAIID generator. It includes 6 sets of 4Kb scan

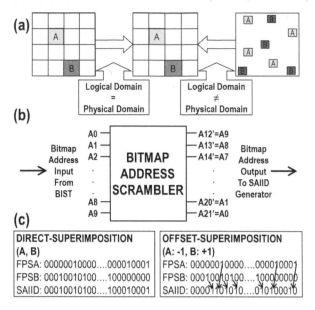

Figure 6.30
(a) Methods for creating domains from large memory blocks. (b) Example of bitmap address mirror scrambler used to create 1K (10 bits) domains from 4Mb memory (22 bits) with logical ? physical domain (c) SAIID generation by encryption of two fingerprint strings (FPS) from domains A and B.

chain registers. The BIST includes a 128b scan chain which outputs bitmap information of an entire 4Mb memory array as a pipeline with the scan clock (CLK). The BIST also generates binary address bits A0–A9 while scanning the bitmap, where bits A0–A9 are used for scrambling into binary address bits A21′–A12′ as shown in Figure 6.30(b). Each 4Kb shift register stores the corresponding 4Kb selected by the domain selection signals d0–d5. These signals are controlled by ANDing of 10 XOR outputs for address match detection, each coupling the corresponding 10-bit address (A12′–A21′ after scrambling) with the OEM domain address selection. The OEM address selection is provided during SAIID generation and is unique to each of the 6 sets of chains. This results in selecting 6 out of 1K domains while scanning the 4Mb FPS to the corresponding 4Kb chain with CLK.

Support for an offset is realized by inclusion of a loop function in the scan chain, which is enabled after the 4Mb FPS has been transferred to the corresponding 6 sets of 4Kb registers. Finally, the ordered bits from each register are coupled to corresponding OR logic such that the resulting 4Kb SAIID is a scrambled, offset, and superimposed binary vector. Any additional round of SAIID generation can be done using the 4Mb FPS kept in the 4Mb memory.

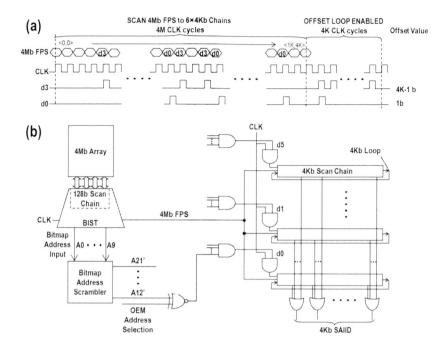

Figure 6.31
(a) Timing diagram of SAIID generator. (b) Example of circuit implementation of 4Kb SAIID generator for 6 domains out of 1K in 4Mb array.

The estimated silicon area needed to implement this SAIID of Figure 6.31 is less than 0.2mm^2 using the 22nm node, which is negligible for a larger chip.

Generation of the 4Mb FPS using a VWL binary search takes <1s for a target of ~32K fails. The 4Mb memory then keeps the generated FPS bits after the final 32K fails have been found. The time required per SAIID generation is subsequently determined by the scan of the 4Mb binary string, which for a 1GHz clock results in 4ms (4M×1ns).

6.9.9.7 SAIID hardware emulation

A self-authenticating intrinsic chip architecture was emulated using an embedded DRAM prototype in 22nm SOI technology [303]. A micrograph of the test site is shown in Figure 6.32. It contains 5 independently testable 20Mb embedded DRAM chips, and 4 independently testable 4Kb eFUSE chips. All encryption and pattern recognition analysis was performed off-chip for this emulation. Although available, the eFUSE was not used in the emulation of the architecture.

Retention bitmaps were generated with a method utilizing VWL to accelerate array device leakage. First, a background of logical 1s was written to

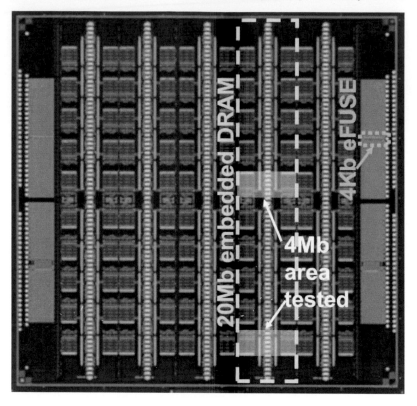

Figure 6.32
Prototype of embedded DRAM with eFUSE in 22nm SOI. The 4Mb area selected for test within one 20Mb embedded DRAM chip is highlighted in white.

all cells in the 4Mb area at nominal supply voltage. After a predetermined pause time, a binary search algorithm varied VWL until a target of 32K fails was met with a tolerance of ±10%. A total of 57 chips were used for this work. All retention bit fail maps exhibited single cell signatures as the dominant pattern for all the parts.

A total of 1168 domains were created by distributing cells from every memory subarray segment into strings of exactly 4Kb length. The average number of fails per domain was 27.4. For each part, 6 random domains were chosen associated their respective random offsets. The offset-superimposed 4Kb strings were recorded as the local SAIIDs. Assuming low collision between the fail locations of different domains, the expected number of fails per overlay is ~164. Figure 6.33(a) shows the distribution of number of fails in the 4Mb original bitmap and number of fails in the SAIIDs, confirming the expectation. Due to possible clustering of fails in the hardware, the number of fails in the 4Mb strings is distributed in a non-uniform manner in the

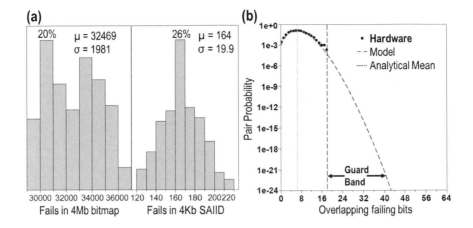

Figure 6.33
(a) Distribution of number of fails in 4Mb bitmaps (left) and 4Kb SAIIDs generated from 6 domains with offset (right). (b) Interchip collision rate in 22nm SOI hardware for 4Kb SAIIDs with 164 failing bits on average, obtained at 25C and supply voltage with $\pm 10\%$ variation. Extrapolation of the model provides a minimum Guard Band (GB) between the maximum number of observed collisions (18) in 57 parts and collision avoidance. The raw value of 43 overlapping fbits provides >99.9999% success for 1M parts.

target window of $32K \pm 10\%$ and is clipped by the test algorithm within this fail count window. The 4Kb strings, on the other hand, are normally distributed due to the successive randomization steps to the domains chosen and their string offsets. Each part was retested at 10% lower and 10% higher supply voltage than the nominal target voltage used for generation, emulating a typical authentication specification requirement for field operation at room temperature. New 4Kb SAIIDs were then created using the same random domain combinations and FPS offsets applied when generating the local SAIIDs of each corresponding part.

The rate of collision between local and regenerated SAIIDs is analyzed by the number of overlapping failing bit addresses (fbits), as shown in Figure 6.33(b). The number of comparisons is $\sim 1.5 \times 10^4$ and includes comparisons between local SAIIDs from different parts. The accompanying model uses the mean number of fails of 164 and peaks at the predicted analytical value $164^2/4096 = 6.6$ according to [122]. The model of Section 6 predicts a success rate >99.9999% ($P_{th} = 10^{-24}$) in 1M parts at 43 fails. This threshold provides an appropriate guard band (GB) relative to the maximum number of collisions observed in the hardware, seen as 18 in Figure 6.33(b).

Figure 6.34(a) shows the number of fbits from the SAIIDs regenerated at different voltages that were authenticated by the local SAIID generated at

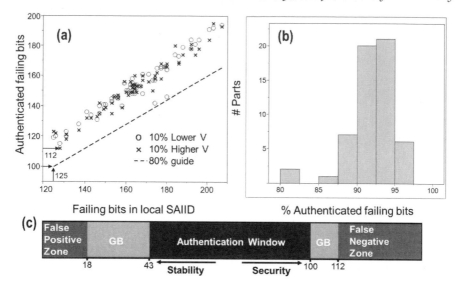

Figure 6.34
(a) Number of authenticated failing bits during pattern recognition using a different voltage relative to that used for generating the local SAIID copy. A guide to the eye at the 80% level is shown. (b) Distribution of minimum percentage (worst case) of authenticated bits per SAIID for all 57 parts. (c) Diagram illustrating relationship between number of authenticated failing bits in hardware and authentication window including guard bands (GB).

nominal voltage. While the lowest number of fbits stored in the local SAIID is 125, the lowest number of authenticated fbits is 112. The distribution of the lowest percentage of fbits required for authentication of each part is shown in Figure 6.34(b). While *most* parts can be authenticated with 92% of fbits recognized, an 80% successful fbit authentication rate is necessary to accommodate *all* parts, suggesting there are no less than 80% stable fbits in the stored SAIIDs. More details can be found in Ref. [352].

The hardware results demonstrate the existence of an authentication window for all parts extending from 43 fbits (false positive threshold of Figure 6.34(b) with GB) to 100 fbits (false negative threshold of Figure 6.34(a) with GB), as illustrated in Figure 6.34(c). Therefore, screening out hardware SAIIDs with less than 100 fbits is more than sufficient to authenticate all available parts, allowing for a worst-case scenario of 43% stable fbits.

6.9.10   Summary

In this chapter, intrinsic ID generation and authentication using a VLSI chip has been covered at length. As discussed, the approach may preferably use an intrinsic feature relating to retention fails using embedded Dynamic Random Access Memory (DRAM). The DRAM cell allows for retention fails to be controlled by changing the sub-threshold leakage. This results in generation of a skewed binary string, having more 0 bits than 1 bits, where 1 is associated with the memory array cell location causing a fail in the bitmap. The challenge of employing retention fails for the generation of intrinsic ID binary strings is how to generate a stable retention fail. This requires operating the device in the sub-threshold domain, and can be realized by controlling the wordline low voltage with fail count feedback during the retention test. One consequence of this fail count feedback method is that the intrinsic ID can be changed dynamically as the fail count target changes at each authentication request, resulting in enhanced hardware security. The discussion further includes a method to improve successful authentication rates in a field using a fuzzy match detection approach [377, 380], while further improving the dynamic micro-ID approach with superimposed BITMAPs. To conclude the chapter, a Self-Authenticating Intrinsic ID (SAIID) chip architecture and hardware implementation method have been proposed. This method is further compatible with hashing algorithms for added security.

Most of the ideas discussed in this chapter use intrinsic ID for chip identification in order to protect the product from counterfeiting. However, the applications of intrinsic IDs are not limited to chip identification, in which intrinsic bitmaps using embedded memory are used as a key. Since an intrinsic bitmap can create various random bit patterns, and cannot be easily copied, one of its most technologically relevant applications is as a cryptographic key [432, 433]. Unlike the application for chip identification, cryptographic keys require 100% stable and secure ID bit string generation, which in turn requires innovations such as advanced circuit design for ID generation and authentication, while integrating the error correction code within the chip [323]. In order to further improve security, implementation of a reconfigurable PUF [203] which limits multiple use of the secret key may also be of special relevance.

In addition to the cryptographic key challenge, a standardization of the intrinsic ID should be strongly encouraged in order to disseminate the use of intrinsic ID in various products. This requires the development of an intrinsic ID engine which can be embedded into any VLSI chip. This engine must take into consideration the system architecture in order to maximize the hardware security advantage [352]. Intrinsic bitmap chip ID using embedded memory is an emerging technology which can greatly benefit from tight interactions among technologists, circuit and system designers, and security engineers [216].

6.10 Conclusion

PUFs are gaining significant attention in the hardware security community because of the well-known vulnerabilities posed by classical cryptography. Apart from security applications, PUFs can also be used as a tool for exploring the impacts of process and environmental variations on timing, power, and performance of general CMOS circuits because of their lightweight nature. In this chapter, an overview of some of the most popular CMOS PUF circuits in the literature is provided. The performance metrics used for analyzing PUF circuits are clearly identified. We also presented a unified design flow for a silicon PUF instantiation. Simulation techniques for measuring the performance metrics of a PUF are also presented. Measurement data observed from the arbiter PUF circuits implemented in 45nm SOI technology node [253] validate the design techniques. Even though the focus of this chapter was on arbiter PUFs, the techniques presented can be applied for other PUF architectures. A class of arbiter PUFs operating in sub-threshold mode called as sub-V_{th} PUF was also presented. These devices were designed, fabricated, and tested based on the methodology presented in this chapter. We also presented a detailed discussion on self-authenticating intrinsic ID architecture and hardware implementation using embedded DRAM.

7

Random Number Generators for Cryptography

Viktor Fischer

Jean Monnet University Saint-Etienne

Patrick Haddad

ST Microelectronics

CONTENTS

7.1 Introduction

Random number generators (RNGs) are computational or physical functions generating a sequence of bits or symbols (e.g., groups of bits – numbers) that do not feature any pattern – generated bits or symbols (numbers) are independent and uniformly distributed. Random number generators have many applications in modern technologies. They are widely used in cryptography, but also in Monte Carlo simulations of complex systems, as noise generators in telecommunication systems, in games, slot machines, etc.

Many methods of generating random numbers exist, starting from ancient methods using dice and coin flipping, sophisticated methods using quantum mechanics, up to modern principles that can be implemented in electronic systems and in particular in logic devices.

Random numbers are crucial in cryptography: they are used as confidential keys, padding data, initialization vectors, nonces in challenge-response protocols, but also as random masks in side channel attack countermeasures. Since the era of Kerckhoff, the confidentiality of data is based on encryption keys: it is supposed that cryptographic algorithm is known to the adversary and that confidentiality is guaranteed only by a confidential key unknown to him.

It is interesting to note that one of the simplest and the most secure encryption techniques called one-time pad is entirely based on a direct use of random numbers. It can be proved that the one-time pad cannot be broken if used correctly, i.e., if each of the generated random sequences is used only once. However, this information-theoretic guarantee only holds if these sequences come from a high quality random source with high entropy.

Compared to other application areas, except for good statistical quality and knowledge of the distribution of generated random numbers, cryptography applications have strong security requirements – random number generators must be cryptographically secure. Cryptographically secure random number generators must generate random numbers that have good statistical quality, and the generated sequences must not be predictable and manipulable.

In cryptographic applications, for security reasons, cryptographic keys and other security critical data must be generated inside cryptographic modules and in particular inside semiconductor devices if the cryptographic module is implemented as a cryptographic system on a chip (e.g., in smartcards). For this reason, we will deal only with generators that can be implemented inside digital devices.

7.2 Random Number Generators and Security Requirements

Practical random number generators are grouped into two main categories: deterministic random number generators (DRNGs), also called pseudo-random number generators (PRGNs), and true random number generators (TRNGs). Each category has different characteristics, advantages, and disadvantages. Therefore, most of the random number generators used in practice are hybrid. Depending on the level of compromise in the choice of parameters, we distinguish between hybrid true random number generators and hybrid pseudo-random number generators.

Deterministic random number generators use deterministic, mostly cryptographic algorithms for generating numbers that seem to be random. Algorithms are selected in such a way that the resulting generator is very fast and gives perfect statistical results (e.g., uniform distribution of generated numbers). However, it is clear that compared to an ideal random number generator, the requirement of independence of output numbers cannot be fulfilled in the case of deterministic generators.

On the other hand, generators of truly random numbers rely on some random process that cannot be controlled. Depending on the source of randomness, we recognize physical (PTRNGs) and non-physical (NPTRNG) random number generators. Physical random number generators extract randomness from some physical, in electronic devices mostly electric phenomena, such as thermal noise, metastability, metastable oscillations, chaos, random initialization of bi-stable circuits, etc. Non-physical random number generators extract randomness from unpredictable human-machine interactions, such as electronic mouse movements, frequency of keystrokes, etc.

Hybrid random number generators (HRNGs) are usually composed of a DRNG preceded by a TRNG, and they take advantage of both building blocks: statistical quality and speed of DRNGs and unpredictability of TRNGs. Characteristics of the hybrid true random number generator (HTRNG) are essentially determined by those of the first true random number generator block and the second deterministic block should ensure computational complexity in the case of entropy failure in the first block. On the other hand, characteristics of the hybrid deterministic random number generator are determined by the second deterministic block, which is seeded regularly by the source of entropy situated in the first block.

When evaluating different RNG principles, security is the main criteria. Security in RNG design is related to the statistical quality and unpredictability of the RNG output. Output of a good generator aimed at cryptographic applications must have statistical properties that are indistinguishable from those of an ideal RNG. Unpredictability means that, knowing current a generator's output (or internal state in the case of DRNG), no preceding and following outputs can be guessed with nonnegligible probability.

The ideal RNG is a mathematical construct [210], which generates numbers that are independent and uniformly distributed. Here, it is important to note that the evaluation of random number generator output by statistical tests, is a necessary but not sufficient condition for a generator to be secure, since statistical tests are not able to assess unpredictability.

Unpredictability of DRNG is related to the computational complexity of the underlying algorithm, the length of the period, and the way the DRNG is initialized (entropy of the seed). Unpredictability of TRNG is related to the entropy rate in generated numbers: entropy rate per output bit equal to one guarantees that the generator output cannot be predicted.

Security of the RNG is determined by its principle, but also by the way the generator is implemented. Several threats are related to the RNG design:

- Hardware-related threats: Failure in generating good random numbers can be caused by component aging, variations in the manufacturing process, and unstable and/or manipulable entropy sources.

- Data leakage: An attacker might use data leaked from the random number generation process to compromise security of the whole system.

- Temporary failure in provisioning random output: Some design can output weak numbers before enough entropy is accumulated.

7.3   Deterministic Random Number Generators

Deterministic random number generators (DRNGs) generate number (bit) sequences that have statistical properties close to an ideal random number generator. The statistical quality of their output is evaluated using dedicated statistical tests, and their security level is assessed using cryptanalytical methods. This class of generators is very easy to implement in logic devices, since they are algorithmic. They offer a very high output bit rate and their output features a very good statistical quality. DRNGs are perfectly adapted to applications that do not require unpredictability, but they are used in some context also as cryptographic primitives, e.g., as a basis of stream ciphers, for cryptographic post-processing, etc.

7.3.1 DRNG Basics

Random number sequences generated in DRNGs are not truly random as they are completely determined by the initial state of the generator (called the seed), by the underlying algorithm, and eventually by its parameters, such as cryptographic keys. Security of the DRNG is linked to its period: the period is the maximum number of random numbers observed at the generator's output before their sequence starts to repeat. For a good (secure) DRNG, the period should be as large as possible and always much larger than the longest practically used number sequence.

However, even generators featuring a sufficiently large period can be cryptographically weak. For example, the behavior of linear congruential generators or linear feedback shift registers can be completely characterized from the knowledge of the generator's successive states (outputs), e.g., by solving a set of linear equations.

Cryptographically secure (strong) DRNGs (CSDRNG) constitute a family of pseudo-random number generators that are suitable for cryptographic applications. The cryptographically secure DRNGs must satisfy two basic requirements: 1) their output must pass statistical tests; 2) they must withstand attacks tending to predict their behavior from a partial or complete knowledge of their initial or current state.

As shown in [282], the first of the aforementioned conditions is fulfilled if the DRNG output can pass all polynomial-time statistical tests, i.e., if no polynomial-time algorithm can correctly distinguish between an output sequence of the generator and a truly random sequence (an output sequence of an ideal RNG) of the same length with probability significantly greater than 0.5.

The second of the above-mentioned conditions is true if: a) by knowing k successive output bits of the generator, the adversary is not able to predict, in polynomial time, the next $(k+1)$-th bit with a probability significantly higher than 0.5; b) by having a partial or complete knowledge of the generator's internal state and/or its output, the adversary is not able to predict the RNG output with a nonnegligible probability in forward and backward direction (depending on the required security level).

A typical DRNG design aimed at cryptographic applications is depicted in Figure 7.1. The DRNG core consists of the internal state register and the next state logic. The seeding procedure is aimed at distilling entropy from the source of true randomness. The extracted random value can then be used to seed the generator's core. The internal numbers available at the core output can be optionally processed using additional generator of output numbers, which is aimed at isolating the RNG's core from the output – thus, the internal state cannot be observed.

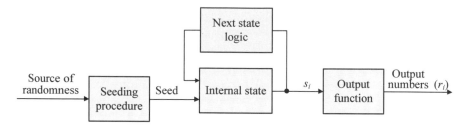

Figure 7.1
General DRNG design

Cryptographically secure DRNGs use well established cryptographic mechanisms or they are based on some difficult number theory problem. Any block cipher can serve as a core of the cryptographically secure DRNG. For instance, this can be a cipher working in a counter mode, in which it encrypts successive counter values using a randomly chosen key. The period of the generator is 2^n, where n is the size of the block. Note that the starting value of the counter and the key must be kept secret otherwise the generator will not be cryptographically secure. For example, if the last output of the generator, the encryption algorithm, and the random key are known, the generated numbers can be entirely predicted in both forward and backward directions.

A hash value of successive counter values generated using some approved one-way (e.g., cryptographic hash) algorithm can be used in cryptographic applications. Although such a method has not been proven to be cryptographically secure, it appears to be sufficient for most applications. Note that in all circumstances, the initial value of the counter (the internal state of the generator) must be kept secret.

One of the most common algorithms based on the difficult number theory problems is the Blum Blum Shub algorithm [47], which is based on a well-known integer factoring problem. Another DRNG algorithm from this family is the Blum-Micali algorithm [48] based on the discrete logarithm problem. The Yarrow algorithm [208] and its successor Fortuna [123] are other examples of commonly used hybrid general purpose DRNG algorithms that are aimed to be implemented in software.

Several cryptographically secure DRNGs were standardized up until now. We can cite the algorithms proposed in Appendix 3.1 and 3.2 of the digital signature standard (DSA) – FIPS 186-2 [337] (note that FIPS 186-2 has been superseded by FIPS 186-3 and FIPS 186-4), or that supplied in Appendix A.2.4 of the ANSI X9.31 standard [205]. The standard NIST SP 800-90A [29] describes three cryptographically secure DRNGs named Hash_DRBG, HMAC_DRBG, and CTR_DRBG, and a fourth DRNG named Dual_EC_DRBG

Table 7.1
Four DRNG classes according to AIS 31 rev. 2.0 [212]

DRNG class	Description
DRG.1	DRNG with forward secrecy (according to ISO 18031)
DRG.2	DRG.1 with backward secrecy (according to ISO 18031)
DRG.3	DRG.2 with enhanced backward secrecy
DRG.4	DRG.3 with enhanced forward secrecy (hybrid DRNG)

which has been shown to be cryptographically weak and probably has a cryptographic back door introduced by the NSA [326].

The AIS 31 methodology revision 2.0 published in 2011 by German BSI (Bundesamt fur Sicherheit in der Informationstechnique) [212], describes four DRNG classes depending on the security level. The four classes named DRG.1 to DRG.4 are listed in Table 7.1.

The classification criteria of these four DRNG groups are based on two properties that should be evaluated during algorithm security analysis: forward and backward secrecy. Forward secrecy means that even if the current generator output and its preceding values are known, it is impossible to determine, i.e., to compute or to estimate with a nonnegligible probability, the future values. Enhanced forward secrecy means that even if the internal state of the generator (and not only its outputs) is known, the future evolution of generated numbers cannot be predicted.

Similarly, backward secrecy means that even if the generator output or its next values are known, it is impossible to determine the preceding values. Finally, enhanced backward secrecy means that the preceding values cannot be determined even if the internal state is known.

7.3.2 Examples of Some DRNGs

In this section, we will present some representative examples of DRNG design featuring different security levels depending on forward and/or backward secrecy guarantees. We suppose that once setup by an initial seed, the DRNG core's next state s_{i+1} is determined by a state transition function ϕ, i.e., $s_{i+1} = \phi(s_i)$. As discussed above, this function should be based on a cryptographically approved algorithm, e.g., a standard block cipher (such as AES) or a standard hash function (such as SHA-3). We will give some examples of the DRNG configuration in the next section. In these examples, $E(x, k)$ represents encryption of the plaintext x using a confidential key k and a block cipher encryption algorithm E.

7.3.2.1 DRNG with backward secrecy

The DRNG with backward secrecy is depicted in Figure 7.2(a). If the state transition function ϕ is realized using an approved one-way cryptographic function (e.g., SHA-3), the behavior of the generator cannot be determined in backward direction because of the one-wayness of the function. However, if the algorithm is known (remember that according to the Kerckhoff principle, this case should be taken into account), all the subsequent states in forward direction can be predicted.

Figure 7.2(b), depicts a DRNG configuration, in which a block cipher of length n (e.g., 128-bit AES cipher) is used as a one-way function. As can be observed, the RNG core state s_i is composed of two n-bit state words s_i' and s_i''. The state of the DRNG in Figure 7.2(b) can be described as follows: $s_{i+1}' = E(s_i', s_i'')$ and $s_{i+1}'' = s_i'$. The DRNG output is $r_i = s_i'$.

Notice that in the given configuration, if the block cipher algorithm and two subsequent outputs r_{i-1} and r_i are known, all future output blocks can be predicted (forward secrecy is not ensured), since $r_{i+1} = s_{i+1}' = E(s_i', s_i'') = E(r_i, r_{i-1})$ and $s_{i+1}'' = r_i$, etc.

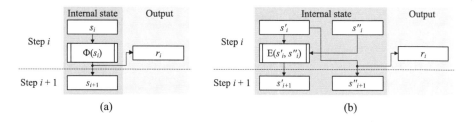

Figure 7.2
DRNG with backward secrecy: (a) DRNG based on a one-way function $\phi(s_i)$; (b) DRNG based on a block cipher behaving as a one-way function

7.3.2.2 DRNG with forward and backward secrecy

The principle of the DRNG with backward and forward secrecy is depicted in Figure 7.3(a). Notice that contrary to the DRNG with backward secrecy, the internal state cannot be directly observed at the generator's output. Instead, the output is given as a bit-wise XOR of two successive internal states. Therefore, if the internal state is not known, the output can neither be predicted in forward direction nor in backward direction.

Figure 7.3(b), depicts a DRNG configuration using a block cipher as a one-way function. Again, the RNG core state s_i is composed of two n-bit state words s_i' and s_i''. As in the previous example, the state of the DRNG in Figure 7.3(b) can be described as follows: $s_{i+1}' = E(s_i', s_i'')$ and $s_{i+1}'' = s_i'$. However, the DRNG output is now $r_i = s_i' \oplus E(s_i', s_i'')$.

As required, if the internal state of the generator is not known, its output can neither be predicted in forward direction nor in backward direction.

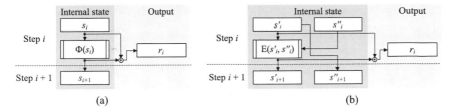

Figure 7.3
DRNG with forward and backward secrecy: (a) principle idea; (b) one-way
configuration

7.3.2.3 DRNG with forward and enhanced backward secrecy

The principle of the DRNG with forward and enhanced backward secrecy
is depicted in Figure 7.4(a). It guarantees that even if the output AND the
internal state of the generator is known, its output cannot be determined
in backward direction. However, if both the generator's output and internal
state are known, its future outputs can be predicted. If the internal state is
not known, the generator's future outputs are unpredictable.

The DRNG with forward and enhanced backward secrecy is at least two
times more expensive (in term of execution time or area) than the generator
with simple backward secrecy – it needs at least two one-way functions to be
implemented: one is used for generating the generator's next state and the
other one for generating the output from the internal state and for isolating
the output from the internal state.

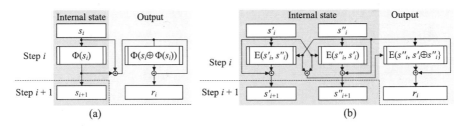

Figure 7.4
DRNG with forward and enhanced backward secrecy: (a) principle idea; (b)
implementation

Implementation of the DRNG with forward and enhanced backward secrecy using a block cipher needs three instantiations of the block cipher (see
Figure 7.4(b)): two for generating the next state and one for isolating the output from the internal state. The state of this generator can be described as
follows: $s'_{i+1} = s'_i \oplus E(s'_i, s''_i)$ and $s''_{i+1} = s''_i \oplus E(s''_i, s'_i)$. The DRNG output is
generated as follows: $r_i = s''_i \oplus E(s''_i, s'_i \oplus s''_i)$.

7.3.3   Evaluation of the DRNG Design and Implementation

The DRNG design and its implementation must be evaluated in several steps. First, the underlying algorithm must be proven to be cryptographically sure. It is therefore preferable to use some approved algorithm. Depending on the targeted security level, backward and forward secrecy must also be demonstrated. The DRNG correct implementation must be verified using test vectors. Finally, the generated sequences must be evaluated using standard statistical tests (see Section 7.4).

7.3.4   Conclusion

Many different DRNG designs exist. Depending on selected application, in some designs, the speed takes priority. In this case, extremely fast DRNGs as those based on LFSRs can be used at the cost of security. If the security cannot be compromised, some cryptographically secure DRNG should be used at the cost of speed and complexity (area). Some generators are intended for a particular purpose (e.g., FIPS-186) some are general purpose DRNGs (e.g., Yarrow-160). A lot of different and possibly insecure modifications of DRNGs are used in many applications. When designing a new DRNG, the designer should reuse existing cryptographically proven algorithms as much as possible. When considering security, he must expect that in the worst case, the secret internal state of the generator can be compromised. In this case, an efficient countermeasure for this dangerous case is to mix entropy, as frequently as possible, from several sources into the new secret state.

7.4   Statistical Testing

The statistical quality of the generator's output must be thoroughly evaluated by statistical tests before the generator is exploited in data security systems. However, no set of statistical tests can absolutely certify a generator as appropriate for usage in a particular application, i.e., statistical testing cannot serve as a substitute for cryptanalysis in the case of PRNGs or analysis of the entropy rate in the case of the TRNGs.

The role of the statistical testing is to evaluate the hypothesis that the generator is indeed random, i.e., that it behaves as an ideal RNG. This hypothesis is called the *null hypothesis* (H_0). The statistical tests should validate this hypothesis. If the generated sequence does not pass tests, the hypothesis is rejected, meaning that the generator does not behave as a good random number generator.

The number of possible statistical tests is practically unlimited. Consequently, no specific finite set of tests can be considered to be complete. For each existing test, some randomness statistic is chosen and used to make a decision about the null hypothesis: accept or reject. For an ideal RNG, such a statistic has a distribution of possible values. A theoretical reference distribution of this statistic is determined through the use of mathematical methods. From this reference distribution, a *critical value* is determined. A statistic value is then computed from the generated sequence and compared to the critical value. If the test statistic value exceeds the critical value, the null hypothesis for randomness is rejected, otherwise, it is accepted.

It is important to note that the results of statistical testing must be interpreted with some care and caution to avoid incorrect conclusions about a specific generator. A good random number generator may produce a sequence that does not appear to be random while a bad RNG may produce an apparently random sequence. Two kinds of errors can occur:

- *Type 1 error* occurs when the statistical test rejects a sequence that is truly random (false reject).

- *Type 2 error* occurs when the statistical test accepts a sequence that is not in fact random (false accept).

The probability of a *Type 1* error is denoted α, and the probability of the *Type 2* error is denoted β. Because of these errors, no statistical test can prove whether a sequence is random or not. Two approaches of statistical testing exist: one based on a so called level of significance of the test and the other one based on a *P-value* evaluation.

- *Statistical testing using a predefined significance level* – the significance level represents the probability of the *Type 1* error (α), i.e., it is the probability that the null hypothesis is rejected although it is true. This approach is used, for example, in the AIS 31 recommendations [212]. For cryptographic applications, α is usually chosen in the interval of $[0.001, 0.01]$. An α of 0.001 indicates that one would expect about one sequence in 1000 sequences to be rejected by the test if the sequence was random.

- *Statistical testing using a P-value approach* – for each statistical test a *P-value* is computed. This *P-value* is the probability that a perfect (ideal) random number generator would have produced a sequence seemingly less random than the sequence that was tested. If the *P-value* is greater than α, then the null hypothesis is accepted. Otherwise, it is rejected. This approach is used in the NIST SP800-22 test suite [361].

7.4.1 Tests FIPS 140-1 and FIPS 140-2

Cryptographic modules that implement a random number generator shall incorporate the capability to perform statistical tests for randomness depending on security level. In the first version of the FIPS 140 standard from 1994

Table 7.2
List of statistical tests FIPS 140-1 and their short description [335]

	Test name	Analyzed feature
T1	Monobit test	Tests the bias of a bitstream
T2	Poker test	Tests the distribution of non-overlapping 4-bit blocks in the bitstream
T3	Runs test	Counts the number of runs of zeros (and of ones) of length $i < 6$ and $i \geq 6$
T4	Long runs test	Searches for runs of length $i \geq 34$ – no such long runs are allowed

(FIPS 140-1 [335]), four statistical tests were recommended. For security levels 1 and 2 defined in the standard, the tests are not required. For level 3, the tests shall be callable upon demand. For level 4, the tests shall be performed at power-up and shall also be callable upon demand. The four tests specified below are recommended. However, alternative tests which provide equivalent or superior randomness checking (e.g., tests NIST SP 800-22 [361]) may be substituted.

Testing strategy

A single bitstream of 20,000 consecutive bits from the generator is tested using the four tests presented in Table 7.2. If any of the four tests fails, then the module shall enter an error state. Note that although the RNG statistical evaluation following the AIS 31 criteria [212] uses the same tests, the testing strategy is not the same (see Section 7.5.5.1).

The first revision of the second version of the FIPS 140 standard (FIPS 140-2) recommended the same statistical tests, but with different security thresholds. However, these four statistical tests were removed from the last revision of the FIPS 140-2 [336] and the only test that is still required is a simple *continuous random number generator test* in which each subsequent generation of an n-bit block shall be compared with the previously generated block of the same size. The test shall fail if any two compared n-bit blocks are equal.

7.4.2   Tests NIST SP 800-22

A section of the National Institute of Standards and Technology (NIST) in US called *Random Number Generation Technical Working Group* (RNG-TWG) has worked since 1997 on the description of a suite of statistical tests. The last update (NIST SP 800-22, rev. 1a) was done in 2010. It is a package of 15 different tests described in [361]. These tests focus on a variety of different types of non-randomness that could exist in a sequence (see Table 7.3).

Each test is based on a calculated test statistic value, which is a function of the generated random data. The test statistic is used to calculate a *P-value*

Table 7.3
List of statistical tests NIST SP 800-22 and their short description

	Test name	Short description
1	Frequency (Monobit) test	Tests proportion of zeros and ones
2	Frequency test within a block	Tests proportion of ones within M-bit blocks
3	Runs test	Tests total number of sequences of identical bits
4	Test for the longest run of ones in a block	Searches for the longest run on ones within M-bit blocks
5	Binary matrix rank test	Tests the rank of disjoint sub-matrices
6	Discrete Fourier Transform (spectral) test	Observes peak heights in the DFT
7	Non-overlapping template matching test	Counts number of occurrences of the pre-specified target strings
8	Overlapping template matching test	Test number of occurrences of the pre-specified target strings
9	Maurers 'Universal statistical' test	Detects whether the sequence can be significantly compressed
10	Linear complexity test	Determines if the sequence is complex enough or not
11	Serial Test	Tests the frequency of all possible overlapping m-bit patterns
12	Approximate entropy test	Compares the frequency of overlapping blocks
13	Cumulative sums (Cusum) test	Tests the cumulative sum of the partial sequences
14	Random excursions test	Tests the number of visits to a particular state within a cycle
15	Random excursions variant test	Counts total number of times that a particular state occurs

that summarizes the strength of the evidence against the null hypothesis. In practice, the *P-value* is compared with a probability of the *Type 1* error α, which is also called the *level of the significance* of the test.

The significance level is fixed in advance by the user. The standard recommends to fix α to 0.001 and to run each test on at least 1000 non-overlapping sequences of one million bits. For each statistical test, a set of *P-values* (corresponding to the set of sequences) is produced.

For a fixed significance level, a certain percentage of *P-values* are expected to indicate failure. For example, if the significance level is chosen to be 0.001 (i.e., $\alpha = 0.001$), then about 0.1 % of the sequences are expected to fail. A random sequence passes a statistical test whenever the *P-value* > α and fails otherwise. A *P-value* > 0.001 would mean that the sequence would be considered to be random with a confidence of 99.9%. A *P-value* $\leq$ 0.001 would mean that the conclusion was that the sequence is non-random with a confidence of 99.9%.

According to the NIST SP 800-22 testing strategy, having *P-value* > α is a necessary, but not sufficient condition for accepting the null hypothesis. The additional requirement is that the *P-values* of individual tested sequences must be uniformly distributed.

Individual tests of the NIST test suite do not have the same weight and do not need the same quantity of statistical data. The most important test is the "Frequency Test". If the generated sequence does not pass this test, the following tests will probably fail, too. While several tests can be applied on sequences as short as 100 bits, the Maurer's "Universal Test" requires a bitstream of at least 387,840 bits as input.

7.4.3   Tests DIEHARD and DIEHARDER

The DIEHARD test suite was proposed in 1995 by George Marsaglia for testing PRNGs aimed at stochastic simulations and not for cryptography. Until the creation of the NIST SP 800-22 test suite, it was the most advanced test suite. However, nowadays, NIST tests are considered to be better adapted for cryptographic applications that those of DIEHARD.

The DIEHARD test suite consists of 12 statistical tests described in [271]. It computes the *P-values*, which should be uniformly distributed. This approach is similar to that of the NIST SP 800-22 test suite. However, unlike in the NIST strategy, a generated random sequence passes the tests from the DIEHARD test suite if the *P-value* is in interval $[0 + \alpha/2, 1 - \alpha/2]$. The original test suite of George Marsaglia was later extended and implemented in the DIEHARDER C library.

7.4.4   Conclusion

Statistical testing represents the first necessary step in evaluation of random number generators. However, statistical tests are not able to evaluate unpredictability that is the most important RNG security requirement. Unpredictability of the PRNGs must be evaluated using cryptanalysis, while unpredictability of the TRNGs can only be guaranteed by a sufficient entropy rate at the generator's output.

7.5   True Random Number Generators

TRNGs rely on some physical source of entropy and on the way the entropy is harvested. Both the source of randomness and the entropy extraction principle are very dependent on the selected technology and therefore a standard or even recommended TRNG does not exist. Recently, German BSI proposed a TRNG evaluation criteria that draws a framework for a good modern and secure TRNG design [212]. Note that NIST also works on such criteria [30].

7.5.1 TRNG Design Basics

The TRNG is typically composed of a digital noise source and an optional entropy conditioning block (see Figure 7.5). Depending on characteristics of the source of randomness and quality of the digital noise, designers select the entropy conditioning (also called post-processing) method aimed at enhancement of statistical properties of generated numbers.

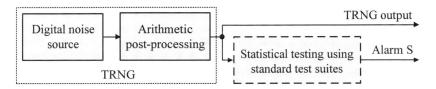

Figure 7.5
Classical TRNG design and evaluation approach

The TRNG principle and its implementation are then evaluated statistically: generated numbers are tested using standard test suites such as FIPS 140-1, NIST SP 800-22, DIEHARD, and DIEHARDER presented in Section 7.4. The generator must not be used if the statistical tests did not succeed (i.e., if Alarm S from Figure 7.5 was triggered).

This approach is not suitable for modern data security systems for several reasons: 1) the post-processing can mask considerable weaknesses of the source of randomness; 2) generic statistical tests can evaluate only the statistical quality of generated numbers and not their entropy (which should guarantee unpredictability as the main security parameter); 3) high-end standard statistical tests are complex and thus expensive and slow, needing huge data sets. Consequently, they are only executed occasionally or on demand and only on selected sets of data.

For these reasons, NIST is working on new recommendations for the entropy sources used for random bit generation – NIST SP 800-90B [30] (currently, only a draft version of the document is available). According to this document, the TRNG design should also include three levels of so-called health tests: startup test (testing all modules of the TRNG), continuous test (essentially testing the noise source), and on-demand test (tests that are more thorough and more complex than continuous test).

German Federal Office for Information Security (BSI) recently proposed an evaluation methodology for physical random number generators (AIS 31) [212], which should help designers to better consider security aspects in their design and which should help evaluators of generators in the evaluation process. The AIS 31-compliant design and evaluation is depicted in Figure 7.6.

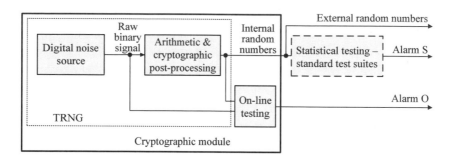

Figure 7.6
AIS 31 compliant TRNG design and evaluation approach

Compared to the classical approach and similarly to the NIST SP 800-90B approach, the AIS 31:

1. redefines the role of an optional arithmetic post-processing – it should correct some occasional small statistical imperfections of the raw binary signal (digital noise) and eventually increase entropy per bit (usually by some data compression method);

2. for the highest security levels, adds mandatory cryptographic post-processing to ensure unpredictability of generated numbers in forward and/or backward direction during a permanent or temporary failure of the entropy source;

3. defines on-line testing strategy by permanently executing a simple and fast total failure test and by executing at initialization and on demand an on-line test aimed at testing raw binary signal (preferably) or internal random numbers;

4. requires either construction of a statistical model aimed at entropy estimation and management or proof of robustness of the entropy source against variations of environmental conditions.

Recently, a new enhanced AIS 31 compliant approach has been proposed in [126]. This extended security approach is depicted in Figure 7.7.

Compared to the AIS 31 technique, the new extended security approach adds evaluation of the source of randomness inside the source of digital noise (before the entropy extraction). This extension of the AIS 31 approach brings several benefits:

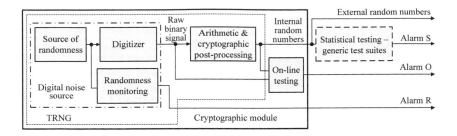

Figure 7.7
Extended security TRNG design and evaluation approach

1. The entropy source testing represents a dedicated statistical test that is better suited to the generator's principle. It can thus be simpler, faster, and, at the same time, more efficient than general purpose statistical tests, since it can better detect generator-specific weaknesses.

2. The entropy source testing can be realized as an embedded measurement of a random physical parameter (e.g., thermal noise or phase jitter of a clock signal). The measured value can then be used as an input parameter of the stochastic model of the source of entropy and can serve for entropy estimation.

3. The parameterized stochastic model of the source of entropy together with the description of the entropy extraction and arithmetic post-processing algorithms can be used to build a stochastic model of the complete TRNG, which can serve for precise entropy management ensuring highest security level while maintaining the highest available bit rate.

The role of the designer would then be as follows:

1. Depending on performance and security requirements, implement one TRNG principle based on a source of digital noise in selected technology.

2. Evaluate the amount of randomness in the given technology by existing embedded tests of entropy (note that all pre-selected TRNG principles must have suitable embedded tests/randomness measurement available).

3. Using the measured parameter, estimate the entropy per bit of the raw binary signal using an existing stochastic model of the source of randomness and select a suitable entropy extraction and algorithmic post-processing method (from a set of options).

4. Depending on the stochastic model of the source of randomness, the entropy extraction and the post-processing algorithm selected, estimate the entropy per bit (or per bit vector) of the generator's output.

5. Evaluate the generator, e.g., using the AIS 31 methodology and required entropy rate.

7.5.2 Sources of the Digital Noise

The entropy rate at the output of the digital noise source is determined by the quality of the physical source of randomness and by the entropy harvesting mechanism. The quality of the physical source of randomness in terms of its use in cryptographic applications is related to statistical parameters and spectrum of the analogue (noisy) signal, but also to its stability in time, independence on manufacturing process, and environmental conditions. The objective of the entropy harvesting mechanism (if necessary) is to extract as much entropy, from the analogue source, as possible and convert it to a digital form – digits.

The source of the digital noise should not be in any way programmable, it should not contain any memory element able to replay previously generated numbers, and it should only contain circuitries reduced to a strict minimum, which is necessary for generating stable digital noise. Any inputs besides ground and DC power are not allowed.

Things that are allowable and even desirable include power filtering and shielding, tamper resistant packaging, memory elements for lot numbering, revision and certification codes, etc.

Many physical entropy sources in digital electronic devices such as random jitter of the clock signal, metastability, temporary (metastable) oscillations, random initialization of bi-stable structures, and violation of the setup and hold time in bi-stable structures can be exploited. Direct utilization of analogue noise sources like noisy diodes, chaos, etc. necessitate analogue components and are therefore out of the scope of our study.

7.5.2.1 Clock jitter as a source of randomness

The most frequently used source of randomness in digital devices is the clock jitter. It is defined as a short-term variation of an event (e.g., a rising or falling edge) from its ideal position in time. The instability of the clock signal in time domain is therefore called the timing jitter. There are several possibilities of how to generate a jittery clock signal in logic devices. These include but are not limited to: inverter ring oscillators, self-timed rings, voltage-controlled oscillator, RC oscillators, etc.

To avoid the confusion, let us first define the clock jitter and its common measurements over time intervals. The instantaneous output voltage of an oscillator can be expressed in the frequency or time domain as [452]

$$V(t) = V_0 sin\left(\omega_0 t + \varphi(t)\right) = V_0 sin\left(\frac{2\pi}{T_0}\left(t + \delta(t)\right)\right) \qquad (7.1)$$

where $\varphi(t)$ is the phase deviation in the frequency domain expressed in radians, and $\delta(t) = \varphi(t)/\omega_0$ is the time deviation (the jitter) expressed in seconds. Relationship between time domain and frequency domain representation of the clock phase/timing instability is depicted in Figure 7.8. For clock applications, time-domain measurements are preferable, since most specifications of concern involve time domain values.

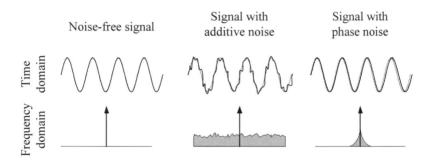

Figure 7.8
Noise-free and noisy signals in time domain and frequency domain

There are three basic measurements of the jitter in the time domain: phase jitter, period jitter, and cycle-to-cycle jitter. The *phase jitter* is defined as the phase advance of the observed clock from an ideal clock with period T_0. The phase jitter is unbounded, and it is difficult to measure. The *period jitter* δ'_n is defined as the difference between measured adjacent clock periods and the ideal clock period T_0. It can be considered the first difference function of the phase jitter. The *cycle-to-cycle jitter* is the measured difference between two successive clock periods. Thus, it contains information about the short-term dynamics of the jitter evolution, and it can be obtained by applying the first order difference to the period jitter or the second order difference to the phase jitter.

The jitter in clock generators is caused by global noise sources and local noise sources [431]. The use of the global jitter sources such as the noise of the power supply, electro-magnetic interference, etc. as a source of randomness should be avoided, because these sources are in general manipulable.

The clock jitter always has two components: random component and deterministic component. Deterministic component of the jitter such as inter-symbol interference and duty cycle distortion is data dependent and usually comes from cross-talks and from power supply loops. It is therefore manipulable and should also be avoided.

Besides the origin of the jitter, two important parameters should be considered: the size of the jitter and its power spectrum. The high frequency range of the spectrum determines the maximum bit rate at the generator's output, while noise in the low frequency range, such as flicker noise, is known to be autocorrelated. The analysis of the jittery clock frequency spectrum can also help in searching for a deterministic component of the jitter, which can be observed as spurious noise peaks at both sides of the signal spectrum (see Figure 7.9).

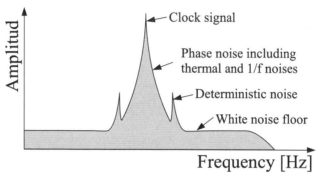

Figure 7.9
Spectrum of a jittery clock signal featuring a deterministic jitter component

In conclusion, the only jitter component that should be freely used as a source of randomness should be that coming from the local Gaussian noise sources such as thermal noise. Unfortunately, in practice, it is difficult to determine the proportion of the jitter coming from local thermal noise (the "good" jitter) on the global jitter which also contains other unwanted components coming from global, deterministic, and low frequency noise sources (the "bad" jitter).

7.5.2.2   Metastability as a source of randomness

Several TRNG designs claim to use metastability as a source of randomness. Metastability can be defined as the ability of an unstable equilibrium electronic state to persist for an indefinite period in a digital system. A synchronous flip-flop can enter the metastable state if the binary input data signal is sampled during its rising or falling edge. In fact, entering metastable state is rare, but since it can have disastrous impact on the synchronous system behavior, manufacturers of logic devices reduce by the flip-flop design the probability of entering a metastable state to a minimum.

Metastability is characterized by a so-called Mean Time Between Failures (MTBF). This time is in order of tens of years in current integrated circuits technologies. It is therefore surprising that some TRNG designs, in which authors claim to use metastability, reach an output bit rate of several Mbits/s (e.g., see [266]). Clearly, it is possible that these TRNGs reach announced bit rates, but it is impossible that the claimed randomness comes from metastability. Using a false source of randomness in cryptography applications is very dangerous, since the true source of randomness is not characterized and consequently the entropy rate cannot be correctly assessed.

7.5.2.3   Other sources of randomness in logic devices

Authors in [438] showed that inverter rings containing an odd number of inverters oscillate temporarily. Duration of this metastable oscillatory state depends on the asymmetry of the two ring branches, which varies dynamically due to the intrinsic noise present in the oscillating ring.

Several TRNGs used random initialization of bi-stable circuits and volatile semiconductor memories. For example, Intel in its last TRNG [422] used a two-inverter bi-stable circuit. At the beginning, both inverters are forced to the logical 1 state using two additional transistors. Then, if the transistors are switched off, bi-stable circuit flips randomly because of an intrinsic noise in two inverters to one of stable states: 1-0 or 0-1. To keep the inverters in balance, a feedback loop aimed at reducing the output bias is implemented.

Another source of random values was shown to be collisions appearing during a simultaneous writing of different data to true dual port memories via two data ports [148]. This principle is suitable for FPGAs, where large dual port memory blocks are available "for free". However, the setup of the collisions in memory writing is very tricky, technology dependent, and can be unstable because of aging and variations in operational environment.

7.5.2.4   Analog to digital conversion: Digitization

Since the source of randomness is an analog phenomenon, it must be converted to random digits. Of course, the conversion methods depend on the source of randomness used. In the case of logic devices, the conversion is made from time domain (analog value jitter) to digits. The most commonly used elements that transform intrinsic noise converted to a clock jitter are synchronous and asynchronous flip-flops as D flip-flop or latches and counters.

Figure 7.10 depicts the principle of the conversion of the clock jitter to random bits. The jittery clock signal (CLJ) is sampled on the rising edge of the reference clock signal (CLK) using D flip-flop.

Depending on the frequency and phase relationship, some samples (output signal Q) can be: 1) Constant – equal to one (samples at triangular markers) or zero (samples at square markers); 2) Random – equal to one or to zero depending on the jitter (samples at circular markers).

Another principle used to convert the clock jitter to random bits exploits synchronous or asynchronous counters. The principle used in [431] is pre-

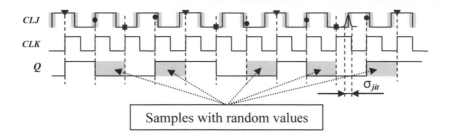

Figure 7.10
Conversion of the clock jitter to random bits – sampling a jittery clock signal using a reference clock

sented in Figure 7.11. The counter counts rising or falling edges of a jittery clock signal during a time window obtained from N reference clock periods. Depending on the jitter size and the measurement time (N periods of the reference clock), few last significant bits of the counter are random.

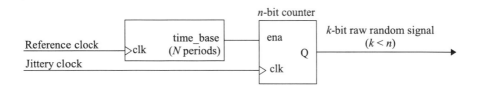

Figure 7.11
Conversion of the clock jitter to random bits – counting edges of a jittery clock signal during a time window generated from a reference clock

Note that some sources of digital noise do not need analog to digital conversion, since the conversion is an integral part of the source of randomness (e.g., random initialization of flip-flops) [422].

7.5.3 Entropy and Its Estimation

In information theory, entropy is a measure of the uncertainty of a random variable. In the context of random number generation for cryptography, the entropy is a measure of guesswork and unpredictability.

The most general definition of entropy is the *Rényi entropy*. The Rényi entropy of a distribution of a discrete variable X with possible values $x_1, ..., x_n$ is defined as:

$$H_\alpha(X) = \frac{1}{1 - \alpha} \log_2 \left(\sum_{i=1}^{n} (p_i)^\alpha \right) \tag{7.2}$$

where $\alpha \geq 0$, $\alpha \neq 1$ and $p_i = \Pr[X = x_i]$.

Unfortunately, the Rényi entropy is difficult to estimate in practice. There are two other entropy measures that are easier to handle: the min-entropy and Shanon entropy.

It can be shown that, for a fixed random variable X, the Rényi entropy $H_\alpha(X)$ decreases monotonically for $\alpha \in [0, \infty)$. So, the most conservative entropy measure is that for $\alpha \to \infty$, which is called the min-entropy. The *min-entropy* is then defined as follows:

$$H_\infty(X) = \inf_{i=1..n} (-\log_2(p_i)) = -\log_2 \sup_{i=1..n} p_i \qquad (7.3)$$

A distribution of a discrete variable X has min-entropy k if for any element x_i of its range, $\Pr[X = x_i] \leq 2^{-k}$. Consequently, if U_d is a uniform distribution over $\{0, 1\}^d$, then the min-entropy of this distribution is $H_\infty(U_d) = d$, and the so-called entropy rate (per bit) is $\delta = 1$. For an n-bit distribution X with min-entropy k and entropy rate $\delta = k/n$, we say that X is an (n, k) distribution.

The most common entropy measure is *Shanon's entropy*, which is the special case of Rényi's entropy for $\alpha \to 1$:

$$H(X) = H_1(X) = -\sum_{i=1}^{n} p_i \log_2 p_i \qquad (7.4)$$

Because of the monotonicity of the Rényi entropy, $H_\infty(X) \leq H(X)$. For any random variable X with uniformly distributed possible values $x_1, ..., x_n$ the Shannon entropy is equal to the min-entropy, i.e., $H(X) = H_1(X) = H_\infty(X)$.

For a binary-valued variable with $\Pr[X = 1] = p$ and $p \in [0, 1]$, it can be shown that $H_1(X) = -(p \log_2(p) + (1 - p) \log_2(1 - p))$ and $H_\infty(X) = \min\{-\log_2(p), -\log_2(1 - p)\} = -\log_2(\max\{p, 1 - p\})$. For the uniform distribution ($p = 0.5$), we obtain $H(X) = H_1(X) = H_\infty(X) = 1$.

Note that entropy cannot be measured. It can only be estimated from the known distribution of possible values of random variables.

7.5.3.1 Stochastic models

Recall that the final goal is to determine the average entropy per generated random number or at least its lower bound. This entropy can be estimated only from the known distribution of random numbers. Ideally, the stochastic model of the digital noise source should quantify this distribution of random numbers and thus serve to calculate the entropy rate at generator's output, or at least to determine lower entropy bounds.

We would like to stress that the role of a stochastic model (i.e., specification of distribution of random variables) is very different from that of a physical model of the noise source, which should model the physical (and not stochastic) behavior of the source. Fortunately, stochastic models of the source of randomness are easier to construct than the physical models that are often very complex.

In the case of binary-valued variables, the stochastic model should give probability of ones $\Pr[X = 1]$ (or probability of an n-bit pattern $\Pr[X_1 = x_1, ..., X_n = x_n]$), depending on random physical parameters (e.g., the size of the clock jitter) and so-called auxiliary random parameters (parameters that can influence, indirectly, the generation of random numbers, such as temperature, power supply, tolerances in the manufacturing process, aging, etc.). Using the model, the entropy can be expressed as a function of these physical parameters that can be determined by experiments.

A comprehensive example of a stochastic model of the TRNG using two noisy diodes is presented in [211]. The construction of the model is based on a study of physical parameters of the generator based on the selected principle and on the used hardware components. Although the model is proposed for the generator using some analog components, it is shown that it can be applied to some extent in other TRNG designs.

Another example of a stochastic model of an oscillator-based TRNG that can be easily implemented in logic devices was published by Baudet et al. in [34]. In this model, the phase $\varphi(t)$ of the output signal $s(t)$ of a ring oscillator is considered to be a Wiener process with drift $\mu > 0$ and volatility σ^2.

Following this assumption, Baudet et al. propose a simple physical TRNG composed of two ring oscillators (RO1 and RO2) and a sampler (D flip-flop – DFF) as presented in Figure 7.12. The output of RO1 is sampled at rising edges of the clock signal at the output of RO2. Note that the model gives no restriction on frequencies f_1 and f_2.

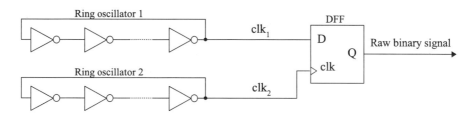

Figure 7.12
Ring oscillator based RNG from [34]

Using the model, the probability of sampling 1, the probability of outputting an n-bit vector, and the entropy rate of this vector of the raw binary signal at the output of the generator are defined as follows:

1. The *probability of sampling 1* at time $t \geq 0$ that is conditioned by $\varphi(0)$ (the phase at time 0) is given as:

$$\Pr[s(t) = 1 | \varphi(0) = x] \approx \frac{1}{2} - \frac{2}{\pi} \sin(2\pi(\mu t + x))e^{-2\pi^2\sigma^2 t} \quad (7.5)$$

2. The *probability of outputting a vector* $b = (b_1, ..., b_n) \in \{0,1\}^n$ at sampling times $0, \Delta t, ..., (n-1)\Delta t$ is given as:

$$p(b) = \Pr[s(0) = b_1, ..., s((n-1)\Delta t) = b_n]$$

$$\approx \frac{1}{2^n} + \frac{8}{2^n \pi^2} \left(\sum_{j=1}^{n-1} (-1)^{b_j + b_{j+1}} \right) \cos(2\pi v) e^{-2\pi^2 Q} \quad (7.6)$$

where $Q = \sigma^2 \Delta t = \frac{\sigma_{jit}^2 T_2}{T_1^3}$ is the *quality factor* and $v = \mu \Delta t = \frac{T_2}{T1}$ is the *frequency ratio* between the sampled and the sampling signal.

3. The *entropy of an n-bit output vector* b is given as:

$$H_n = \sum_{b \in \{0,1\}^n} p(b) \log p(b)$$

$$\approx n - \frac{32(n-1)}{\pi^4 \ln(2)} \cos^2(2\pi v) e^{-4\pi^2 Q} \quad (7.7)$$

Using equation (7.7), the lower bound of the Shanon entropy rate per bit at the generator output is given as follows:

$$H_{min} \approx 1 - \frac{4}{\pi^2 \ln(2)} e^{-4\pi^2 Q} = 1 - \frac{4}{\pi^2 \ln(2)} e^{\frac{-4\pi^2 \sigma_{jit}^2 T_2}{T_1^3}} \quad (7.8)$$

As can be seen, the lower entropy bound is determined by the (mean) frequencies of the two ring oscillators and the jitter variance per period T_1. All of these parameters are easy to measure, and the entropy bound is therefore very simple to assess.

If different realizations of the random variable are not independent, the stochastic model should rather specify a conditional probability $\Pr[X_n | X_1, ..., X_{n-1}]$ depending on the characteristics of the source of randomness. From this probability a conditional entropy rate $H(X_n | X_1, ..., X_{n-1})$ or at least its lower bound, should be computed.

7.5.4   Post-processing

Physical and nonphysical sources of true randomness in random number generators are often imperfect and consequently, the raw binary signal can be biased, generated numbers can be correlated, or the entropy rate per generated number can be insufficient. These statistical weaknesses should be removed in an algorithmic post-processing block [212]. Note that the algorithmic post-processing block is sometimes called conditioner [30] or entropy extractor [195]. The role of this block is, essentially, to make generated numbers statistically and computationally indistinguishable from the output of an ideal TRNG.

The closeness of the generated output to that of an ideal one can be evaluated using a statistical distance. The statistical distance between distributions of random variables X and Y taking values in subset T of the set S, is defined as: $\Delta(X, Y) = \max_{T \subset S} |\Pr[X \in T] - \Pr[Y \in T]|$. In other words, X and Y are ε-*close* if $\Delta(X, Y) \leq \varepsilon$.

A deterministic randomness extractor Ext: $\{0,1\}^n \rightarrow \{0,1\}^m$ (called (k, ε)-*extractor*) is a procedure that takes an n-bit sample from a weak random source (any of all possible (n, k) distributions of variable X) as an input and gives an m-bit output that is statistically ε-close to the uniform distribution U_m. Note that the deterministic randomness extractors are suitable only for independent stochastic sources.

One of the simplest and commonly used deterministic entropy extractors is the parity filter. It breaks the source n-bit string into m blocks of length $r = \lfloor n/m \rfloor$ and outputs the parity of each block. It can dramatically reduce the bias on the generator output at the cost of reducing r-times its bit rate. However, the bias of the output bit-stream is reduced only if the original bits are independent.

The main advantages of this entropy extractor are its simplicity and the possibility to maintain a constant output bit-rate. In [97], the author shows that the relation between the probability that the output bit is equal to one ($\Pr[Y = 1]$), probability of the input bit being equal to one ($\Pr[X = 1]$), and compression factor r is given by equation (7.9), in which B represent the bias of the output bit-stream. As can be seen, if $\Pr[X = 1]$ is different from 0 or 1, then the bias B converges towards 0 when r tends to infinity.

$$\Pr[Y = 1] = 0.5 - 2^{r-1} \left(\Pr[X = 1] - 0.5\right)^r = 0.5 - B \qquad (7.9)$$

Another frequently used entropy extractor is that of von Neumann [443]. First, the biased output bit stream is broken in pairs of bits. Then, for each pair, the output bit is equal to 0 if the input pair was 01, to 1 if the pair was 10, and the pair is skipped if it was 00 or 11. The output bit rate is data dependent and the process will yield an unbiased random bit after $1/(2\Pr[X = 1](1 - \Pr[X = 1]))$ pairs on average.

If the input stream is stationary, and possibly biased, the output is unbiased. However, if the original stream is auto-correlated, the output may remain auto-correlated. Also note that the von Neumann's entropy extractor will produce a biased output if the input stream features a cycle with period 2. If the entropy extractor is implemented in hardware, it may interfere with the generator and produce exactly this type of occurrence.

Using bijective functions (e.g., block ciphers) as bias correctors is not recommended because they just transform one weakness (the bias) to another (correlation) [212].

In order to enable designers to find the tradeoff between security and cost (in terms of area or execution time), according to AIS 31 recommendations [212], algorithmic post-processing can be separated into arithmetic post-processing and cryptographic post-processing, both of which remain optional (however, note that for the class PTG.3, the cryptographic post-processing is required).

For good-quality (and possibly more expensive) sources of randomness, complex post-processing is not necessary and the raw binary signal can be used directly as the TRNG output. Consequently, the final cost of the generator will be reduced. If the output signal of a simpler (and possibly cheaper) source of digital raw random signal has some statistical weaknesses, they can be corrected by the arithmetic post-processing at the cost of complexity.

The role of the potentially expensive cryptographic post-processing is to temporarily generate computationally secure, high quality pseudo-random numbers when the source of entropy completely fails (note that the total failure test always has some latency).

Of course, the role of the arithmetic and cryptographic post-processing can be jointly ensured, e.g., by an approved cryptographic post-processing function. A typical approach is to collect weakly random data and feed it into a cryptographic hash function. The output of the hash function is then used as if it were a sequence of random bits. It is important to note that no theoretical justification for this way of using a fixed cryptographic hash function to do deterministic extraction exists [428].

Finally, remember that one can always use a DRNG for cryptographic post-processing, but the DRNG must fulfill requirements of the class DRG.3 as presented in Table 7.1.

7.5.5 TRNG Design Evaluation and Testing

Security of a TRNG design must be thoroughly evaluated [126]. Namely, two security requirements are essential:

- *The statistical quality of generated numbers* guarantees that attacks can only succeed by using an exhaustive search for the secret.

- *Unpredictability* means that even knowing the last generator's output, no other output can be predicted with non-negligible probability in a forward or backward direction.

While the statistical quality of the generated numbers is relatively easy to verify, evaluating unpredictability is not straightforward, since it cannot be measured or tested. The entropy (and thus unpredictability) can only be estimated using a *stochastic model*.

A perfect generator should be *robust against environmental fluctuations, aging, and attacks*. In practice, perfect and permanent robustness against attacks and manipulations cannot be reached. Even a generator that is robust to all known attacks may be vulnerable to new attacks in the future. The only way to ensure long-term resistance against attacks is to *permanently execute dedicated on-line tests* able to detect, quickly and reliably, even temporary reduction of the entropy rate. Embedded tests must be based on existing stochastic model having, as an input parameter, the size of the physical phenomenon that is used as a source of entropy (e.g., the clock jitter).

A thorough security evaluation is not trivial since it necessitates expertise in several scientific fields, such as microelectronics and physics for understanding random physical processes in electronic circuits, mathematics and statistics for constructing simple, but sufficiently precise stochastic models, information processing and information theory for entropy estimation and entropy management, and cryptography for dealing with data security and cryptographic post-processing.

7.5.5.1 TRNG evaluation and testing for a security certificate

Because evaluation of the entropy rate is not trivial, official institutions responsible for the security certificate delivery help designers and evaluators by proposing standard evaluation approaches. Two different approaches currently exist: the earlier one proposed by the German BSI in the AIS 31 recommendations [212] and the later one (only a draft version is available up to now) proposed by American NIST in cooperation with the NSA [30]. Next, we will describe basic features of the two approaches, their common characteristics, and their differences.

AIS 31 approach to entropy evaluation

According to the AIS 31 standard, the designer must deliver a stochastic model that should be used for entropy estimation. Internal random numbers for class PTG.2 and PTG.3 must have an entropy rate of at least 0.997. Note that according to AIS 31, internal random numbers are present as the output of the post-processing block (see Figure 7.7).

Besides the requirements on the stochastic model, AIS 31 proposes a set of nine statistical tests and the strategy of testing the target of evaluation (the TRNG). Depending on the targeted security level (PTG.1, PTG.2, or PTG.3) two procedures are proposed: Procedure A and Procedure B.

AIS 31 testing strategy for Procedure A is clearly defined: what kind of tests, how much data, how many test repetitions, and how many rejections allowed. Procedure A consist in the application of six statistical tests T0 to T5 aimed at testing internal random numbers (see Table 7.4). For an ideal RNG the probability that test procedure A fails is almost 0.

The disjointness test (test T0) compares 2^{16} successive 48-bit blocks of random data. All these blocks must be different. Probability of a collision for an ideal RNG is 10^{-17}. Tests T1 to T4 are the four tests of the standard FIPS 140-1 (see Section 7.4.1). Test T5 is the correlation test from [282].

Test procedure B is usually applied to the raw binary signal of TRNGs (if the raw binary signal is available). The goal is to ensure that the entropy rate is sufficiently large. A small bias and slight one-step dependencies are permitted, but no significant more-step dependencies. This procedure uses three statistical tests T6 to T8 (see Table 7.5). For independent and identically

Table 7.4
Tests of the AIS 31 – Procedure A [212]

Test name	Analyzed feature
T0 Disjointness test	2^{16} 48-bit random blocks must be different
T1 Monobit test	Tests the bias of a bitstream
T2 Poker test	Tests the distribution of non-overlapping 4-bit blocks in the bitstream
T3 Runs test	Counts the number of runs of zeros (and of ones) of length $i < 6$ and $i \geq 6$
T4 Long runs test	Searches for runs of length $i \geq 34$ – no such long runs are allowed
T5 Autocorrelation test	tests correlation between successive bits

Table 7.5
Tests of the AIS 31 – Procedure B [212]

Test name	Analyzed feature
T6 Uniform distribution test	Special case for contingency tables
T7 Comparative test for multinomial distribution	Test for homogeneity
T8 Coron's entropy test	Modified Maurer's 'Universal Entropy Test' blocks in the bitstream

distributed (IID) random variables, the test T8 gives estimation of the entropy rate. Unlike test procedure A, the size of the input data block is not fixed – it depends on the input sequence. Again, for an ideal RNG the probability that test procedure B fails is almost 0.

NIST SP 800-90B approach to entropy evaluation

In contrast with the BSI approach from AIS 31, the NIST approach does not explicitly require existence of a stochastic model. Instead of using the model, entropy evaluation is based on groups of selected generic tests. Since only a draft version of the standard is available today, we will present the general testing strategy rather than the role of individual tests.

The aim of the first group of eight tests is to determine whether the generated data is IID. Two cases are possible: 1) if the data is IID, then the min-entropy is estimated from the number of observations of the most common output value (note that this approach is not meaningful if data is not IID); 2) if the data is not IID, a conservative set of entropy tests is used.

If the process is not IID, a set of five tests (collision, collection, compression, Markov, and frequency) should be used in order to estimate the entropy rate in the raw binary signal. The claimed confidence level of these tests is

95%. Although it is accepted that entropy evaluation is not precise in this case, it is claimed that the entropy is underestimated rather than overestimated.

7.5.5.2   Embedded statistical tests

The use of reliable and precise generic statistical tests for embedded TRNG testing is not practical because these tests are slow and expensive. The solution is to use some simplified embedded statistical tests which should be faster, cheaper, and possibly less accurate. Embedded tests should thus detect important deviation from TRNG normal operation (small deviations are accepted). Three different kind of tests are required: the start-up test, the on-line test, and the total failure test. Note that in the NIST SP 800-90B standard, the embedded tests are called Health Tests.

According to the AIS 31 standard, the designer has the freedom to chose the best startup, on-line, and total failure tests. However, the NIST SP800-90B standard specifies minimum requirements on the total failure tests by proposing two tests: the Repetition Count Test and the Adaptive Proportion Test. The designer can propose other types of tests, but he must show that the alternative tests are able to detect the same failure conditions. In the following paragraphs, we will present requirements on different embedded tests following the AIS 31 standard.

Start-up test

The start-up test consists of testing all elements of the TRNG. It must be performed after power-up, after an error message (e.g., from the total failure test or on-line test), after system reset or generator reset, or after a pause during which the generator did not operate, for example in order to reduce the power consumption.

The source of entropy must be tested during the start-up test by calling the on-line test presented in the following paragraph. The testing of algorithmic post-processing blocks (if some are used) must be performed using a known answer test. The generator may not output any random bits before the start-up test ends with success.

On-line and on-demand test

The on-line test shall detect intolerable statistical defects of the raw random number sequence while the RNG is in operation. The generator must not output any random numbers when a defect has been detected. Therefore, the test must be as fast as possible.

Depending on the TRNG application, the on-line test should always be triggered in the framework of the start-up procedure, externally (on demand), and for PTG.3 at regular intervals. It can also run continuously.

Total entropy failure test

The total failure test shall run continuously when TRNG is in operation. It must be fast and must have small probability of false alarm. It can test the raw binary signal, but is preferable if it measures the size of physical random phenomenon used as a randomness source, i.e., inside the noise source. This will make it faster and more sensitive to weaknesses of the source.

When the total failure test fails, the generator may output temporarily internal random numbers with the post-processing algorithm of class DRG.2, if the memory of the internal state is big enough to deliver temporarily enough entropy. The latency of the total failure test and the size of the internal memory (state register or FIFO) are thus closely related to each other.

7.6 Example of an AIS 31 Compliant TRNG

In this section, we present an AIS 31 compliant design of a TRNG. Our aim is to show a good design strategy instead of a "good" TRNG. We decided to use an elementary ring oscillator based TRNG from [34] as an example, since it is very simple and at the same time, it clearly demonstrates the complexity of a robust TRNG design.

7.6.1 Elementary Ring Oscillator Based RNG

Our elementary ring oscillator-based RNG is depicted in Figure 7.12. It is composed of two inverter-based ring oscillators. The output clock of one oscillator is sampled periodically after K rising edges of the second oscillator output using a D flip-flop (DFF). The length of the time interval determined by K periods of the second oscillator clock defines the jitter accumulation time and thus determines the size of the accumulated jitter and consequently the entropy rate at the generator's output. The T flip-flop at the output of ring oscillator 1 guarantees the duty cycle close to 50/50 even at high oscillator frequencies and thus reduces the bias at the generator's output.

This RNG has the advantage that it is very simple to implement in logic devices, since it is composed of standard logic gates available in all integrated circuit technologies. Moreover, the design of the elementary ring oscillator based RNG does not need any circuit level optimization. In most cases, such RNGs do not even need manual placement and routing. However, such optimization can reduce variation of ring oscillator frequencies due to process variations.

The source of randomness exploited in this RNG is the clock jitter produced in both ring oscillators. Due to the noisy current that occurs in transistors instantiated in ring oscillators, the generated clock signal features un-

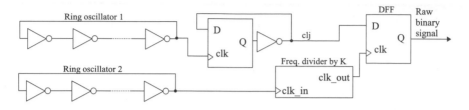

Figure 7.13
Elementary ring oscillator based RNG

predictable jitter. Authors in [101] showed that if the jitter comes only from the thermal noise, the timing of edges produced by an inverter-based ring oscillator drifts around the average half period and the drift has a Brownian motion characteristic.

Because of the difference in frequencies of the two generated clocks, the bit stream produced by the circuitry from Figure 7.12 contains some bits that are derived from samples, which are close to the rising or falling edges of the sampled ring oscillator clock. Since positions of rising and falling edges of the sampled signal are unstable, because of the random jitter, the corresponding samples contain some randomness. Figure 7.14 shows an example of a sampling process for the division factor $K = 3$.

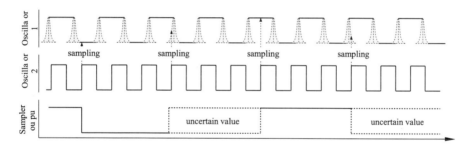

Figure 7.14
Example of randomness extraction from the jitter in the elementary ring oscillator based RNG for $K = 3$

7.6.2 Stochastic Model and Entropy Estimation

As explained in previous sections, the statistical modeling of the generator is one of the most important phases in the random number generator design, since it is essential for proving that the entropy rate at the generator's output is sufficient and consequently that the generated bit stream is unpredictable.

Before presenting our modeling strategy, we recall that the stochastic model should describe the distribution of random variables based on measurements of the underlying physical process (behavior of the raw random analog signal). These measurements are then used as input parameters in the parameterized stochastic model of the source of randomness. This stochastic model is then combined with a (deterministic) model of the digitization process in order to provide a stochastic description of the raw binary signal. Finally, this description is used for computing the theoretical value of the entropy per bit at the generator's output before the post-processing. This general approach of the raw binary signal generator modeling is depicted in the top panel of Figure 7.15.

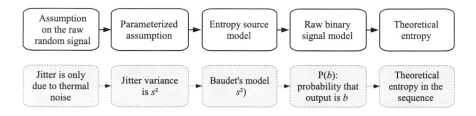

Figure 7.15
General approach in P-TRNG stochastic modeling (top) and its translation to an elementary ring oscillator TRNG design (bottom)

One of advantages of the selected TRNG is that we can directly use the model of Baudet et al. presented in Section 7.5.3. The bottom panel of Figure 7.15 represents application of the general stochastic modeling approach described in the previous paragraph, to the ring oscillator based RNG model of Baudet et al.

We recall that the selected model can be used for estimating the probability of ones (equation (7.5)), the probability of a vector of bits (equation (7.6)), the entropy of a bit vector (equation (7.7)), and the lower entropy bound (equation (7.8)) at the generator's output depending on the size of the jitter coming from the thermal noise and periods of generated clocks, which are the input parameters of the model. We also recall that the authors consider only the component of the jitter coming from the thermal noise, because it is not manipulable.

The most important contribution of the approach of Baudet et al. is their modeling of random processes in the frequency domain, depending on the two clock frequencies (f_1 and f_2) and the variance of the Gaussian jitter.

Equation (7.8) giving the lower entropy bound can be adapted to our elementary TRNG from Figure 7.14 to the following form:

$$H_{min} = 1 - \frac{4}{\pi^2 ln(2)} e^{\frac{-\pi^2 \sigma_{th}^2 KT_2}{T_1^3}} \tag{7.10}$$

where σ_{th}^2 is the variance of the jitter due to thermal noise, K is the frequency division factor, and T_1, T_2 are periods of the clock signals generated by oscillator 1 and 2, respectively.

It is clear that for entropy estimation using this model, σ_{th}^2, K, T_1, and T_2 must be known. Two approaches are possible:

1. The designer evaluates the parameters in all possible operating conditions (including border and corner conditions) and it takes the worst case values of the model input parameters for entropy estimation.

2. Using the underlying stochastic model, the designer defines the security thresholds for all the above-mentioned parameters. Then the real values of these parameters are measured continuously in the TRNG module in order to ensure that they do not drop under the defined threshold guaranteeing sufficient entropy rate at the generator's output.

The second approach is preferable, because it can detect degradations and other processes (e.g., attacks) that were unknown at the time of generator's design. Next, we will illustrate how to construct model-based embedded statistical tests based on these measurements.

7.6.3 Dedicated Embedded Statistical Tests

As presented in Section 7.5.5, according to the AIS 31 recommendations, three kinds of tests should be proposed: the start-up test, the on-line test, and the total failure test. Next, we will discuss these kind of tests in the context of our generator.

7.6.3.1 Start-up test

We recall here, that the start-up test consists of testing all elements of the TRNG after power-up, after an error message (e.g., from the total failure test), or after generator or system reset. The source of entropy must be tested during the start-up test by calling the on-line test presented in the following section. The algorithmic post-processing (if any) must be tested using a known answer test. Since our objective is to propose a high-quality raw binary signal that does not necessitate any post-processing, our start-up test will be the same as the on-line test, which is presented in the next paragraph.

7.6.3.2 On-line test

We have seen in Section 7.6.2 that the RNG model links the entropy contained in the generated sequence with periods T_1 and T_2, division factor K, and the variance of the jitter due to thermal noise. In this section, we introduce the on-line test,which is based on this link.

First, we must determine the division factor K in conjunction with the value of σ_{th}^2 for known (measured) T_1 and T_2, which will guarantee sufficient entropy rate. We call this phase of the TRNG design a calibration process.

Then the embedded test will consist in checking on-line the size of the jitter variance in relationship with T_1 and T_2. This approach is completely compliant with the AIS 31 standard, which defines the on-line test as a quality check of the generated random number while the PTRNG is in operation. Furthermore, according to AIS 31, such test should be realized by a physical measurement of the source of randomness, what is also true in our case.

Let us now introduce the principle of an on-line measurement of the jitter. Our objective is to propose a simple and comprehensible method that is easy to implement in logic devices. It is important to note that because of the differential principle of the elementary ring oscillator based TRNG, in which both the sampled and the sampling signals are generated with the same structure (ring oscillator), global noise sources are suppressed [127]. Therefore, only local noise sources are considered in the model.

Authors in [154] showed that the variance of the jitter coming from independent noise sources is proportional to the variance of its cumulative sums. This property is very useful for the embedded jitter measurement. In fact, the resolution of the jitter measurement, which should at least be as precise as the jitter standard deviation, can be obtained by accumulating enough jitter for measuring the variance of the cumulative sum.

Let J_i be the i-th realization of the jitter, whose variance is σ^2, and then let us consider the following cumulative sum:

$$s(i) = \sum_{k=N}^{k=2N-1} J(k+i) - \sum_{k=0}^{k=N-1} J(k+i) \tag{7.11}$$

It is shown in [153] that the variance of the cumulative sum $s(i)$ is proportional to the variance $\sigma^2(N)$ defined as follows:

$$\sigma^2(N) = 2 \cdot N \cdot \sigma_{th}^2 + N^2 \cdot \sigma_{fl}^2 \tag{7.12}$$

where σ_{th}^2 is the variance of the jitter coming from the thermal noise and σ_{fl}^2 is the variance of the jitter coming from the flicker noise.

The jitter measurement is then realized in two steps. In the first step, a synchronous counter counts the number of rising edges Q_i^N of the output clock of RO1 during a timing window defined by the frequency division by N of the clock of RO2.

The second step consists in computing the variance $V(N)$, which is the variance of $Q_{i+1}^N - Q_i^N$, i.e., the variance of the difference between two successive measured number of edges.

Finally, it is possible to show that the variances $V(N)$ and $\sigma^2(N)$ are linked by the following equation:

$$V_N = \frac{4\sigma^2(N)}{T_1^2} \tag{7.13}$$

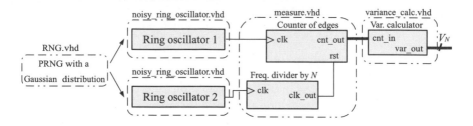

Figure 7.16
Simulation setup of the jitter variance measurement

It is therefore clear that from a simple execution of the two above-mentioned steps, $\sigma^2(N)$ can be easily obtained.

Before implementing the jitter measurement method in hardware, it is useful to validate the principle by simulations. The simulations should confirm the validity of the method, but also give an estimate of the measurement precision. The objective of the simulation is to recover the same jitter size at the output of the jitter measurement circuitry as the size of the jitter injected to the device under test.

Our simulation setup is depicted in Figure 7.16. Since the generation of the flicker noise for simulations is not straightforward, we did not include the flicker noise in our simulations.

The jitter with a Gaussian distribution was generated separately for each oscillator, using a PRNG from the VHDL package RNG.vhd [418], and it was then injected to corresponding oscillator during VHDL simulations. The rising edges of clock generated in oscillator 1 are counted during the time window determined by N periods of the reference clock coming from oscillator 2. Results of the simulations for different sizes of the injected jitter and different accumulation times (number of reference clock periods N) are presented in Table 7.6.

Table 7.6
Results of the simulation of the jitter variance measurement

$\sigma_{th}^2\,[\mathrm{s}^2]$	N	Expected V_N	Measured V_N	Relative error
$3.5 \cdot 10^{-21}$	16384	5.71	7	17 %
$3.5 \cdot 10^{-21}$	24576	8.57	9	5 %
$3.5 \cdot 10^{-21}$	32768	11.42	11	4 %
$1.4 \cdot 10^{-20}$	16384	22.84	28	18 %
$1.4 \cdot 10^{-20}$	24576	34.27	38	10 %
$1.4 \cdot 10^{-20}$	32768	45.69	45	2 %

The results show that the relative error of the variance measurement can be as small as 2 or 4% depending on the size of the jitter. As could be expected, in order to reduce the measurement error, N must be sufficiently big (many thousands). Reduction of this measurement error is important because it will impact reliability of the embedded on-line tests that will be built on these measurements.

Once the method has been validated, we can consider its implementation in hardware and/or software. Figure 7.17 presents the structure of a model-based dedicated test based on the proposed jitter measurement. Certain parts of this system can be implemented in a dedicated hardware and for some parts of the measurement, such as the variance calculator, it is more reasonable to implement them in software running on an embedded processor (although their implementation in hardware is still possible).

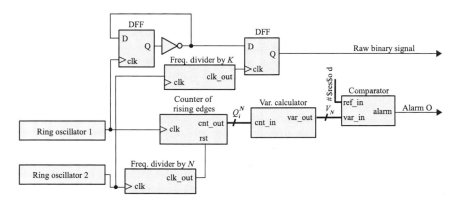

Figure 7.17
Structure of an elementary ring oscillator based RNG including its model-based dedicated statistical test

However, we must recall a very important fact: the stochastic model of Baudet et al. used in our example is based on the jitter coming only from the thermal noise (the flicker noise is neglected). Unfortunately, the proposed embedded method of jitter measurement evaluates the total jitter, which also includes the jitter coming from the flicker noise. Two strategies can be used to solve this problem.

The first one consists of computing the variance $Q_{i+1}^N - Q_i^N$ for several timing windows (several values of N) and then, by a curve fitting, identifying the parts due to the flicker noise and to the thermal noise from the measured $\sigma^2(N)$ using equation (7.12). Finally, only the part due to the thermal noise should be considered in the statistical test triggering the alarm. This strategy is very expensive to be implemented in hardware or in an embedded system in general because of the polynomial curve fitting needed in this measurement method.

The second strategy can directly use the structure depicted in Figure 7.17. This strategy consists of two steps. A preliminary step is a calibration step, which consists of observing the size of σ_{th}^2 and σ_{fl}^2 and their proportion in a safe environment. Such observation can be done using the strategy defined in the previous paragraph (the curve fitting). Then, knowing σ_{th}^2, the lower entropy bound can be estimated. The second step of this strategy consists again of counting the number of edges of the first oscillator during a timing window defined by the frequency division by N of the second oscillator and then in computing $\sigma^2(N)$. However, in this second case, $\sigma^2(N)$ is measured for some single value of N, and it does not give any information about the proportion of the jitter components coming from flicker noise and thermal noise.

However, in this second strategy, we can use the fact that for a constant jitter accumulation window, the proportion between σ_{th}^2 and σ_{fl}^2 remains constant. Indeed, electronic noise models presented in [262] showed that the thermal noise and flicker noise cannot change independently. Thus, reduction of $\sigma^2(N)$ means that the noise source changed its behavior and that the value of σ_{th}^2 that guarantees sufficient entropy rate is certainly reduced too.

Because it does not need complex curve fitting, this second strategy can be exploited as a simple on-line test, in which the measured $\sigma^2(N)$ value is compared with a reference threshold guaranteeing sufficient entropy rate. This reference value is obtained in the so called RNG calibration phase presented later in this section.

7.6.3.3   Total failure test

The goal of the total failure test is to detect quickly and reliably when the entropy rate at generator's output falls to zero. It should preferably test some TRNG parameter, which indicates the total failure of entropy. Reliability of the test means that probabilities of the Type 1 and Type 2 error must be very small, because each false alarm would necessitate execution of a long start-up test. Frequent generation of false alarms could even cause the total unavailability of the generator, because the generated numbers must not be output before the start-up test was not finished. Because the generic statistical tests are not adapted to the generator's principle, they are not fast enough and not sufficiently sensitive to generator's weaknesses.

In the case of the elementary ring oscillator based TRNG, the total failure of entropy can only be due to a catastrophic reduction of the relative jitter between the two oscillators. The total failure test can therefore verify that the relative jitter variance is not null by comparing successive Q_{i+1}^N and Q_i^N, which should always be different. The advantage of this solution for the total failure test is that the counter of rising edges from the embedded on-line test (see Figure 7.17) can be re-used.

In order to reduce the probability of the false alarm triggered by this test to a negligible value, we propose to shift the output of the comparator to an

M-bit shift register and to trigger the total failure alarm only if M successive counter values were identical. The number of bits of the shift register must be chosen so that the total failure test is faster than the generation of one random bit. This choice (the value of M) is made during the TRNG calibration phase. The complete structure of the proposed TRNG including the on-line test and the total failure test is depicted in Figure 7.18.

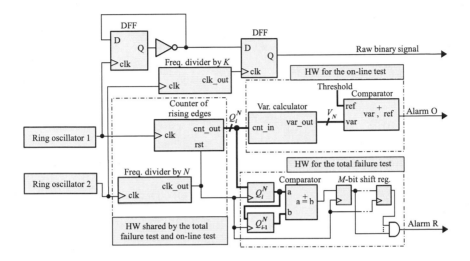

Figure 7.18
Structure of a complete elementary ring oscillator based RNG including the on-line test and the total failure test

We recall that the size (area) and power consumption of the total failure test should also be considered during the design, since it must run continuously.

7.6.4 RNG Calibration

The calibration is a process of setting up parameters of the RNG for guaranteeing sufficient entropy rate of the raw binary signal and sensitivity of the embedded tests. In the context of the elementary ring oscillator based RNG, this calibration consists of determining K (the division factor of the frequency divider determining jitter accumulation time), N (accumulation time for the variance measurement), and M (the speed of the total failure test and the probability of the false alarm).

The calibration process of the TRNG is realized in two steps. The first one is a measurement of parameters σ_{th}^2/T_1^2 and T_2/T_1. These parameters can be obtained using the measurement procedure described in the previous section.

In the second step, the measured RNG parameters are used as inputs in the stochastic model in order to determine the entropy rate in the raw binary signal as a function of K.

Figure 7.19 shows results of a practical experiment – implementation of the elementary ring oscillator based TRNG including jitter measurement circuitry in Altera Cyclone III FPGA in a hardware module dedicated to TRNG testing. It can be seen that experimental results of measuring the variance $\sigma^2(N)/T_1^2$ as a function of N fit well with the curve $2.46 \cdot 10^{-6}N + 9.96 \cdot 10^{-9}N^2$. Moreover, using the same experimental setup, we were able to measure the frequency ratio $T_2/T_1 = 0.702$.

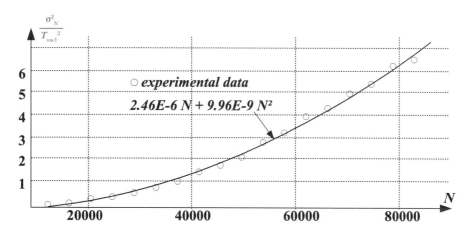

Figure 7.19
Variance $\sigma^2(N)/T_1^2$ as a function of N for an elementary ring oscillator based TRNG implemented in Altera Cyclone III FPGA

Knowing the variance and the frequency ratio, we can determine the entropy rate of the raw binary signal as a function of K (see Figure 7.20). From this function, we can observe that in order to obtain entropy rate higher than 0.9, K has to be bigger than 206 500, and to have entropy rate over 0.997, K has to be bigger than 616 300.

The curve presented in Figure 7.19 was obtained for N varying from $3000_{(16)}$ to $15000_{(16)}$ in steps of $1000_{(16)}$. As presented in Section 7.6.3, the on-line test should compute the variance $\sigma^2(N)/T_1^2$ just for one value of N. We selected $N = 3000_{(16)} = 12\,288_{(10)}$, since for smaller accumulation times the impact of the flicker noise on the total jitter is smaller.

For selected value of $N = 12\,288$ and for selected value of $K = 616\,500$, we obtain the threshold variance $V_{thr}(N) = 4\sigma^2(N)/T_1^2 = 1.534$. If the measured variance $V(N)$ is bigger than this threshold value, the entropy rate at generator's output does not fall under 0.997.

Next, we can determine the length of the on-line test. We recall that for computing one Q_i^N value, we need N periods of the reference clock. For computing the variance $V(N)$, we need $L = 262\,134$ values of Q_i^N. So the total length of the on-line test is $t_{olin} = L \cdot N = 3\,221\,102\,592$ periods of the reference clock. During this time, about $5\,225$ random bits were generated, that is less than $20\,000$ random bits needed for a simple generic test.

Finally, we can determine the length of the total failure test. As explained in Section 7.5.5, it should be faster than the generation of one random bit. Since we need $K = 616\,500$ periods of the reference clock for generating one random bit and $N = 12\,288$ periods of the reference clock for computing each value of Q_i^N, the shift register should have $M < 50$ bits. We selected $M = 10$ as a good compromise so the total length of the total failure test is $t_{tot} = M \cdot N = 122\,880$ periods of the reference clock. In this case, if ten successive values of Q_i^N are identical, the total failure alarm is triggered. The probability of the false alarm is practically zero and it is very fast.

We underline that for selected K, the generated bit stream of more than 1 Gbit random data could pass without any problems the NIST SP 800-22 test suite with the significance level 0.001. Note that our entropy estimation is much more rigorous than just a statistical testing using even the most sophisticated generic statistical tests.

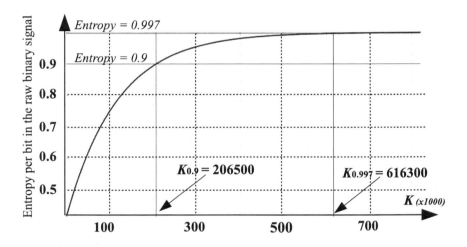

Figure 7.20
Entropy per bit of the raw binary signal as a function of K

In order to observe, what value of K would be sufficient for letting output data pass the tests, we generated one megabyte of random data for several K values, and we submitted the generated sequences to the procedure B of the AIS 31 test suite. The results are depicted in Figure 7.21. As can be seen, the minimal value of the division factor K necessary for the success of the test is $23\,040$.

The value of K determined by statistical tests is thus 30 times smaller than the value required to guarantee entropy rate over 0.997 just from the thermal noise. This important difference is probably be due to the fact that the model is very conservative and that it does not take into account the flicker noise, which is considered to be auto-correlated.

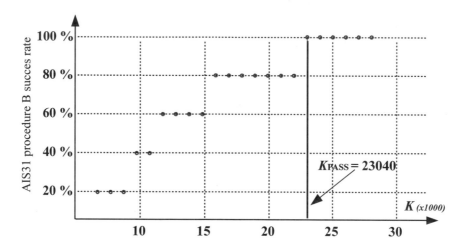

Figure 7.21
Success rate of the AIS 31 procedure B as a function of K

7.7   Conclusion

Random number generation is the Achilles' heel of cryptographic systems, it is therefore vital we get it right. Consequently, we should keep the design as simple as possible in order to be able to perfectly understand where the randomness comes from and what physical parameters can have impact on it.

We showed that evaluation of the TRNG using just a standard test suite is definitely not a good way of designing a secure generator. Instead, we showed a good approach based on entropy estimation from a carefully designed stochastic model that was built after detailed analysis of the underlying physical phenomena.

8

Side-Channel Attacks

Josep Balasch

K. U. Leaven

Oscar Reparaz

K. U. Leaven

Ingrid Verbauwhede

K. U. Leaven

Sorin A. Huss

Technical University of Darmstadt

Kai Rhode

Technical University of Darmstadt

Marc Stöttinger

Nanyang Technological University

Michael Zohner

Center for Advanced Security Research

CONTENTS

8.1   Introduction

Modern cryptographic algorithms are typically designed around computationally hard problems. Generally speaking, breaking a cryptographic system implies finding its secret key. In classical cryptanalysis, an adversary that knows the algorithm specification and has access to its input/output messages tries to overcome these computationally hard problems by employing abstract mathematical models. If the computational effort necessary to break the algorithm is beyond the ability of any adversary, then it can be considered secure.

In practical scenarios however such a security model does not always hold. In fact, proving the mathematical security of an algorithm does not imply that an *implementation* of such algorithm is secure. Let us consider an adversary that wants to extract the secret key used by a smart card implementing a signature generation algorithm such as RSA [347]. Rather than trying to come up with a way to solve the computationally hard problem of factoring very large numbers, an adversary will most likely try to extract the secret key by exploiting the physical characteristics of the implementation. For instance, he might try to extract the key by reading the memory contents or even corrupting the firmware.

8.2 Taxonomy of Physical Attacks

Physical attacks encompass all those attacks in which an adversary tries to tamper with a cryptographic device in order to extract its secret keys. The wide range of existing physical attacks gives rise to multiple categorization options. Time, cost, or expertise, are just a few examples of factors enabling different space partitions. In the following, and as commonly accepted in the field, we categorize physical attacks according to two orthogonal criteria [269].

A first criterion divides physical attacks depending on the way an adversary manipulates the device under attack:

- **Passive attacks** encompass all attacks in which a target device is operated according to its design specifications. During a passive attack the adversary collects data by simply observing physical properties of the device during the normal execution of some operation. These data, referred to as *side channel information*, is exploited by using *side channel analysis*. Passive attacks are often referred to as *side channel attacks*.

- **Active attacks** include all attacks in which the normal functioning of the device is intentionally manipulated. The aim of the adversary consists in inducing an erroneous behavior on the operation being executed by the device. The *faulty information* collected as a result of the manipulation is later exploited by using *fault analysis*. Active attacks are often referred to as *fault attacks*.

Independently of the technique used the end goal of the attacker is to process the collected information, either side channel or fault, in order to learn secret or sensitive information of the device. Recall that in the most typical scenario an adversary performs physical attacks during the execution of some cryptographic algorithm with the objective of learning information about the secret cryptographic key(s) employed.

A second criterion further subdivides physical attacks into three groups according to the level of intrusion performed on the device.

- **Non-invasive attacks** do not require physical alterations on the target device, i.e., the attacks can be carried out without need of accessing internal elements of the chip. Side channel attacks based on monitoring execution time [229], power consumption [230], or electromagnetic emanations [134,338] fall into this category. Fault attacks based on glitch injection in external signals [24], e.g., clock, voltage, etc., are common examples of active non-invasive attacks.

- **Semi-invasive attacks** require a certain degree of intrusion on the device conditioned so that the passivation layer is not damaged. In practice attacks that require chip depackaging fall in this category. Exemplary passive semi-invasive attacks are probing attacks employed for instance to read out memory cell contents [366]. Notice that the quality of side channel leakages such as electromagnetic radiation can be enhanced by depackaging the chip. Active semi-invasive attacks include those attacks exploiting the photoelectric effects caused by lasers or white light [402].

- **Invasive attacks** involve full access to the interior of the device. Passive invasive attacks are based on probing techniques, whereas active invasive attacks focus on altering the layout of the target device.

Besides providing a more granular taxonomy, the latter classification gives an intuition of the cost of the attack, i.e., the higher the degree of intrusion, the larger the cost of the equipment required to mount the attack. Notice that semi-invasive and, particularly, invasive attacks require expensive specialized equipment such as laser cutters, probing stations, and/or focused ion beams to access the inside of the chip.

In the rest of this chapter we will focus on non-invasive side channel attacks, as they require (relatively) inexpensive equipment and thus pose a threat to the security of embedded devices. Side channel analysis is a very active research field as indicated by the growing number of articles on the topic. Even today, more than 15 years after the seminal publications, commercial devices with security functionalities are shown to be vulnerable to side channel attacks. KeeLoq-based remote keyless entry systems [111, 200], the contactless MIFARE DESFire MF3ICD40 card [320], the FPGA bitstream encryption used by Xilinx Virtex FPGAs [295], or the secure Atmel CryptoMemory EEPROMs [26], are just some examples of devices that have been recently broken.

8.3   Side Channel Analysis

Side channel analysis exploits physically observable information leaked by cryptographic devices when performing some operation. Contrary to theoretical models employed in cryptanalysis in which an adversary tries to break cryptographic constructions using mathematical models and input/output message pairs, practical realizations of cryptography are susceptible to more dangerous attacks enabled by the existence of side channels. Time, power consumption, or electromagnetic emanations are the most common examples of observable phenomenons that an attacker can exploit to gain knowledge of secret cryptographic keys.

In the rest of this section we give some basics on side channel analysis. We first introduce the origin of side channel leakage in circuits, and briefly enumerate the capabilities of an adversary to collect this information. We then enumerate different types of side channel attacks, with particular emphasis on the seminal Simple Power Analysis (SPA) and Differential Power Analysis (DPA) attacks. Finally, we give an overview of currently known countermeasures.

8.3.1 Capturing Leakage in CMOS Devices

Complementary Metal Oxide Semiconductor (CMOS) is the predominant technology used nowadays in integrated circuits. One of the characteristics of CMOS, which partially justifies its dominance in the field, is its convenient energy usage. In fact most of the energy consumption of CMOS devices occurs dynamically as the circuit changes its state, while the static power consumption of the device remains rather low. Logic cells in a circuit are typically implemented using complementary CMOS, i.e., there exist a pull-up network composed of P-transistors connected to V_{DD} and a pull-down network of N-transistors connected to GND.

For illustrative purposes let us consider the case of one of the simplest CMOS cells: an inverter, whose input/output transition table is depicted in Figure 8.1 (left). When there is no input transition the inverter circuit remains idle, and only the (rather low) static component of the current flows through the conducting transistor. When there is an input transition from high to low $(1 \rightarrow 0)$, the pull-up network of the circuit (P transistor) becomes active. A dynamic current flows through the circuit in order to charge the load capacitance, as shown in Figure 8.1 (center), effectively generating an output transition from low to high $(0 \rightarrow 1)$. Finally, when the input transition goes from low to high $(0 \rightarrow 1)$ the pull-down network becomes active. In this case there appears a short-circuit current that causes the load capacitance to be discharged, producing an output transition from high to low $(1 \rightarrow 0)$.

This example highlights the inherent leakage behavior of CMOS devices — the dynamic current drawn by a circuit leaks information about its inner transitions. How to measure this leakage is the first issue an adversary needs to tackle. Typically, the power consumption of a circuit can be monitored by placing a (shunt) resistor in the current path of the target device and measuring its voltage drop over time. Electromagnetic radiation carries the same information as the power consumption and can be monitored by placing a coil over the surface of the target device. A clear advantage of the latter approach is that it provides access to more localized leakage, e.g., an adversary can target the leakage of a specific element rather than the whole circuit. As

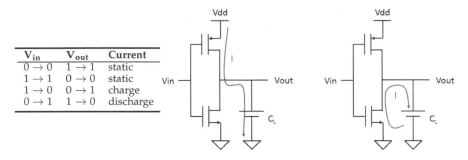

V_{in}	V_{out}	Current
$0 \to 0$	$1 \to 1$	static
$1 \to 1$	$0 \to 0$	static
$1 \to 0$	$0 \to 1$	charge
$0 \to 1$	$1 \to 0$	discharge

Figure 8.1
CMOS inverter. Transition table (left), charge circuit (center), and discharge circuit (right)

a downside, electromagnetic radiation is often more noisy than power signals. Independently of the leakage source chosen, an adversary requires an acquisition device to digitize the leakage signal and some auxiliary equipment such as power supplies or signal generators to run the target device. The minimum cost for building such a setup is clearly rather low, particularly when compared to invasive attacks.

One of the first objectives of an adversary is to quantify the data-dependent leakage captured as this can help enhancing attacks at a latter stage. One of the most employed leakage models in the literature is the *Hamming Weight Model*. It basically assumes that the power consumption of a CMOS device is directly correlated to the number of low to high transitions. In other words, the power consumption of a device when switching to a value Y is given by $H_W(Y)$. Other models, such as the *Hamming Distance Model*, take also into account the low to high output transitions. In this case the power consumption of a device when switching from a value X to a value Y is given by $H_D(X, Y) = H_W(X \oplus Y)$. Figure 8.2 depicts a series of power measurements obtained from a CMOS device performing transitions in its 8-bit data bus. For this device in particular the leakage model is quite close to a Hamming distance model, as the overall number of transitions determines the peak amplitude of the power measured in that cycle. In other words, the same amount of current is drawn whether there is an X to Y transition or an Y to X transition in the data bus.

Further leakage models proposed in the literature aim to consider different transition leakages [325] or to assign different weights to specific elements of the circuit. The branch of profiled attacks, which will be discussed later in the chapter, uses advanced statistical tools to fully characterize a device's leakage [69, 373].

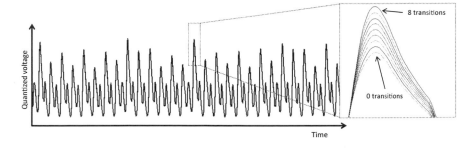

Figure 8.2
Transitions on the data bus of a CMOS device

8.3.2 Inspection of Power Traces

A trace captured during the execution of a cryptographic operation carries several types of information of interest to an adversary: the existence of *patterns* within the measurement is related to the structure of its implementation; the *amplitude* of the curve peaks carries information about the type of operation being performed and the data being processed; and the *timing* distinguishes variations in the program flow.

In order to illustrate these effects, let us assume an adversary that possesses a cryptographic device that performs symmetric-key encryption and whose specifications are not publicly available. Using a suitable measurement setup, the adversary captures a power measurement trace as shown in Figure 8.3. The trace contains a series of patterns (highlighted by dotted lines) that stem from the structure of the algorithm's implementation. In particular, one can easily identify nine identical patterns followed by a shorter, yet rather similar, one. Such a construction is consistent with the 128-bit version of the Advanced Encryption Standard (AES-128), consisting of 10 encryption rounds with the particularity that the last round omits one operation.

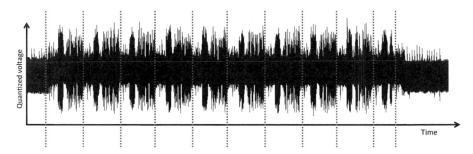

Figure 8.3
Power curve of a full AES-128 encryption, rounds highlighted

Zooming into one of the rounds, as illustrated in Figure 8.4, allows one to visualize their inner structure. The round operations of AES — SubBytes, ShiftRows, MixColumns, and KeyAddition — give rise to four distinguishable patterns. Notice that the amplitude of some peaks within the SubBytes and MixColumns computations is slightly higher than the rest. As these operations are often implemented in software by means of look-up tables in memory, one can infer that memory access operations consume more power than, e.g., atomic operations in the Arithmetic Logic Unit (ALU).

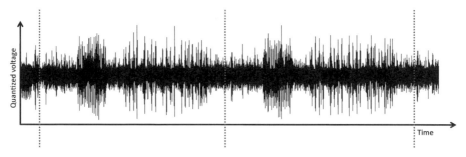

Figure 8.4
Power curve of two rounds of an AES-128 encryption, rounds highlighted

In the following section we focus on attacks that exploit leakage information contained in the power consumption and/or electromagnetic emanations to recover secret cryptographic keys. As a final remark, note that timing side channels are captured by using the same experimental setups, as the acquisition device is inherently measuring over time.

8.3.3 Simple Power Analysis (SPA)

Simple power analysis (SPA), introduced by Kocher et al. in [230], tries to exploit key-dependent patterns present in one (or very few) power traces, often by simply visual inspection. While it is a powerful attack, this level of interpretation of the curves typically requires some expertise and/or detailed knowledge on the target device and the implementation.

The existence of patterns that can be exploited by an adversary arises due to key-bit dependent operations or branches. This is particularly troubling in some public key cryptosystems whose algorithms contain conditional statements directly dependent on secret key bits. A typical example of such constructions is given by modular arithmetic algorithms such as the right-to-left binary modular exponentiation [285] employed for instance in RSA [347]. This algorithm iterates over each bit of the exponent and always performs at least a modular squaring. If the scanned bit equals one an additional modular multiplication is executed.

It is not difficult to grasp where the leakage in this algorithm lies — the sequence of multiplications and squarings in a *single* power trace directly yields the value of the exponent employed in the modular exponentiation. In the case the operation carried out by the device is the generation of an RSA signature, then the exponent corresponds to the device's private RSA key. An example of an SPA attack on such an implementation can be seen in Figure 8.5.

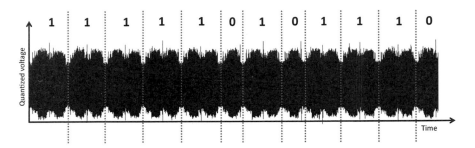

Figure 8.5
SPA attack on square-and-multiply exponentiation algorithm

While the effects of the attack previously described are devastating from a security point of view, it is rather difficult and uncommon to break cryptographic implementations using a single power trace. Often, SPA attacks are based on combining the leakage of multiple power traces with other information about the implementation and/or the target device. Mangard in [268] and VanLaven et al. in [437] show how to break insecure implementations of the AES by exploiting leakages in the key schedule routine. Similar attacks exist for other block ciphers such as Serpent in [82] or Camellia in [458]. SPA can also be used to detect more granular timing leakages and thus enhance timing attacks such as the one by Koeune and Quisquater [231] against non-constant implementations of the MixColumns step of the AES. Using SPA an attacker can detect the time leakages *inside* the rounds rather than aggregated leakages at the end of the encryption process. Other SPA attacks target implementations of algorithms to generate prime numbers.

8.3.4   Differential Power Analysis (DPA)

Differential Power Analysis (DPA) is a powerful statistical technique that jointly processes a (generally large) set of *power measurements* corresponding to the execution of some cryptographic implementation to retrieve secret material. DPA was originally introduced in [230]. Due to its powerful nature, DPA stimulated a fruitful research and development effort both in countermeasures and new refined attacks in industrial and academic circles.

DPA differs from SPA in some key aspects. Most importantly, and in contrast to SPA, DPA does not require a detailed knowledge of the device characteristics or the specific implementation of the cryptographic algorithm running on the chip. The attacker should know that, for example, the particular chip implements AES-128 encryption (i.e., she knows which algorithm is running) and could carry out the attack even if she did not know the exact underlying hardware architecture or software implementation (for example, the datapath width, the technology employed, or the clock speed). The fact that DPA even works with mild assumptions regarding the target device greatly increases the risk exposition of embedded devices to this attack technique.

Quite importantly, the principles that make DPA work are relatively general and do not depend on the specific cryptographic algorithm implemented (for example RSA [347], AES [94], or SHA-1 [109], just to name a few) or the particular platform used (for example, a software implementation running on a 8051 microcontroller, a hardware implementation on an FPGA prototype, or a full-custom design on an ASIC). As a consequence, DPA can be applied to any implementation of a cryptographic algorithm that does not feature specific countermeasures against it.

Another important difference between SPA and DPA is that DPA can improve its effectiveness by collecting additional traces. SPA relies on visual inspection of the power consumption traces, meaning that if there is enough noise present in the traces, the information that SPA extracts will be corrupted by noise and as a consequence the outcome of the SPA will be incorrect. On the contrary, an important strength of DPA is that it employs statistical techniques that can cancel out noise if there is a sufficiently large number of power measurement traces. DPA is a robust technique in the sense that in addition to being able to cancel out noise, it tolerates deviations from the initial assumptions about the device and works relatively well even with rough approximations to the device behavior.

A first high-level description of a DPA attack is as follows. Suppose that some particular chip is performing an AES-128 encryption. On the one hand, we record the instantaneous power consumption of the chip while performing multiple encryptions. During each encryption, the chip handles some intermediate values that depend on a small chunk of the key. (Let us assume that the instantaneous power consumption depends on the intermediate value handled by the chip. This dependency may not be visible in a single trace, due to noise present, but should be detectable on average.) On the other hand, we can predict the power consumption of a particular intermediate value appearing on the execution for each hypothesized subkey, for each encryption. Next, DPA compares the set of actual power measurements with the predictions of some intermediate value, and repeats this process for each subkey. For the correct subkey, the match between predictions and actual power measurements will be higher. In this way, the correct subkey can be distinguished among all others.

Let us have a closer look at the previous process with a concrete example. Suppose that the chosen intermediate value corresponds to the first Sbox lookup $S(p_i + k)$[1]. Here p_i is an 8-bit chunk of the plaintext, and k is an 8-bit chunk of the key we want to recover. We record a number of measurements with different plaintexts p_i corresponding to the instant where the Sbox output is handled. On the other hand, we predict the power consumption of the intermediate value $S(p_i + k)$. Let us focus on the least significant bit b_0 of the Sbox output $S(p_i + k) = b_7 \ldots b_0$. Our prediction is that values with $b_0 = 1$ will consume more than values with $b_0 = 0$. We partition the set of power measurement traces into two subsets based on the value of b_0. In other words, we form two sets of traces S_0 and S_1, corresponding to power measurement traces where $b_0 = 0$ or $b_0 = 1$, respectively. The critical observation of DPA is that, on average, the power consumption of traces in S_0 will be slightly smaller than traces in S_1. This is because, according to our assumptions, the device tends to consume a little bit less when the handled value $S(p + k)$ satisfies $b_0 = 0$ than in the case $b_0 = 1$. This effect can be detected by simply taking the mean value of each set, $\bar{S}_i$ for $i = 0, 1$, and checking that the difference of the means is non-zero, $|\bar{S}_0 - \bar{S}_1| > 0$.

The previous observation does not directly recover the key. Nevertheless, it can be easily turned into a key-recovery attack as follows. For each subkey hypothesis k, we compute the corresponding intermediate values $S(p_i + k)$ and partition accordingly the traces into two sets S_0 and S_1. Notice that for two different keys, the hypothesized intermediate values for each trace will differ, implying that the partition of the set of traces into the two sets S_0 and S_1 will be different for each key. The key that produces the larger difference between means of these sets, $|\bar{S}_0 - \bar{S}_1|$ will likely be the correct one. Note that often the attacker does not know the exact time sample corresponding to the handling of the intermediate value $S(p_i + k)$, so the entire process is normally repeated for each time sample in some suitable time window.

The process of a typical DPA attack is illustrated in Figure 8.6. The device under attack is an 8-bit CMOS RISC microcontroller clocked at 3.58 MHz implementing AES-128 encryption. The voltage drop across a shunt resistor of 50 Ω was sampled at 2 GS/s using an off-the-shelf oscilloscope. In total, 1000 instantaneous power consumption curves were measured. The plot on top in Figure 8.6 depicts one instantaneous power consumption of the microcontroller during a part of an AES-128 encryption. The plot below represents, in light grey, the difference $|\bar{S}_0 - \bar{S}_1|$, for each time sample, for each subkey hypothesis. These are called *differential curves*. For the correct subkey, the differential curve is plotted in solid black. Notice that only for the case of correct subkey, the differential curve has characteristic peaks, i.e., time samples where the difference $|\bar{S}_0 - \bar{S}_1|$ is relatively big. This noticeable difference precisely tells that the partitioning into two sets of curves according to the (key hypothesis dependant) value of b_0 actually leads to two sets with different mean power consumption. This fact is compatible with the initial hypothesis about the device consuming more when a value with $b_0 = 1$ is handled,

---

[1]Neglecting key whitening.

and thus the correct key is revealed. Note that for an incorrect key, the value $|\bar{S}_0 - \bar{S}_1|$ will be close to 0.

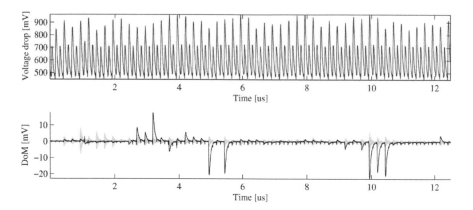

Figure 8.6
Exemplary DPA attack. Top: exemplary curve. Bottom: differential curves

A great deal of research effort has been put in the last decade in the field of DPA. A first line of research is the efficiency improvement on the number of traces required for successfully performing a DPA. Normally, this is achieved by exploiting some additional information about the device behavior and building a leakage model that approximates the behavior. A notable example of this direction is the CPA distinguisher [55]. This kind of attack is crafted to extract keys in a more efficient way at the cost of targeting a narrow family of devices. On the contrary side, significant effort has been put into generic DPA techniques that are able to defeat devices with possibly unknown leakage behaviors [137]. This is usually achieved at the cost of more traces.

8.3.5 Internal Collision Attacks

Collision attacks were introduced by Schramm et al. [376] in 2003. They exploit the fact that two runs of a cryptographic algorithm with different inputs can yield the same intermediate result at a certain point in time. More formally, two executions with different inputs x_1 and x_2 and same secret key k, can produce the same intermediate result $z = f(x_1, k) = f(x_2, k)$. Collisions are an inherent feature of cryptographic algorithms with an iterative structure in which the same transformations are repeatedly executed. In this type of algorithm an intermediate value can even collide at multiple points in time during a single encryption process.

In terms of power analysis a collision can be found when parts of power traces have a high similarity during a time span. The length (and shape) of such an interval depends on multiple aspects, e.g., algorithm, implementation, target device, etc., ranging from single cycle operations to a sequence covering multiple cycles. An adversary capable of detecting such patterns is able to exploit the side channel information provided by the collision. In particular, collisions help an adversary to reduce the search space for the secret key.

8.3.6 Profiled Attacks

The techniques described so far attempt key extraction without any (or much) prior knowledge about the side channel leakage of an implementation. Profiled attacks on the other hand make use of a learning step to obtain a priori information about the device's leakage. The better an adversary is able to characterize the leakage of a device, the higher her success rate will be. The profiling stage relies on the assumption that an adversary has access to a device identical to the one under attack. In some cases, a further requirement is that the adversary can change secret keys on the training device. While these conditions might be difficult to meet in real scenarios, profiled attacks are amongst the most powerful side channel techniques.

Template attacks, introduced by Chari et al. [69], have been shown to be the strongest side channel attack possible from an information theoretical point of view. During profiling an adversary produces a multivariate characterization of the noise in the instantaneous leakage signal. This characterization, referred to as *template*, consists of a mean vector and a covariance matrix $(\mathbf{m}, \mathbf{C})$. During the profiling phase, the adversary uses the training device to collect different power traces for different input data x_i and keys k_i. Then, he groups the traces corresponding to a certain pair (x_i, k_i) and calculates its corresponding template $(\mathbf{m}, \mathbf{C})_{(x_i, k_i)}$. During the attack stage, the adversary obtains a power trace $\mathbf{t}$ from the device under attack. He then evaluates the probability density function of the multivariate normal distribution using each template $(\mathbf{m}, \mathbf{C})_{(x_i, k_i)}$ and the measurement $\mathbf{t}$. More formally, he computes the following probability:

$$p(\mathbf{t}; (\mathbf{m}, \mathbf{C})) = \frac{1}{\sqrt{(2\pi)^T \cdot det(\mathbf{C})}} \cdot \exp\left(-\frac{1}{2} \cdot (\mathbf{t} - \mathbf{m})' \cdot \mathbf{C}^{-1} \cdot (\mathbf{t} - \mathbf{m})\right) \quad (8.1)$$

This probability indicates how much each template fits to the measurement, i.e., the higher the probability, the better match between both. Once all probabilities are computed, the adversary applies a maximum-likelihood (ML) decision rule to find the correct template, which at the same time yields the value of the correct key.

In practice there are several issues that need to be dealt with when using templates. First of all, notice that the size of the covariance matrix $\mathbf{C}$ grows quadratically with the number of points in the power trace. For the attack to be computationally feasible it is thus important to reduce the number of points, and only consider those which carry useful information. Finding points of interest (POI) inside power traces is key to successfully carry out a template attack, but unfortunately no optimal mechanism exists to do this.

In [69] the sum of pairwise differences of the average measurements is used to highlight and select interesting points, but other alternatives such as the sum of absolute pairwise differences or even the variance have been used in related works. In [339], Rechberger and Oswald use DPA as means to identify suitable points for building templates. Other tricks often found when dealing with templates include ways to efficiently compute equation (8.1). For instance, it is typical to apply logarithms to avoid the exponentiation, or to use the identity matrix instead of the covariance matrix to avoid computing the matrix inversion.

Stochastic attacks, originally proposed in [373], classifies samples by using a single accurate multivariate characterization of the probabilistic noise together with an approximation of the deterministic, data-dependent component in a vector subspace. More formally, the stochastic attack models each point in the power curve $I_t(x,k)$ as a stochastic variable composed of two parts such that $I_t(x,k) = h_t(x,k) + R_t$. The data-dependent part $h_t(x,k)$ is a function depending on the input (known) data d and subkey k, while R_t is assumed to be a stochastic variable representing zero mean noise. A full description of the steps followed in stochastic attacks is out of the scope of this book — for details and proofs see [373].

The fact that the deterministic component $h_t(x,k)$ is approximated inherently makes the success rate of stochastic attacks lower or equal to that of template attacks, which perfectly estimates the multivariate noise. As an advantage, the profiling step of stochastic attacks require the collection and processing of less traces than template attacks.

8.3.7   Countermeasures

Even though there is a rich and vast literature on side channel countermeasures, there is currently no mechanism capable of providing perfect security. Formally proving resistance against side channel attacks is in fact an open research problem. Due to this, a general approach to protect implementations consists in adding multiple known countermeasures with the goal of minimizing the options of a successful attack. Note that adding countermeasures, particularly in commercial products, is not a straightforward process and requires management of sensitive trade offs, e.g., between security, efficiency, or cost, among others.

Side channel countermeasures can be implemented in different abstraction layers. In the following we describe in detail one such countermeasure framework that allows for synthesis of side channel resistant synthesis of cryptoprocessors.

8.4 A Framework for Side-Channel Resistant Synthesis of Crypto-Processors

8.4.1 Introduction

Nowadays, embedded devices find application in an increasing part of everyday life. Due to the ongoing progress of the fabrication technology integrated devices get more and more powerful in terms of computational throughput as well as in terms of complexity. However, the increasing number of embedded devices also gives rise to potential unintended exploitations. In order to secure the devices against misuse and create a secure communication environment, various protocols and cryptographic schemes have been developed.

In the last two decades another attack type emerged, targeting directly at implementations that process sensitive data. This kind of assault is called *side-channel analysis attacks* and exploit physical observables that depend upon the processed data. Side-channel attacks are especially dangerous due to their general applicability to cryptographic schemes and their non-invasive nature. Thus, a side-channel attack can be performed on various implementations and a compromised secret is very difficult to be detected. Examples for devices targeted are smart cards, USB security tokens, RFID tags, or even complex cryptographic co-processors. The most prominent examples of successfully mounted side-channel attacks on daily products are the attacks on the access system KeeLoq [111] and the RFID card MIFARE [320].

Various countermeasures have been suggested in order to harden the resistance of embedded devices against such attacks. These countermeasures are often specifically adapted to an algorithm, implementation, or device, thus limiting their general applicability, cf. [324] Also, it is nearly impossible for a designer to predict the side-channel resistance of an embedded device before its implementation and a subsequent in-depth analysis. Thus, countermeasures against side-channel attacks are often developed and implemented in an additional step at the end of the design process of a device. At this point in time, however, the available resources on the device are rather limited, forcing the designer to either consider less efficient countermeasures or to perform a major redesign of the device. Figure 8.7 depicts the usual hardware design flow of an embedded system commonly used in a security-sensitive application.

Usually, the design of a side-channel resistant hardware module is done in two separate phases, which may lead to a cyclic process. First, the module is specified in a hardware description language (HDL), for instance VHDL, and its correctness is tested by functional simulation. After passing the functional tests, which only ensure an error free behavior in the normal operation mode within the specification parameters of the device, the implementation phase starts. This phase makes use of CAE tools that assist the designer in translating a high-level specification in a more hardware-related description. If the designer has an appropriate model of the target technology platform, a security engineer can then perform a first side-channel evaluation of the hardware design. Usually, this side-channel analysis is done after the implementation phase of the design and is based on a prototype for the physical verification test.

During the side-channel evaluation, the hardware designer provides all necessary information to the security engineer. For the evaluation phase, the security analyst has to select an appropriate attack method that depends on the threat scenario of the circuit's application. Additionally, he has to construct a physical model of the device that is as precise as possible and describes the expected exploitable information leakage. Both the model and the prototype are then used in the evaluation phase. This phase aims at identifying the observable side-channel leakage and, if possible, also its origin. If a successful side-channel attack is identified, this information can then be used to restart the design cycle. If, on the other hand, the exploitable side-channel leakage is too low to reveal the circuit secret such as the cryptographic keys, the target device is deemed side-channel resistant. Such a design flow requires at least a designer, who is skilled in both side-channel analysis and hardware design. However, designers who have expertise in both areas are hard to find.

In this section we investigate the question how to introduce an automated and supportive side-channel evaluation to the conventional CAE toolset to the hardware design. We thereby propose a framework that supports both the security engineer as well as the hardware engineer in the construction of side-channel hardened devices. The main contributions of the side-channel resistant synthesis framework AMASIVE to the state of the art are:

1. Automatic analysis of given HDL design descriptions of cryptographic modules in terms of information and data flow, and

2. Automatic insertion of several first-order countermeasures into the given HDL code before logic synthesis.

By manipulating the HDL description of the circuit, this method can be applied to various implementation platforms such as full custom ICs, ASICs, or FPGAs. This chapter is structured as follows. Section 8.4.2 provides some background to the topic of side-channel analysis in terms of attacks and well-known countermeasures. Then, Section 8.4.3 gives an overview of related work. Section 8.4.4 introduces the architecture of the AMASIVE framework for the supportive design of side-channel hardened devices. The evaluation sub-component of the AMASIVE framework is described in Section 8.4.5. Section 8.4.9 then introduces the reshaping method, which may be viewed as a major step towards a secure synthesis methodology that automatically embeds countermeasures in a given design. In Section 8.4.10 we demonstrate

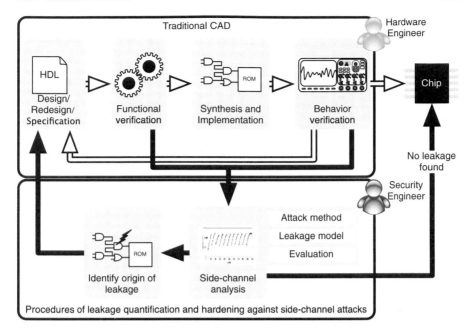

Figure 8.7
Traditional side-channel aware hardware design flow

the advantages of the proposed framework during the analysis and the subsequent hardening of a PRESENT block cipher implementation on top of a Xilinx Virtex 5 FPGA platform.

8.4.2 Background

In the civil cryptographic community, implementation attacks were first introduced in the nineties. Such attacks use weaknesses in the implementation of a cryptographic algorithm or application that handles security-sensitive information. One class of implementation attacks are the so-called side-channel attacks, which we further detail in this section.

8.4.2.1 Side-channel attacks

Side-channel attacks (SCA) are a class of non-invasive implementation attacks, which exploit physical observables of an implementation of a cryptographic algorithm in order to extract secret information, for instance the secret key of an encryption scheme. They are especially threatening since it is hard to notice a successful side-channel attack. These kinds of attacks are mounted during the operation of a device, i.e., runtime, while the implementation is running in normal operation mode without any violation to its operational specifications, cf. [269]. The most commonly exploited observables are the run-time of the execution, the power consumption, and the electromagnetic emission, cf. [324]. This additional information may enable an adversary to break a cryptographic system, even if the algorithm was mathematically proven to be secure. The focus of this chapter is on side-channel attacks that exploit the power consumption of a device, the so called power analysis attacks. We refer to [269] for a more detailed introduction to the basic side-channel analysis material.

8.4.2.2 Power analysis attack methods

Power analysis attacks exploit the data-dependent switching activity of a cryptographic implementation in order to extract information of intermediate values of internal operations or of the current internal state of the algorithm. The unintended information leakage from the measured power consumption is then exploited by making assumptions about the processed data during an execution, computing a hypothesis about the power consumption based on the assumptions, and verifying them using statistical methods. The hypotheses are the expected values of the observed physical quantity caused by the adversary's suggested intermediate value changes or state transitions. The construction of theses hypotheses does not require detailed information about the implementation in general and focuses only on parts of the secret key[2] for feasibility reasons by applying the *divide and conquer* principle.

Since the introduction of power analysis attacks a lot of progress has taken place on both sides, for attacks as well as for countermeasures. Table 8.1 lists several fundamental power analysis attack methods, classified into two main groups and annotated with the according references in the literature. Note that besides the ones listed in the table, many derivative attacks exist in literature.

The simple power analysis (SPA) attack, for instance, exploits the unbalanced control flow of public cryptography schemes like RSA, ECC, or McEliece, cf. [230, 292, 391] or direct power magnitude differences for the amount of Hamming weight or distance [287]. The different power consumption values of a captured trace are directly interpretable due to the known control flow decisions in the algorithm. In contrast, the template attack and the stochastic approach, listed in the table in the section profiling, exploit the intermediate values of internal operations using multiple measurements. The stochastic approach differs from the template attack by using a more advanced power model based on subspaces and linear regression.

---

[2]So-called subkeys

Table 8.1
Power analysis attack methods

Type	Name	Comment	Reference
Non-profiling	Differential power analysis	First attack on intermediate values	[230]
	Correlation power analysis	Most popular attack method	[55]
	Mutual information analysis	Considers non-linearity	[137]
	Linear regression analysis	Utilizes linear regression	[105]
	Power analysis collision	Exploits internal collisions	[296]
	Power amount analysis	Considers time intervals	[425]
Profiling	Simple power analysis	First control flow attack	[230]
	Template attack	Utilizes power distribution	[69]
	Stochastic approach	Exploits subspace representation	[372]

8.4.2.3 Countermeasures

Since the introduction of power analysis attacks, various countermeasures have been studied and developed. An intuitive approach for a countermeasure is to decrease the exploitable power consumption of the device. The exploitable information leakage due to the data-dependent power consumption can be reduced mainly by two different basic approaches. The first approach, *hiding*, focuses on the physical behavior of the implementation.

The principles of hiding are to decouple the data specific power consumption from the actual internal processed intermediate values. Increasing the noise of the unwanted power consumption or dissipation is one possibility to decrease the signal-to-noise ratio and thus decouple the power consumption from the internal processed values. The unmodified intermediate values, which rely on the input values, can be decoupled from their characteristic power consumptions by scrambling them in two domains: the time domain and the amplitude domain. An overview over the different hiding methods is provided in Table 8.2.

The second approach, *masking*, reduces the degree of exploiting the data-dependent power consumption by internally randomizing the processed intermediate values by using auxiliary or masked values instead. On an abstract level, masking uses the principle of secret sharing to increase the effort of extracting useful information from the data-dependent power consumption. The incoming public values as well as the intermediate values are masked by a randomly chosen value, the so-called masking value. Thus, the

Table 8.2
Classification of countermeasure methods

Type	Name	Level	Reference
Masking	Masked values	Algorithm	[62, 167]
	Masking logic	Cell	[128, 429]
	Masked dual-rail pre-charge logic (MDPL)	Cell	[330, 331]
	Threshold implementation	Algorithm	[297, 315]
Hiding	Shuffling	Algorithm	[269, 412]
	Random delay, clock	Cell	[260, 261]
	Dual-rail logic	Cell	[166, 439]
	Current load logic	Cell	[67]
	Noise, random switching	Cell	[149, 417]
	Register pre-charging	Algorithm	[260]

internal operation process is unknown and not directly derivable from intermediate values from the known input or output data. In order to compute the correct output, the intermediate results have to be corrected at the end of the cryptographic algorithm run. Without knowing the mask value the adversary cannot construct correct hypotheses for the data-dependent power consumption, thereby making a side-channel attack more complex. An overview over the different masking techniques is provided in Table 8.2.

8.4.3   Related Work

Nowadays in hardware design, the resistance of the device to side channel attacks is often considered at the end of the design process, which makes this important step more dependent on the available resources than on the actual security requirements, cf. [35]. The late consideration is especially disadvantageous as it forces the designer to repeat the overall design process in order to include side channel countermeasures. The reasons for a rather late consideration of side channel vulnerabilities are manifold such as the short time to market, unpredictability of side channel properties during the design process, or an unidentified side-channel information leakage of the algorithm. Thus, side-channel vulnerabilities are detected too late to be taken into account when deciding on the design architecture. This situation is rather unfortunate, because the designer has access to an enormous amount of knowledge about the device. So, she or he is in the position to perform a much more accurate side-channel analysis than an outside attacker ever could. However, identifying, combining, and utilizing this knowledge is a difficult task and requires experience in side-channel analysis next to hardware design skills. Recently, there have been contributions that aim at timely support of the designer in securing a device against side-channel attacks in both software and hardware implementations.

In [407] Standaert et al. proposed a unified framework to cope with side channel issues of implementations. This work introduces various formalisms and methods to identify side-channel leakage in order to compare the vulnerability of several design variants. Countermeasure implementation for side channel attacks in software implementations is tackled by two approaches, which provide automated code analysis and countermeasure integration, cf. [35, 299, 300]. The work in [35] applies a side channel attack to the assembled code and utilizes an estimation of the mutual information [407] to evaluate and to highlight vulnerabilities of a code section. The second approach, proposed by Moss et al. [299, 300], features an automatic insertion of masks by evaluating the secrecy requirements of an intermediate value in the program code.

The first step towards an intelligent insertion of side-channel related countermeasures during the design phase of a hardware implementation was proposed in [340]. Here, Regazzoni et al. introduced a tool set that automatically transforms each element of the design netlist into the side-channel resistant logic style MCML. The exchange of netlist elements is based on an evaluation aimed at identifying security sensitive parts, which need to be hardened by means of this logic style. In two of these three approaches the designer has to formulate the leakage assumption for the side-channel vulnerabilities, which presumes a designer with considerable expertise in such attacks. So, the designer has to decide which attacks a potential adversary would conduct on this design. The proposal in [35] overcomes this problem by utilizing an estimation of the mutual information as side-channel distinguisher. Since the selected leakage model has the greatest impact on the success of the attack, cf. [114, 321], it would be reasonable to select the best model for each attack scenario. In addition, the leakage model relies on an adequate hypothesis function in order to identify the vulnerabilities in the design. Hence the best exploitable hypothesis functions of intermediate states or intermediate value transitions of the implementation is selected in order to use an appropriate leakage mode for the evaluation.

In this chapter we introduce the AMASIVE framework, which supports the designer in a comprehensive way when developing side-channel resistant devices. The AMASIVE framework can be used to automatically identify side channel vulnerabilities at different stages in the design flow of embedded systems, to suggest countermeasures, and to verify the effectivity of the proposed countermeasures. Thereby, the designer is notified early about potential vulnerabilities and can assess the costs of implementing possible countermeasures in the design. To the best of our knowledge, the AMASIVE framework is the first approach to automatically identifying vulnerabilities and suggesting various countermeasures early in the hardware design process. The idea of the AMASIVE framework is rather to support the designer in developing an overall side-channel resistant design than to enforce specific countermeasures. Thus, the designer is supported by a much more flexible design process, which enables her or him to directly adapt the design to given security requirements. The strength of this framework is the adaptability to various implementations by exploiting several attacking procedures, leakage models, and side-channel distinguishers. Similar to [407], we consider an attack model for our analysis in order to secure an implementation specifically with respect to its threat scenario.

8.4.4   The AMASIVE Concept

The basic idea of the AMASIVE framework relies on a multistage approach of security analysis and countermeasure suggestions (cf. Figure 8.8). Both, the security analysis and countermeasure proposals are intended to interact with each other. The security analysis can further be subdivided into the *information collection*, the *graph representation*, and the *vulnerability analysis* phases, respectively. In summary, the information collection phase combines all relevant design information that are then used to construct an intermediate graph representation, which in turn forms the basis for the vulnerability analysis. Its output is then forwarded to both the designer and the countermeasure suggestion module.

The countermeasure suggestion utilizes the data gained from the security analysis as well as information taken directly from the hardware design to suggest various countermeasures, which are taken from the countermeasure database. The countermeasures are then automatically embedded in the original VHDL code by extending the involved modules of the hardware design and by automatically generating the interface between the original and new module in the VHDL code. In the current version the countermeasures are stored as VHDL code entities aimed to be directly embedded in the original code, but this approach is easily extendable by using dedicated hard-macros in order to maintain a deterministic physical behavior of the embedded security modules. After the insertion of the security module the complete adapted VHDL code is resynthesized again. In Section 8.4.9 we will detail this secure (re-)synthesis step.

Depending on the stage of the design process, more detailed information about the design can be included in the graph representation and considered in the subsequent security analysis. Thus, while potential side-channel vulnerabilities can be vague at the beginning of the design process, they can become more concrete at later stages, and can automatically be verified as soon as there is an actual implementation available. Additionally, the designer is informed about potential vulnerabilities and can thus take them into account at an early stage in the design process.

In the following we first detail the security analysis, which is used to extract leakage hypotheses. Then describe the countermeasure suggestion method in detail and their automatic insertion into the original design. Since the AMASIVE concept is general and can be applied in the context of both

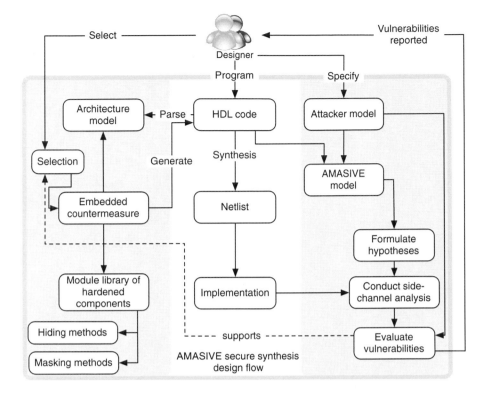

Figure 8.8
Overview on the AMASIVE workflow

different architectures and design tools, we first discuss it from a high-level point of view as depicted in Figure 8.8 and then present a detailed instantiation using VHDL code as the information source.

8.4.5   Automated Side Channel Analysis

AMASIVE identifies side channel vulnerabilities from three steps as depicted in Figure 8.9. First, various information is collected from an arbitrary set of sources, such as the implementation of a cryptosystem or its designer (Figure 8.9(a)). Secondly, this information is combined into a graph, which represents both the design and an attacker model, which specifies the attacker's capabilities (Figure 8.9(b)). Lastly, both the graph and the attacker model, i.e., a specification of the assumed capabilities of an attacker, are used to identify and to highlight side-channel vulnerabilities directly in the design (Figure 8.9(c)). The results from this analysis phase can be directly used either for an actual side-channel attack or in the countermeasure selection phase.

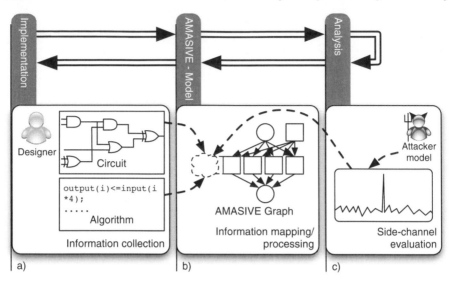

Figure 8.9
Architecture of the AMASIVE framework

The graph and the attacker model are the basic parts of the security analysis and represent the relation between the device and the attacker in the real world. In the following, we start with a general overview on the information collection, the graph and the attacker model, and the concept of the security analysis. A more detailed presentation on the realization of these phases is given in Section 8.4.8.

8.4.6   Information Collection

In the information collection phase the basic data that are required for the vulnerability analysis are gathered from various sources. The sources, from which the information are gathered, depend on the type of the implementation and the design tools being used. For example, for a VHDL design the required information can be obtained from the VHDL source code, the simulated design, and the synthesised netlist. Additionally, we require the designer to specify the envisaged capabilities of the attacker, such as his available side-channel attacks, the number of traces that can be measured, and the computational power at hand. Templates for attacker models featuring default values for such capabilities are provided too.

We structure the type of information that we require for the subsequent analysis into two categories: structural and functional information. Structural information describes the information flow of the algorithm as well as the interconnection of the modules in a hardware design. The structural information is used to construct the graph representation of the algorithm during the graph representation phase. In contrast, functional information describes the operations that are performed when processing the data. The functional information is used to evaluate the complexity and to generate an adequate side-channel hypothesis during the vulnerability analysis. Note that both types of information can be provided at a different granularity, thus allowing the vulnerability analysis to vary in precision. For instance, if the VHDL code is already written but not directly synthesizable in full detail, at least the interconnection of the components can already be read and potential information leaks can be highlighted. When the design is synthesizable and a target platform is available, the design can be implemented and a complete side-channel analysis, using the pre-determined information leaks, can be performed.

8.4.6.1 Models

In order to make the security analysis independent of the underlying device, we use the collected information to construct two abstract models. The first model specifies the capabilities of the attacker and can be changed independently of the envisaged implementation. The second model is a graph representation of the design that models the control and data-flow of the implementation at hand. Both models interact during the subsequent security analysis in order to determine potential side-channel vulnerabilities.

8.4.6.2 Graph representation model

The AMASIVE graph as outlined in Figure 8.9(b) models the device in terms of its architectural and algorithmic features and is introduced as the basis for both the security analysis and the communication with the designer.

The graph G expresses a relation between a set of nodes V, which represent execution modules, and a set of edges E, which connect nodes. Nodes can have input edges that specify the origin of the processed data and output edges that specify the resulting values. Since the AMASIVE graph is used as the foundation for a later side-channel evaluation, we define it such that it contains all relevant information for the analysis. That is, we model how the values are processed, at which steps in the algorithm a secret is being processed, and at which instructions during the execution secret values are stored. Accordingly, we distinguish between three kinds of nodes: *operations*, *entropy sources*, and *registers*. Operations describe modules, which modify the processed data, such as an S-box module or an XOR gate. Entropy sources model secret values, inserted during the execution of the algorithm, i.e., keys or random numbers. Lastly, registers are elements in an architecture that store data and thus pose a potential point of data leakage. Additionally, the start nodes and the end nodes of the graph are marked in order to identify the input to and the output of an algorithm implemented in hardware.

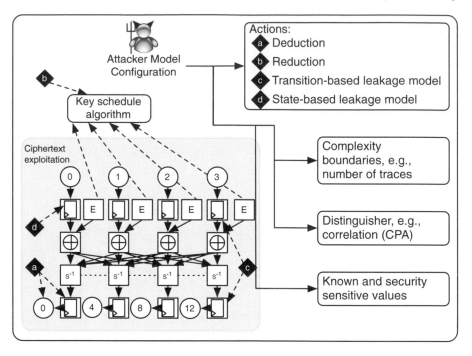

Figure 8.10
Scheme of the attacker model configuration

8.4.6.3   Attacker model

We introduce an attacker model in order to evaluate the security of a device.
   This model represents an attacker, who aims at recovering security sensitive information. Various features can be assigned to such an attacker model.
Figure 8.10 depicts an example for the attacker model as well as its assignable
features. We assign a local view $\mathbb{V} \subseteq (V \times \mathbb{R})$ to the attacker model that sets
all nodes of the previously introduced graph $G = (V, E)$ in relation to an
entropy level, which in turn specifies the uncertainty of the attacker regard-
ing the value of the node. Furthermore, we create a set $\mathbb{S} \subseteq \mathcal{P}(V)$, consisting
of sets of security sensitive nodes, which may lead to a key recovery when
they are known to the attacker. With the known and security sensitive nodes
marked, the attacker model is aware of both the starting points and the target
points in the graph. However, the attacker model still requires a set of actions
$\mathbb{A}$ for traversing the graph. These actions represent the performable attacks
of the attacker model. An action $a \in \mathbb{A}$ is defined as a function $a : \mathbb{V} \mapsto \mathbb{V}$.
A requirement function $R_G : (\mathbb{V}, \mathbb{A}) \mapsto \{0, 1\}$ checks if the requirements
for performing an action $a \in \mathbb{A}$ in the attacker view $\mathbb{V}$ are satisfied. R_G is

defined as

$$R_G(\mathbb{V}, a) = \begin{cases} 1 & \text{, if requirements hold for } \mathbb{V} \\ 0 & \text{, else.} \end{cases} \tag{8.2}$$

The strength of the attacker can be increased or decreased by adding or removing known nodes and actions, or by changing the complexity boundaries. Note that the attacker model is independent of the constructed graph and can be varied without requiring any change in the graph.

8.4.7 Security Analysis

We express the security analysis as a game where the goal of the attacker is to recover a set of security sensitive nodes from the graph $G = (V, E)$ according to our preliminary work [464]. The attacker is given a set of initially known nodes $\mathfrak{I}$ by adding these nodes with an entropy of zero to his view $\mathbb{V}$:

$$\mathbb{V} = \bigcup_{\forall n \in \mathfrak{I}} \{(n, 0)\} \tag{8.3}$$

Every node n, contained in the set $(V \backslash \mathfrak{I})$, is added to $\mathbb{V}^+$ together with its respective number of bits n_b as entropy:

$$\mathbb{V}^+ = \mathbb{V} \cup \bigcup_{\forall n \in (V \backslash \mathfrak{I})} \{(n, n_b)\} \tag{8.4}$$

Subsequently, the attacker tries to apply actions in order to recover information about further nodes. Algorithm 8.1 details the security analysis game. The attacker wins the game if the complexity for guessing a group of security sensitive nodes is lower than his pre-defined computational complexity bound. If the attacker wins the game, the designer is notified by highlighting the recovered nodes and by visualizing the corresponding sequence of actions that led to the node's recovery.

8.4.8 Diagnosis Framework

In the following we detail an implementation of AMASIVE, which analyzes FPGA designs represented in VHDL code. First, we describe the sources from which we collect the required information for the analysis. Then we outline the construction of the graph. Lastly, we detail the attacker model and the subsequent steps performed in the security analysis.

8.4.8.1 Information collection

The information we need from the hardware design is the structure of the algorithm and a functional description of the operations as denoted in the VHDL code of the module at hand. Also, in order to specify the attacker, we

---

Algorithm 8.1 AMASIVE security analysis.

Require: all nodes $n \in V$ of the AMASIVE graph $G = (V, E)$ are contained in the view $\mathbb{V}$
Require: set of security sensitive nodes $S \subseteq \mathcal{P}(V)$
Require: ordered list of N actions to traverse the graph $\mathbb{A} = \{a_1^G, a_2^G, ..., a_N^G\}$ with corresponding requirement function $R_G : (\mathbb{V}, \mathbb{A}) \mapsto \{0, 1\}$
Require: computational complexity boundary C
  1: $i = 1$
  2: **repeat**
  3:     **if** $R(\mathbb{V}, a_i^G) = 1$ **then**
  4:         $\mathbb{V} = a_i^G(\mathbb{V})$
  5:         $i = 1$
  6:     **else**
  7:         $i = i + 1$
  8:     **end if**
  9: **until** $i > N$
10: Attacker wins if $\exists \; \{s_1, s_2, ..., s_p\} \in S$ with $(s_j, \epsilon_j) \in \mathbb{V}$ such that $(\sum_{j=1}^p \epsilon_j) < C$

---

need information about the attacker model. In order to obtain this information, we require the following sources: the VHDL source code, its simulation results, and the designer.

The VHDL source code is parsed in order to extract the structure of the algorithm. Furthermore, we define constraints for the VHDL source code, in order to increase the readability of the graph. Such constraints are, for instance, that each operation node in the graph is described in a separate VHDL entity. We also introduce comment-based markers aimed for a direct communication of the designer with the VHDL parser. By using such a marker the designer can, e.g., enter names for nodes or define the type of a VHDL entity.

The AMASIVE framework is able to map the complete functionality of the investigated algorithm to the graph model. Therefore, simulation procedures and the tools from the Xilinx, Inc. design suite ISE are utilized to generate a lookup table, which describes the functionality of each entity. After the functionality of the VHDL entity is provided for each node in the graph, the graph is able to mimic the complete implemented algorithm. Thus, the framework is able to directly simulate the complete algorithm on top of commercial VHDL simulation tools such as ISIM at a functional level even without detailed information about the current implementation.

We only query the designer if information cannot be obtained autonomously for the given code. In the current version, we ask the designer to specify the attacker model by stating the public and security sensitive values, by specifying the leakage model, and by defining the security constraints. After collecting the information we compose it using an XML scheme. In this scheme we define four different elements: a register, an operation, an entropy source, and a channel. Each element is attributed by general features such as a name, an identifier, a bit size, and an entropy level.

8.4.8.2 Constructing the models

In order to construct the models, we first need to instantiate all elements of the XML description as nodes. The instantiation ensures that each implemented module is also represented in the graph. When each entity is instantiated, loops in the VHDL code are unrolled in order to ease the subsequent synthesis step. From the instantiation of the entities we obtain a node for every execution of a module on the FPGA or on the ASIC envisaged as the implementation platform.

In the next step the nodes are connected using the channel elements. The connected nodes result in a graph that represents the steps in the execution of the algorithm. Subsequently, permutations are dissolved, since on an FPGA they are represented as a permutation of wires in contrast to ASIC platforms. The dissolving is done by using the permutation lookup table, gained from the simulation of the VHDL design, and connecting the outputs of the previous and the inputs of the next elements of the permutation.

The last step in the process of constructing the graph is the assignment of non-trivial features to the nodes. These attributes describe features that have to be computed from a combination of directly readable features. Non-trivial features are, for instance, the invertibility of functions, the bijectivity of functions, and the vulnerability of a function with respect to side-channel attacks.

8.4.8.3 Graph analysis

After the construction of the graph, the security analysis is performed. We provide the CPA (correlation power analysis) distinguisher as a first example from the class of side-channel distinguishers. An essential step of the CPA is to make appropriate leakage predictions by formulating hypotheses about exploitable intermediate values or register transitions in the algorithm, cf. [114]. The identification of a suitable hypothesis function can currently be performed for the Hamming weight (HW) model and for distance-based leakage models such as the Hamming distance (HD) model. Note that additional side-channel attacks, such as the template attack, can easily be integrated into the analysis phase of AMASIVE.

In order to start the security analysis of the graph, the attacker model has to be defined using the specifications of the information collection phase. The view of the attacker is initialized by assigning to all public nodes the entropy value zero and to all non-public nodes their corresponding bit-size as entropy values. The security sensitive nodes are assembled into groups and are then added to the set of security sensitive nodes. Subsequently, the actions for the attacker model are assigned. Currently, we have defined four actions: deduc-

---

Algorithm 8.2 Hypothesis function identification for HW model.

Require: view $\mathbb{V}$ for the graph $G = (V, E)$
Require: computational complexity boundary C
  1: Identify operation node op, where node i and entropy node e are input
      nodes, o is the output node, $\{(i, \epsilon_i), (e, \epsilon_e), (o, \epsilon_e)\} \subseteq \mathbb{V}$ holds, and $\epsilon_e > 0$
  2: **if** $\epsilon_i = 0$ and $epsilon_o > 0$ **then**
  3:     hypothesis function $\mathfrak{h} = HW(op(i, e))$
  4: **else if** $\epsilon_i > 0$ and $\epsilon_o = 0$ and op is invertible **then**
  5:     hypothesis function $\mathfrak{h} = HW(op^{-1}(o, e))$
  6: **else**
  7:     return error: no suitable hypothesis found
  8: **end if**
  9: **if** $\epsilon_e < C$ **then**
 10:     return hypothesis function $\mathfrak{h}$
 11: **else**
 12:     return error: no suitable hypothesis found
 13: **end if**

---

tion, reduction, leakage exploitation using the Hamming weight model, and leakage exploitation using the Hamming distance model. The framework exercises these actions in order to automatically analyze the resulting graph.

The simplest action, the deduction, evaluates an operation or propagates a value and sets the entropy value of the corresponding node in the attacker view to zero. The deduction requires either the input or the output of a node to be known and the node to be invertible. The reduction infers information about a node and reduces its entropy. A reduction currently requires an associated node to have a reduced complexity or an operation to have certain features such as non-surjectivity.

A power attack based on the Hamming weight model, cf. [269], sets the entropy of the corresponding node in the attacker view to the value zero. The requirement of a power attack based on the Hamming weight model is a suitable hypothesis function within the correlation power analysis. The process of finding such a function is denoted in Algorithm 8.2. A power attack based on the Hamming distance model sets the entropy values of all nodes, included in the hypothesis function, to zero. However, finding the required hypothesis function for the Hamming distance model is somewhat more complex due to the requirement of a register transition. Algorithm 8.3 depicts the procedure of finding a hypothesis function for the Hamming distance model.

The security analysis of the graph is then performed as stated in Algorithm 8.1 using the defined actions. If the attacker wins the security analysis, then the actions and nodes, which led to the recovery of the secret key, are presented to the designer. Furthermore, in order to determine the practical feasibility of the attack, a CPA is performed for each identified hypothesis

---

Algorithm 8.3 Hypothesis function identification for HD model.

Require: view $\mathbb{V}$ for the graph $G = (V, E)$
Require: computational complexity boundary C

1: Identify register transition from register node $(p, 0) \in \mathbb{V}$ to register node $(r, \epsilon_r) \in \mathbb{V}$ with entropy source $(e, \epsilon_e) \in \mathbb{V}$ along the path and $\epsilon_r, \epsilon_e > 0$
2: **if** a suitable register transition is identified **then**
3:     **for all** edges i from/to r **do**
4:         traverse graph along i until node $(n, 0) \in \mathbb{V}$ is reached
5:         add path from i to n to hypothesis function $\mathfrak{h}_t$
6:     **end for**
7:     hypothesis function $\mathfrak{h} = HD(p, \mathfrak{h}_t)$
8:     compute total entropy E of $\mathfrak{h}$ by summing up all entropy values for all entropy source nodes in $\mathfrak{h}$
9:     **if** $E < C$ **then**
10:         return $\mathfrak{h}$
11:     **end if**
12: **end if**
13: return error: no suitable hypothesis found

---

function. Hence, the CPA is used as an evaluation means to assess whether the key hypothesis via side-channel exploitable intermediate results is feasible. If the correlation of the correct hypothesis key is outstandingly higher than the correlation value of the other keys hypothesis values, then the correct hypothesis function was found. Every hypothesis function is written into a C source file to be subsequently executed by a MATLAB script that implements a CPA. The feasibility of each hypothesis function is then verified by checking whether the required number of measurements is within the complexity boundary. If the CPA succeeds within the pre-defined complexity boundary, then the node is deemed recoverable and its entropy is updated according to the results of the CPA. Otherwise, the node is deemed not recoverable, and all consecutive actions have to be re-evaluated.

When all planned power side-channel attacks have been performed and their feasibility has been evaluated, the designer is notified whether the mounted attack was successful and how high the complexity for recovering each node was. The designer can then evaluate for which nodes he should consider to implement countermeasures. Afterwards, he or she can then rerun the security analysis in order to quantify the achieved security improvement.

8.4.9 Secure (Re-)Synthesis Approach

One of the advanced features of the AMASIVE framework is the autonomous injection of countermeasures in a given VHDL code. This automated, au-

Table 8.3
Types of modules in the AMASIVE architecture graph

Node type	Component	AMASIVE identifier
Register	Register element/flip flop	*register(clk,rst,en)*
Switch	Multiplexer	*mux(sel)*
Permutation	Permutation module	*permutation*
Non-linear function	Substitution module	*non-linear,* *bilinear,invertible*
Data path	Top level module of data path	*data path*

tonomous process is achieved by the concept called *secure (re-)synthesis*, which is detailed in this section. First, the general idea of embedding secure module fragments is introduced, followed by the available countermeasures in the current version of AMASIVE. Please note that this version of the AMSIVE framework focuses on block cipher algorithms, an extension to other classes of encryption algorithms is envisaged in future work. Therefore, the presentation of both the functionality of the framework and the implementation examples is focused on the structure of block ciphers.

Before embedding countermeasures automatically by the outlined framework, the hardware structure of the data path is mapped into a unidirectional graph, which from now on is denoted as the architecture graph $G_A(V, E)$. Compared to the phrasing process for the information analysis in Section 8.4.5, $G_A(V, E)$ represents the structure of the data path and not the information and data flow within the algorithm. Thus, each node V_i of $G_A(V, E)$ represents a component and each edge denotes an interconnection between components. Being more precise, the architecture graph has four different types of nodes that represent different elementary modules of circuits that are used in a cryptographic algorithm. Table 8.3 lists all these nodes and the corresponding modules.

Currently, parameter flags in the VHDL code, the so-called AMASIVE identifiers, are used to provide the parser with the necessary information which type of node the module should be assigned to. Other additional information, provided by the *data path* AMASIVE identifier, is the top-level module in the VHDL code, which contains the complete data path. Thus, the parser can even handle complex systems based on multiple cores with separated data paths. By assigning the AMASIVE identifier to the investigating data path, the framework can autonomously generate the architecture graph of the core part of the system.

$G_A(V, E)$ is essential as a foundation in order to properly extend the circuit with side-channel resistant modules or to exchange modules or circuit components and to replace them by hardened components. After $G_A(V, E)$

Algorithm 8.4 Generating a side-channel hardened circuit.

Require: VHDL code assigned with AMASIVE identifier
 1: Parse the given design code hierarchically beginning with the data path module, which marked with the AMASIVE identifier *data path*
 2: Generate $G_A(V, E)$ based on VHDL code
 3: Manipulating $G_A(V, E)$:
 4: Select a node V, which should be manipulated by embedding a side-channel countermeasure
 5: Manipulate the graph $G_{A,Sec}(V, E) \leftarrow G_A(V, E)$ by establishing new edges E_+ to connect new embedded nodes V_+ in $G_A(V, E)$
 6: Generate new VHDL code (VHDL$_{Sec}$), based on the original VHDL code, secured VHDL macros, and the new $G_{A,Sec}(V, E)$
 7: (Re-)Synthesis the side-channel security circuit based on new VHDL$_{Sec}$

---

is manipulated by an automated countermeasure embedding method, a code generator function uses $G_A(V, E)$ to produce new synthesizable VHDL code for implementing the hardened circuit, see Algorithm 8.4.

There are several functions available in the framework for the automated exchange of the data path. The first two ones are used to insert nodes into $G_A(V, E)$ next to existing nodes or in between existing nodes. The third function is a combination of these insertions that adds more complex structures to the existing $G_A(V, E)$ and thereby manipulates the original to a large extent. The combined application in form of different orders and numbers of iterations can be applied to generate more complex structures such as mirroring the complete data path and thus allowing boolean masking of the whole data path. Algorithm 8.5 and Algorithm 8.6 denote the procedures of the first two basic functions.

These basic manipulation functions of the architecture are essential to embedded various countermeasures. Currently the AMASIVE framework supports the following generic countermeasures:

Algorithm 8.5 Add new i nodes $V_{+s,i}$ sequentially between existing nodes V_{pre} and V_{succ}.

Require: Architecture graph $G_A = (V, E)$ and node V_i that becomes the successor V_{succ} after the $V_{+s,i}$ are placed
 1: $V_{succ} \leftarrow V_i$ and V_{pre} are $\forall V$, whereas V_i is the successor node
 2: Get $\forall E_i$ between V_{succ} and V_{pre}
 3: Reroute E_i to connect V_{pre} and $V_{+s,i}$
 4: Create new edges $E_{+,i}$ to connect $V_{+s,i}$ and V_{succ}
 5: Return $G_{A,+} = (V, E) = (\{V, V_{+s,i}\}, E)$

Algorithm 8.6 Add new i nodes $V_{+p,i}$ in parallel to existing node V_i.

Require: Architecture graph $G = (V, E)$ and node V_i working in parallel to
   node $V_{+p,i}$
  1: Identify successor node V_{succ} and predecessor V_{succ} of V_i
  2: Add component to merge two edges into one edge between V_i and V_{succ}
     by using Algorithm 8.5
  3: Create new edge E_+ to assign the new nodes $V_{+p,i}$ to the graph with
     predecessor nodes V_{pre} and V_{succ}
  4: Return $G_{A,+} = (V, E) = (\{V, V_{+p,i}\}, E)$

- Random register switching/pre-charching

- Component masking

- Boolean masking of the data path.

For instance, Algorithm 8.5 describes a sequential placement for positioning the XOR operations for embedding boolean component masking in the circuit. After the application of Algorithm 8.5 to the two XOR components, Algorithm 8.6 is used to position a copy of the masked function in parallel to the original one for calculating the correction term, see center illustration of Figure 8.11. Other countermeasures can be implemented by using combinations of these two algorithms and side-channel hardened components.

All these countermeasures can be embedded fully automatically in the original data path without manually writing any VHDL code segment. The countermeasures can be applied to the original circuit description at considerably different levels of granularity. For instance, the designer can choose between masking all registers or only every second register and then to evaluate the resulting drop in security. Thus, she or he is able to vary the tradeoff between resource consumption and security level. The transformation from the standard component to the hardened version as supported in the current framework is visualized in Figure 8.11. Each of the countermeasures needs a source of entropy, which can be supported by an external source or by an additional VHDL component. When using an external source, an additional port at the top-level module will be generated, in the other case currently an LFSR is added to the design and properly connected. The hardened modules are saved as VHDL code in the countermeasures data base, thus allowing a more flexible application to different implementation platforms.

8.4.10   Evaluation Results for Block Cipher PRESENT

In this section we demonstrate the analysis and the hardening of the block cipher PRESENT [51] by exercising the AMASIVE framework. For the sake of understandability, we focus on the analysis of the first and the last round only. Figure 8.12(a) depicts the block diagram of a PRESENT implementation, which processes a 64-bit input message and a 80-bit key. In each round this encryption algorithm separates the 64-bit wide input into 4-bit blocks and XORes them with 4-bit blocks of the 64-bit round key. The result of this operation is then stored in a register, from where it is substituted and permutated. In total, PRESENT performs 31 rounds as well as an additional last round, where the last round key is added, and the result is stored in a register.

8.4.10.1 Information collection

The workflow starts with the analysis of PRESENT by parsing its VHDL code and by producing an XML description, which contains the instantiated modules AddRoundKey, S-box, P-layer, and their communication channels. We obtain the following elements in the XML description: two operations (AddRoundKey (ARK) and S-box), a permutation (P-layer), a register (Reg), an entropy source (Entropy), and the channels, which connect these elements.

Next, we simulate the VHDL code of the design in order to yield a functional description of the operations. Thus, we obtain a 16×16 lookup-table

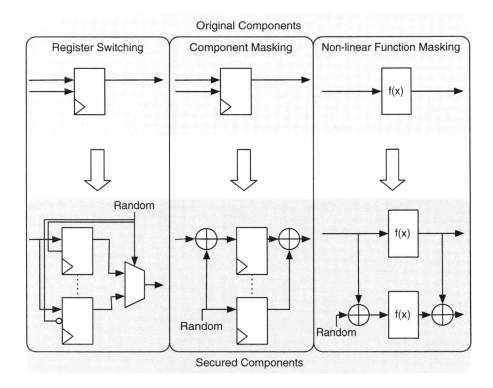

Figure 8.11
Component transformation for secure synthesis

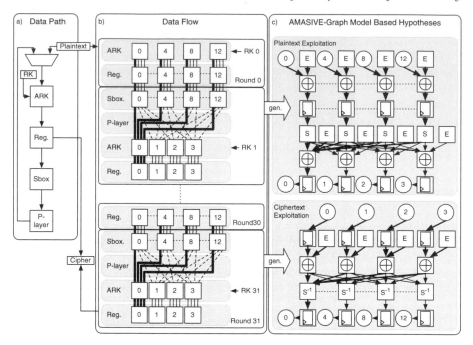

Figure 8.12
PRESENT block cipher represented as an AMASIVE graph

for the S-box and a 256×16 lookup-table for the ARK operation. The permutation P-layer is stored in a 64×2 lookup-table.

Finally, we define our attacker model. We assume the input and output of the algorithm to be known, since this is a common scenario in side-channel analysis. Subsequently, we mark all Entropy elements as security sensitive and assume that the knowledge of all Entropy elements of each round leads to the recovery of the secret key. We allow our attacker to be able to run a brute-force attack, i.e., to measure power for every single possible input value, up to a complexity of 2^{32} and to perform up to 10,000 measurements. Then, we grant the attacker the ability to execute a deduction, a reduction, and a CPA using the Hamming distance model.

8.4.10.2   Model building

In order to construct the graph, we instantiate each XML element the number of times it will be executed. The numbers of instantiations of each element for the PRESENT example are denoted in Table 8.4. Subsequently, we connect the channels to the nodes and dissolve the permutations using the P-layer lookup-table. A fragment of the resulting graph is visualized in Figure 8.12. This figure details the processing of four 4-bit blocks for the first and the last round. Note that the total graph for the first and the last round consists of four times the graph depicted in Figure 8.12(b).

Next, we evaluate non-trivial features of the operations. The S-box is determined to be bijective and invertible. The ARK operation, which is an XOR element between two nodes, is determined to be non-injective and invertible in case that at least one of both input nodes and the output node are known.

8.4.10.3 Security analysis of the unprotected implementation

After generating the graph, we perform the security analysis utilizing the pre-defined attacker model. Figure 8.12(b) depicts the first and last round of the resulting graph and can be referred to for the visualization of the security analysis. Since we have multiple nodes of one type in each round, we denote the i-th node n of round d as n_i^d. In the following we denote the ARK nodes as x, the S-box nodes as s, the Reg nodes as r, the Entropy nodes as e, the Input nodes as i, and the Output nodes as o. Also, we use their corresponding capital letters, together with a round index, to denote the set of all associated nodes of a round.

We begin the analysis by initializing the view $\mathbb{V}$ of the attacker with the input nodes *Input* and output nodes *Output*:

$$\mathbb{V} = \bigcup_{\forall n \in (Input \cup Output)} \{(n, 0)\}$$

The remaining nodes are added to $\mathbb{V}$ with their corresponding bit-size as entropy. Then we add the Entropy nodes to the security sensitive nodes by grouping all nodes $e^d \in V$ of one round into:

$$S = \bigcup_{0 \leq d < 32} \{\{e_0^d, e_1^d, ..., e_{15}^d\}\}$$

Finally, we determine and prioritize the available actions. The priority of the actions is: deduction ded_G, reduction red_G, and the CPA using the Hamming distance leakage model HD_G.

After the initialization we start the security analysis of the AMASIVE graph. The first deduction ded_G on $\mathbb{V}$ yields:

$$\mathbb{V}^+ = ded_G(\mathbb{V}) = \{(n, \epsilon) \in \mathbb{V} | n \notin R^{31}\} \cup (R^{31} \times 0)$$

In other words, we can simply recover the value of the last round's Reg nodes $R^{31} = \{r_0^{31}, r_1^{31}, ..., r_{15}^{31}\}$ by propagating the value of the output. We set

Table 8.4
Numbers of resulting nodes in the AMASIVE graph

XML elements	ARK	S-box	P-layer	Reg	Entropy	Channel
initially	-	-	-	16	-	16
per round	16	16	1	16	16	64
last round	16	-	-	16	16	48
Total	512	496	31	528	512	2048

$\mathbb{V} = \mathbb{V}^+$ and since ded_G successfully recovered some nodes, we can again perform ded_G on the updated $\mathbb{V}$, yielding:

$$\mathbb{V}^+ = ded_G(\mathbb{V}) = \{(n, \epsilon) \in \mathbb{V} | n \notin X^{31}\} \cup (X^{31} \times 0)$$

Thus, the second ded_G yields the value of the ARK nodes of the last round $X^{31} = \{x_0^{31}, x_1^{31}, ..., x_{15}^{31}\}$ by simply propagating the value from the recovered Reg nodes R^{31}.

We again set $\mathbb{V} = \mathbb{V}^+$ and observe that we cannot apply ded_G to $\mathbb{V}$, since no value can be propagated and no operation may be evaluated. Thus, we verify the requirements for the reduction and also observe that it can not be applied to $\mathbb{V}$. Thus, we verify the requirement for the CPA using the Hamming distance model HD_G. Indeed, by applying Algorithm 8.3 to $\mathbb{V}$ we obtain hypotheses for all nodes in E^0 and E^{31}, which are visualized in Figure 8.12(c). In the following we use the short notation $n_{j,b}$ to refer to the b-th most significant bit of n_j.

The hypothesis function $\mathfrak{h}_{e_0^0}$ for element $e_0^0 \in E^0$ is given by:

$$\mathfrak{h}_{e_0^0} = HD(r_0^0, r_0^1)$$
$$= HD(S(i_0 \oplus e_0^0), (x_{0,0}^1 || x_{1,0}^1 || x_{2,0}^1 || x_{3,0}^1))$$

where $x_{j,b}^1 = S(i_j \oplus e_j^0)_b \oplus e_{j,b}^1$ for $0 \leq j < 16$. Analyzing the hypothesis function of e_0^0 shows that we also have to make hypotheses for e_1^0, e_2^0, e_3^0, and e_0^1 too. This raises the complexity of a CPA to $(2^4)^5 = 2^{20}$, which still lies in the predefined complexity boundary of 2^{32}. Thus, we obtain a successful Hamming distance hypothesis and are able to set the entropy values for the nodes e_0^0, e_1^0, e_2^0, e_3^0, and e_0^1 to zero.

The hypothesis function $\mathfrak{h}_{e_0^{31}}$ for $e_0^{31} \in E^{31}$ is:

$$\mathfrak{h}_{e_0^{31}} = HD(r_0^{30}, r_0^{31})$$
$$= HD(S^{-1}(s_{0,0}^{31} || s_{4,0}^{31} || s_{8,0}^{31} || s_{12,0}^{31}), r_0^{31}))$$

where $s_{j,b}^{31} = o_{j,b} \oplus e_{j,b}^{31}$. In contrast to the hypothesis $\mathfrak{h}_{e_0^0}$, $\mathfrak{h}_{e_0^{31}}$ only requires a complexity of 4 bits, making an attack more likely to succeed with a smaller number of measurements.

Since we found a hypothesis for HD_G, we can update $\mathbb{V}$ as follows:

$$\mathbb{E} = E^0 \cup E^1 \cup E^{31}$$
$$\mathbb{V}^+ = HD_G(\mathbb{V}) = \{(n, \epsilon) \in \mathbb{V} | n \notin \mathbb{E}\} \cup (\mathbb{E} \times 0)$$

We use ded_G again on the updated $\mathbb{V}$, which would result in the recovery of X^0 and X^{31}. However, we stop the analysis at this point, since we have collected sufficient information. The current set of $\mathbb{V}$ contains the round keys of the first, second, and last round. Due to the key-schedule of PRESENT, using one round key, we can already recover the actual key by guessing the 2^{16} remaining bits.

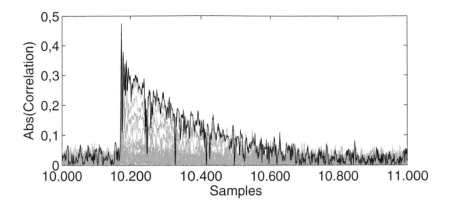

Figure 8.13
CPA on PRESENT with hypothesis $\mathfrak{h}_{e_0^{31}}$

8.4.10.4 Power analysis attack on the unprotected PRESENT

However, in order to verify the practicability of the attack, we performed a CPA using the hypotheses $\mathfrak{h}_{e_i^{31}}$ on an FPGA mounted on a SASEBO-II board, which executes the outlined PRESENT implementation. We exercised 10,000 measurements and used the Pearson correlation coefficient as comparison method between the hypotheses and power traces, cf. [55]. The resulting correlation of the attack on the last round of PRESENT using $\mathfrak{h}_{e_0^{31}}$ is depicted in Figure 8.13, which visualizes the absolute values of the correlation coefficient over the samples, and confirms the correctness of the hypothesis. Samples represent points in time of the monitored power consumption of the PRESENT encryption operations, which are reflected in each of the captured power traces. After 3400 captured traces the complete round key may be recovered and thus the secret key of the block cipher can be revealed with an computational effort of just 2^{16}.

8.4.10.5 Automatic secure synthesis of PRESENT

In order to verify the security gain of the available countermeasures, we generated hardened versions of the PRESENT implementation and repeated the attack. We used the same input parameters (plaintext) for all attacks, as well as the same secret key. In total we recorded 100,000 traces from each side-channel hardened design implemented and running on the SASEBO-GII platform. These traces were then evaluated with the same hypotheses, $\mathfrak{h}_{e_0^{31}}$, which where previously applied to the unprotected version.

We evaluated four different side-channel hardened designs, which were secured by applying different countermeasures to the original design. At first only a simple register pre-charging by random register switching for the

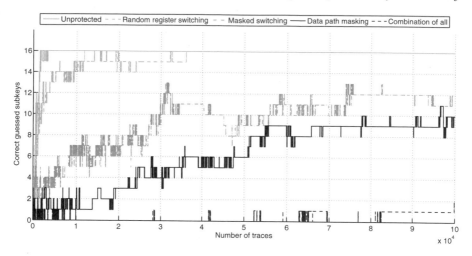

Figure 8.14

CPA on side-channel hardened PRESENT with hypothesis $\mathfrak{h}_{e_0^{31}}$

round state of PRESENT was implemented, as depicted on the left hand side of Figure 8.11. This countermeasure slightly increased the resistance against a CPA attack using $\mathfrak{h}_{e_0^{31}}$. All subkey nibbles were revealed with a ten times higher effort[3]. The second design masks the registers with an XORed random value before the state is stored (cf. centered design in Figure 8.11). By means of this countermeasure, only 12 of 16 nibbles of the subkey were revealed using all 100,000 traces. In the third design, a boolean masking was implemented on the whole data path. Please keep in mind that the embedding of all these countermeasures was done in a completely automatic manner by the advocated framework. Now, the third countermeasure increased the security even more, causing only 10 of 16 nibbles of the subkey to be identified using all 100,000 traces. For the last design we combined all mentioned countermeasures. Now, just 2 nibbles of the subkey were revealed using 100,000 traces. These evaluation results are clearly visible from Figure 8.14, which depicts the effect-effort relationship for the unprotected and the hardened encryption modules, respectively. Here, we quantify the effect by the number of revealed subkeys (0 to 16) and the effort by the amount of traces (up to 10^5) needed to guess them correctly.

This practical evaluation clearly demonstrates that the AMASIVE framework is able to autonomously generate hardened designs. In addition, the resource-security tradeoff for side-channel hardening, given a design as a starting point, can be adjusted according to the security scenario at hand. The security evaluation is based on the algorithmic information flow and thus as-

---

[3]34000 traces were needed instead of just 3400

sures that an appropriate hypothesis for attacking the design is found. Due to the modularized structure and the usage of the (re-)synthesize concept, the AMASIVE framework is a very flexible tool set for securing devices to be eventually implemented on quite different platforms early in the design phase.

8.4.11 Acknowledgements

The work presented in this contribution was supported by the German Federal Ministry of Education and Research (BMBF) in the project RESIST through grant number 01IS10027A. The fourth author is supported by the German Federal Ministry of Education and Research (BMBF) within EC SPRIDE and by the Hessian LOEWE excellence initiative within CASED. We would like to thank André Seffrin for providing us with his VHDL parser, Adeel Israr for his contribution to implementing efficient parsing methods, and Oliver Stein for his substantial contributions to security analysis. We also like to thank Annelie Heuser for fruitful discussions on suitable leakage models developed within the RESIST project.

9

Remarks and Conclusion

Farhana Sheikh

Portland, Oregon

Leonel Sousa

University of Lisbon

CONTENTS

Security and privacy will continue to increase in importance as we move into the next generation of technology innovations where each person will be immersed in a world of sensors, actuators, and mobile connectivity. The collection, aggregation, and communication of large amounts of personal and non-personal data requires increasingly level of protection from malicious attacks.

Security and privacy are important values and developing ways to assure them in an era of electronic devices is paramount, especially in the field of communication and information systems. Cryptography has been the key in supporting the design of secure systems, both at hardware and software levels. As presented in this book, different types of cryptographic primitives and functions have been proposed and used in the development of secure devices and to enable secure sharing and communication of data and information. However, it is also evident that the main challenge in designing secure systems, rather than their cryptographic algorithms, is their implementation. The right design decisions can lead to implementations that are less prone to attacks, which became more and more sophisticated with the advances in technology and the growth of the available computational power.

This book gives a holistic perspective on the design of systems that provide security and privacy. These systems rely on the formulation of the cryptographic algorithms, designed based on dedicated or more general purpose architectures, and implemented on circuits and systems with the current technology. For those implementations, not only figures of merit such as the cost, throughput, and power consumption (especially important for mobile systems) are considered, but also their resistance to attacks such as side-channel attacks. The material presented and discussed in the various chapters are fundamental to unveil the details required for designing secure systems.

The background, algorithms, circuits, and the architectures presented along the chapters of this book provide a solid basis and practical hints to design highly provable secure systems. However, to design secure systems in the future, one should be continuously aware of the evolution of computer systems and cryptographic algorithms. For instance, should quantum computing become a viable option, the classical approaches to asymmetric cryptography such as ECC and RSA would be broken, and consequently the services provided by them. Towards preventing this, alternative several cryptosystems have started being proposed. Lattice-based Cryptography (LBC) is such a system [341]. LBC shows much promise where other post-quantum cryptosystems lack the variety of applications that it can be developed for, while enabling a number of provably secure cryptosystems.

Bibliography

[1] M. Agrawal, N. Kayal, and N. Saxena. Primes is in p. *Annals of Mathematics*, 160(2):781–793, 2004.

[2] Martin Aigner. *A Course in Enumeration*. Springer Verlag, New York, 2007.

[3] A. Al-Rawi, A. Lansari, and F. Bouslama. A new non-recursive algorithm for binary search tree traversal. In *Electronics, Circuits and Systems, 2003. ICECS 2003. Proceedings of the 2003 10th IEEE International Conference on*, volume 2, pages 770–773, 2003.

[4] J. L. Alperin and Rowen Bell. *Groups and Representations*. Springer Verlag, New York, 1995.

[5] P. Amberg, N. Pinckney, and D.M. Harris. Parallel high-radix Montgomery multipliers. In *Signals, Systems and Computers, 2008 42nd Asilomar Conference on*, pages 772–776, 2008.

[6] Fatima Amounas. A novel approach for enciphering data based ECC using catalan numbers. *International Journal of Information & Network Security*, 2(4):339–347, August 2013.

[7] James Anderson and James Bell. *Number Theory with Applications*. Prentice Hall, Upper Saddle River, 1997.

[8] Titu Andreescu and Dorin Andrica. *Number Theory—Structure, Examples, and Problems*. Springer Verlag, New York, 2009.

[9] Titu Andreescu, Dorin Andrica, and Ion Cucurezeanu. *An Introduction To Diophantine Equations*. Springer Verlag, New York, 2010.

[10] S. Antão, J.-C. Bajard, and L. Sousa. Elliptic curve point multiplication on gpus. In *IEEE International Conference on Application-specific Systems Architectures and Processors - ASAP*, pages 192–199. IEEE, 2010.

[11] S. Antão, J.-C. Bajard, and L. Sousa. RNS based elliptic curve point multiplication for massive parallel architectures. *The Computer Journal 2011 - Oxford Journals*, 55(5):629–647, 2011.

[12] S. Antão, R. Chaves, and L. Sousa. AES and ECC cryptography processor with runtime configuration. In *International Conference on Advanced Computing and Comunications — ADCOM*, pages 1–8. IEEE, 2009.

[13] S. Antão, R. Chaves, and L. Sousa. Compact and flexible microcoded elliptic curve processor for reconfigurable devices. In *IEEE Symposium on Field Programmable Custom Computing Machines — FCCM*, pages 193–200. IEEE, 2009.

[14] S. Antão and L. Sousa. The crns framework and its application to programmable and reconfigurable cryptography. *ACM Transactions on Architecture and Code Optimization*, 9(4), 2013.

[15] ARM Architecture Group. ARMv8 instruction set overview. Tech. rep., ARM Limited, 2011.

[16] Mark Armstrong. *Groups and Symmetry*. Springer Verlag, New York, 2010.

[17] Michael Artin. *Algebra*. Addison-Wesley, Boston, 2010.

[18] Benno Artmann. *Euclid: The Creation of Mathematics*. Springer Verlag, New York, 1999.

[19] Avner Ash and Robert Gross. *Elliptic Tales: Curves, Counting, and Number Theory*. Princeton University Press, Princeton, 2012.

[20] Robert Ash. *Information Theory*. Dover Publications, New York, 1991.

[21] Avner Asher and Robert Gross. *Fearless Symmetry: Exposing the Hidden Patterns of Numbers*. Princeton University Press, Princeton, 2006.

[22] Kubilay Atasu, Luca Breveglieri, and Marco Macchetti. Efficient AES implementations for ARM based platforms. In Hisham Haddad, Andrea Omicini, Roger L. Wainwright, and Lorie M. Liebrock, editors, *Proceedings of SAC*, pages 841–845. ACM, 2004.

[23] Jean-Philippe Aumasson, Luca Henzen, Willi Meier, and Raphael C.-W. Phan. SHA-3 proposal BLAKE: Submission to NIST (Round 3). http://ehash.iaik.tugraz.at/wiki/BLAKE, 2010.

[24] Christian Aumüller, Peter Bier, Wieland Fischer, Peter Hofreiter, and Jean-Pierre Seifert. Fault Attacks on RSA with CRT: Concrete Results and Practical Countermeasures. In B. Kaliski, Ç. Koç, and C. Paar, editors, *Proceedings of the 4th International Workshop on Cryptographic Hardware and Embedded Systems: CHES 2002*, volume 2523 of *Lecture Notes in Computer Science*, pages 260–275. Springer-Verlag, 2002.

[25] J.-C. Bajard, L.-S. Didier, and P. Kornerup. Modular multiplication and base extensions in residue number systems. In *IEEE Symposium on Computer Arithmetic - ARITH*, pages 59–65. IEEE, 2001.

[26] Josep Balasch, Benedikt Gierlichs, Roel Verdult, Lejla Batina, and Ingrid Verbauwhede. Power analysis of atmel cryptomemory — recovering keys from secure eeproms. In Orr Dunkelman, editor, *Proceedings of the 12th conference on Topics in Cryptology - CT-RSA 2012*, volume 7178 of *Lecture Notes in Computer Science*, pages 19–34. Springer-Verlag, 2012.

[27] Maria Welleda Baldoni, Ciro Ciliberto, and Giulia Maria Piacentini Cattaneo. *Elementary Number Theory, Cryptography and Codes*. Springer Verlag, New York, 2008.

[28] B. Baldwin, A. Byrne, M. Hamilton, N. Hanley, R. P. McEvoy, W. Pan, and W. P. Marnane. FPGA implementations of SHA-3 candidates: CubeHash, Groestl, LANE, Shabal and Spectral Hash. In *Symposium on Digital Systems Design (DSD 2009)*, pages 783–790. Euromicro, 2009.

[29] E. Barker and J. Kelsey. Recommendation for Random Number Generation Using Deterministic Random Bit Generators, NIST Special Publication 800-90A. Online. Available at: http://csrc.nist.gov/publications/nistpubs/800-90A/SP800-90A.pdf, 2012.

[30] E. Barker and J. Kelsey. Recommendation for the Entropy Sources Used for Random Bit Generation, NIST Special Publication 800-90B. Online. Available at: http://csrc.nist.gov/publications/drafts/800-90/draft-sp800-90b.pdf, 2012.

[31] Paul Barrett. mplementing the rivest shamir and adleman public key encryption algorithm on a standard digital signal processor. In *Advances in Cryptology—CRYPTO '86 Proceedings*, volume 263 of *Lecture Notes in Computer Science*, pages 311–323, 1987.

[32] J. Barth, D. Plass, E. Nelson, C. Hwang, G. Fredeman, M. Sperling, A. Mathews, T. Kirihata, W.R. Reohr, K. Nair, and Nianzheng Caon. A 45 nm SOI embedded DRAM macro for the POWER^{TM} processor 32 MByte on-chip L3 cache. *Solid-State Circuits, IEEE Journal of*, 46(1):64–75, Jan 2011.

[33] D.W. Bauder. An Anti-Counterfeiting Concept for Currency Systems. Research Report PTK-11990, Sandia National Laboratories, Albuquerque, NM, 1983.

[34] M. Baudet, D. Lubicz, J. Micolod, and A. Tassiaux. On the security of oscillator-based random number generators. *Journal of Cryptology*, 24(2):398–425, 2011.

[35] Ali Galip Bayrak, Francesco Regazzoni, Philip Brisk, François-Xavier Standaert, and Paolo Ienne. A first step towards automatic application of power analysis countermeasures. In Leon Stok, Nikil D. Dutt, and Soha Hassoun, editors, *DAC*, pages 230–235. ACM/IEEE, 2011.

[36] A. Bernal and A. Guyot. Design of a modular multiplier based on Montgomery's algorithm. In *Proc. 13th Intl. Conf. Design of Circuits and Integrated Systems*, pages 680–685, 1998.

[37] Daniel J. Bernstein. AES speed reports. http://cr.yp.to/aes-speed.html.

[38] Daniel J. Bernstein and Peter Schwabe. New AES software speed records. In Dipanwita Roy Chowdhury, Vincent Rijmen, and Abhijit Das, editors, *Proceedings of INDOCRYPT*, volume 5365 of *Lecture Notes in Computer Science*, pages 322–336. Springer, 2008.

[39] G. Bertoni, J. Daemen, M. Peeters, and G. V. Assche. On the indifferentiability of the sponge construction. In *Advances in Cryptology — EUROCRYPT 2008*, pages 181–197. Springer, 2008.

[40] G. Bertoni, J. Daemen, M. Peeters, and G. V. Assche. The keccak sha-3 submission: Submission to nist. http://ehash.iaik.tugraz.at/wiki/Keccak, 2010.

[41] Guido Bertoni, Luca Breveglieri, Pasqualina Fragneto, Marco Macchetti, and Stefano Marchesin. Efficient Software Implementation of AES on 32-Bit Platforms. In Burton S. Kaliski Jr., Çetin Kaya Koç, and Christof Paar, editors, *Cryptographic Hardware and Embedded Systems: CHES 2002*, volume 2523 of *Lecture Notes in Computer Science*, pages 159–171. Springer, 2002.

[42] Rabi Bhattacharya and Edward Waymire. *A Basic Course in Probability Theory*. Springer Verlag, New York, 2007.

[43] Swarup Bhunia, Saibal Mukhopadhyay, and Kaushik Roy. Process variations and process-tolerant design. In *International Conference on VLSI Design*, pages 699–704. IEEE, 2007.

[44] Eli Biham and Adi Shamir. Differential Cryptanalysis of DES-like Cryptosystems. In Alfred Menezes and Scott A. Vanstone, editors, *Proceedings of CRYPTO*, volume 537 of *Lecture Notes in Computer Science*, pages 2–21. Springer, 1990.

[45] Robert Bix. *Conics and Cubics*. Springer Verlag, New York, 1998.

[46] David Blaauw, Kaviraj Chopra, Ashish Srivastava, and Lou Scheffer. Statistical timing analysis: From basic principles to state of the art. *IEEE Transactions on Computer-Aided Design of Integrated Circuits and Systems*, 27(4):589–607, 2008.

[47] L. Blum, M. Blum, and M. Shub. A Simple Unpredictable Pseudo-Random Number Generator. *SIAM Journal on Computing*, 15(2):364–383, 1986.

[48] M. Blum and S. Micali. How to Generate Cryptographically Strong Sequences of Pseudo Random Bits. *SIAM Journal on Computing*, 13 (4):850–864, 1984.

[49] T. Blum and C. Paar. High-radix Montgomery modular exponentiation on reconfigurable hardware. *Computers, IEEE Transactions on*, 50(7):759–764, Jul 2001.

[50] Andrey Bogdanov, Dmitry Khovratovich, and Christian Rechberger. Biclique Cryptanalysis of the Full AES. In Dong Hoon Lee and Xiaoyun Wang, editors, *Proceedings of ASIACRYPT*, volume 7073 of *Lecture Notes in Computer Science*, pages 344–371. Springer, 2011.

[51] Andrey Bogdanov, Lars R. Knudsen, Gregor Leander, Christof Paar, Axel Poschmann, Matthew J. B. Robshaw, Yannick Seurin, and C. Vikkelsoe. PRESENT: An Ultra-Lightweight Block Cipher. In Pascal Paillier and Ingrid Verbauwhede, editors, *Cryptographic Hardware and Embedded Systems: CHES 2007*, volume 4727 of *Lecture Notes in Computer Science*, pages 450–466. Springer, 2007.

[52] Monica Borda. *Fundamentals in Information Theory and Coding*. Springer Verlag, New York, 2011.

[53] Antoon Bosselaers, René Govaerts, and Jose Vandewalle. comparison of three modular reduction functions. In *Advances in Cryptology—CRYPTO '93 Proceedings*, volume 773 of *Lecture Notes in Computer Science*, pages 175–186. Springer, 1994.

[54] E. Brier and M. Joye. Weierstraß elliptic curves and side-channel attacks. In *Public Key Cryptography*, volume 2274 of *Lecture Notes in Computer Science*, pages 183–194. Springer, 2002.

[55] Eric Brier, Christophe Clavier, and Francis Olivier. Correlation power analysis with a leakage model. In Marc Joye and Jean-Jacques Quisquater, editors, *Cryptographic Hardware and Embedded Systems: CHES 2004*, volume 3156 of *Lecture Notes in Computer Science*, pages 16–29. Springer, 2004.

[56] G. Bronner, H. Aochi, M. Gall, J. Gambino, S. Gernhardt, E. Hammerl, H. Ho, J. Iba, H. Ishiuchi, M. Jaso, R. Kleinhenz, T. Mii, M. Narita, L. Nesbit, W. Neumueller, A. Nitayama, T. Ohiwa, S. Parke, J. Ryan, T. Sato, H. Takato, and S. Yoshikawa. A fully planarized 0.25μm CMOS technology for 256 Mbit DRAM and beyond. In *VLSI Technology, 1995. Digest of Technical Papers. 1995 Symposium on*, pages 15–16. IEEE, 1995.

[57] Ezra Brown and Bruce Myers. Elliptic curves from mordell to diophantus and back. *American Math. Monthly*, 109(7):639–649, August 2002.

[58] Hannes Brunner, Andreas Curiger, and Max Hofstetter. On Computing Multiplicative Inverses in $GF(2^m)$. *IEEE Trans. Computers*, 42(8):1010–1015, 1993.

[59] James D. R. Buchanan, Russell P. Cowburn, Ana-Vanessa Jausovec, Dorothée Petit, Peter Seem, Gang Xiong, Del Atkinson, Kate Fenton, Dan A. Allwood, and Matthew T. Bryan. Forgery: 'fingerprinting' documents and packaging. *Nature*, (7050):475, 2005.

[60] Christoph Bsch, Jorge Guajardo, Ahmad-Reza Sadeghi, Jamshid Shokrollahi, and Pim Tuyls. Efficient helper data key extractor on fpgas. In *Cryptographic Hardware and Embedded Systems*, volume 5154 of *Lecture Notes in Computer Science*, pages 181–197. Springer Berlin Heidelberg, 2008.

[61] Benton H. Calhoun, Sudhanshu Khanna, Randy Mann, and Jiajing Wang. Sub-threshold circuit design with shrinking cmos devices. In *IEEE International Symposium on Circuits and Systems*, pages 2541–2544. IEEE, 2009.

[62] David Canright and Lejla Batina. A very compact "perfectly masked" S-Box for AES. In Steven M. Bellovin, Rosario Gennaro, Angelos D. Keromytis, and Moti Yung, editors, *ACNS*, volume 5037 of *Lecture Notes in Computer Science*, pages 446–459. Springer, 2008.

[63] Celine Carstensen, Benjamin Fine, and Gerhard Rosenberger. *Abstract Algebra: Applications to Galois Theory, Algebraic Geometry, and Cryptography*. De Gruyter, Berlin, 2011.

[64] J. W. S. Cassels. *An Introduction to the Geometry of Numbers*. Springer Verlag, New York, 1997.

[65] Certicom Research. Standards for efficient cryptography 1 (sec 1) : Elliptic curve cryptography, version 2. Certicom Research, 2009.

[66] Certicom Research. Standards for efficient cryptography 2 (sec 2) : Recommended elliptic curve domain parameters. Certicom Research, 2010.

[67] Alessandro Cevrero, Francesco Regazzoni, Micheal Schwander, Stéphane Badel, Paolo Ienne, and Yusuf Leblebici. Power-gated MOS Current Mode Logic (PG-MCML): A Power Aware DPA-Resistant Standard Cell Library. In Leon Stok, Nikil D. Dutt, and Soha Hassoun, editors, *DAC*, pages 1014–1019. ACM/IEEE, 2011.

[68] S. Chang, Ray Perlner, William E. Burr, Meltem Sönmez Turan, John M. Kelsey, Souradyuti Paul, and Lawrence E. Bassham. Third-round report of the SHA-3 cryptographic hash algorithm competition. NIST Interagency Report 7896, National Institute for Standards and Technology, November 2012.

[69] Suresh Chari, Josyula R. Rao, and Pankaj Rohatgi. Template attacks. In Burton S. Kaliski Jr., Çetin Kaya Koç, and Christof Paar, editors, *Cryptographic Hardware and Embedded Systems: CHES 2002*, volume 2523 of *Lecture Notes in Computer Science*, pages 13–28. Springer, 2002.

[70] Ricardo Chaves, Georgi Kuzmanov, Leonel Sousa, and Stamatis Vassiliadis. Improving sha-2 hardware implementations. In *Cryptographic Hardware and Embedded Systems: CHES 2006*, pages 298–310. Springer, 2006.

[71] Ricardo Chaves, Georgi Kuzmanov, Leonel Sousa, and Stamatis Vassiliadis. Rescheduling for optimized sha-1 calculation. In *Embedded Computer Systems: Architectures, Modeling, and Simulation*, pages 425–434. Springer, 2006.

[72] Ricardo Chaves, Georgi Kuzmanov, Leonel Sousa, and Stamatis Vassiliadis. Cost-efficient sha hardware accelerators. *Very Large Scale Integration (VLSI) Systems, IEEE Transactions on*, 16(8):999–1008, 2008.

[73] S. Chellappa, A. Dey, and L.T. Clark. Improved circuits for microchip identification using SRAM mismatch. In *Custom Integrated Circuits Conference (CICC), 2011 IEEE*, pages 1–4. IEEE, 2011.

[74] W. N. Chelton and M. Benaissa. Fast elliptic curve cryptography on fpga. *IEEE Transactions on Very Large Scale Integration (VLSI) Systems*, 16(2):198–205, 2008.

[75] Lindsay Childs. *A Concrete Introduction to Higher Algebra*. Springer Verlag, New York, 2009.

[76] Benny Chor and Oded Goldreich. An improved parallel algorithm for integer gcd. *Algorithmica*, 5(1–4):1–10, June 1990.

[77] D. Chudnovsky and G. Chudnovsky. Sequences of numbers generated by addition in formal groups and new primality and factorization tests. *Advances in Applied Mathematics*, 7(4):385–434, 1986.

[78] Henri Cohen. *Number Theory: Volume I: Tools and Diophantine Equations*. Springer Verlag, New York, 2007.

[79] Henri Cohen. *Number Theory: Volume II: Analytic and Modern Tools*. Springer Verlag, New York, 2007.

[80] Henri Cohen. *A Course in Computational Algebraic Number Theory*. Springer Verlag, New York, 2010.

[81] Henri Cohen, Gerhard Frey, Roberto Avanzi, Christophe Doche, Tanja Lange, Kim Nguyen, and Frederik Vercauteren. *Handbook of Elliptic and Hyperelliptic Curve Cryptography*. Chapman & Hall/CRC, Boca Raton, 2005.

[82] Kevin J. Compton, Brian Timm, and Joel VanLaven. A simple power analysis attack on the serpent key schedule. *IACR Cryptology ePrint Archive*, 2009:473, 2009.

[83] John Conway and Neal Sloane. *Sphere Packings, Lattices, and Groups*. Springer Verlag, New York, 1999.

[84] W. A. Coppel. *Number Theory: An Introduction to Mathematics*. Springer Verlag, New York, 2009.

[85] M. Cortez, A. Dargar, S. Hamdioui, and G.-J. Schrijen. Modeling SRAM start-up behavior for physical unclonable functions. In *Defect and Fault Tolerance in VLSI and Nanotechnology Systems (DFT), 2012 IEEE International Symposium on*, pages 1–6. IEEE, 2012.

[86] C. Costea, F. Bernard, V. Fischer, and R. Fouquet. Analysis and enhancement of ring oscillators based physical unclonable functions in FPGAs. In *Reconfigurable Computing and FPGAs (ReConFig), 2010 International Conference on*, pages 262–267, 2010.

[87] Thomas Cover and Joy Thomas. *Elements of Information Theory*. John Wiley & Sons, Hoboken, 2006.

[88] David Cox. *Galois Theory*. John Wiley & Sons, Hoboken, 2004.

[89] David Cox, John Little, and Donal O'Shea. *Using Algebraic Geometry*. Springer Verlag, New York, 2005.

[90] David Cox, John Little, and Donal O'Shea. *Ideals, Varieties, and Algorithms: An Introduction to Computational Algebraic Geometry and Commutative Algebra*. Springer Verlag, New York, 2008.

[91] Richard Crandall and Carl Pomerance. *Prime Numbers: A Computational Perspective*. Springer Verlag, New York, 2005.

[92] Paul Cull, Mary Flahive, and Robby Robson. *Difference Equations: From Rabbits to Chaos*. Springer Verlag, New York, 2005.

[93] Luigi Dadda, Marco Macchetti, and Jeff Owen. The design of a high speed asic unit for the hash function sha-256 (384, 512). In *Design, Automation and Test in Europe Conference and Exhibition, 2004. Proceedings*, volume 3, pages 70–75. IEEE, 2004.

[94] Joan Daemen and Vincent Rijmen. *The Design of Rijndael: AES — The Advanced Encryption Standard*. Springer, New York, 2002.

[95] Matthew Darnall and Doug Kuhlman. AES Software Implementations on ARM7TDMI. In Rana Barua and Tanja Lange, editors, *Proceedings of INDOCRYPT*, volume 4329 of *Lecture Notes in Computer Science*, pages 424–435. Springer, 2006.

[96] Abhijit Das. *Computational Number Theory*. CRC Press, Boca Raton, 2013.

[97] R. B. Davies. Exclusive OR (XOR) and hardware random number generators. Online. Available at: http://webnz.com/robert/, February 2002.

[98] J. Delvaux and I. Verbauwhede. Side channel modeling attacks on 65nm arbiter pufs exploiting cmos device noise. In *IEEE Symposium on Hardware-Oriented Security and Trust*, pages 137–142. IEEE, 2013.

[99] Jeroen Delvaux and Ingrid Verbauwhede. Fault injection modeling attacks on 65nm Arbiter and RO Sum PUFs via environmental changes. Cryptology ePrint Archive, Report 2013/619, 2013.

[100] C. Demi, M. Jankowski, and C. Thalmaier. A $0.13\mu m$ 2.125MB 23.5ns embedded flash with 2GB/s read throughput for automotive microcontrollers. In *Solid-State Circuits Conference, 2007. ISSCC 2007. Digest of Technical Papers. IEEE International*, pages 478–617. IEEE, 2007.

[101] A. Demir, A. Mehrotra, and J. Roychowdhury. Phase noise in oscillators: a unifying theory and numerical methods for characterization. *Circuits and Systems I: Fundamental Theory and Applications, IEEE Transactions on*, 47(5):655–674, 2000.

[102] W. Diffie and M.E. Hellman. New directions in cryptography. *Information Theory, IEEE Transactions on*, 22(6):644–654, Nov 1976.

[103] Whitfield Diffie and Martin Hellman. New directions in cryptography. *IEEE Transactions on Information Theory*, IT-22(6):644–654, November 1976.

[104] John Dixon and Brian Mortimer. *Permutation Groups*. Springer Verlag, New York, 1996.

[105] Julien Doget, Emmanuel Prouff, Matthieu Rivain, and François-Xavier Standaert. Univariate Side Channel Attacks and Leakage Modeling. *Journal of Cryptographic Engineering*, 1:123–144, 2011.

[106] Ling Dong, Kefei Chen, Mi Wen, and Yanfei Zheng. Protocol engineering principles for cryptographic protocols design. In *Software Engineering, Artificial Intelligence, Networking, and Parallel/Distributed Computing, 2007. SNPD 2007. Eighth ACIS International Conference on*, volume 3, pages 641–646. IEEE, 2007.

[107] David Dummit and Richard Foote. *Abstract Algebra*. John Wiley & Sons, Hoboken, 2003.

[108] Stephen R. Dussé and Burton S. Kaliski. A cryptographic library for the Motorola DSP56000. In Ivan Bjerre Damgård, editor, *Advances in Cryptology EUROCRYPT 90*, volume 473 of *Lecture Notes in Computer Science*, pages 230–244. Springer Berlin Heidelberg, 1991.

[109] D. Eastlake, 3rd and P. Jones. US secure Hash algorithm 1 (SHA1). RFC Editor, 2001.

[110] Wolfgang Ebeling. *Lattices and Codes*. Springer Spektrum, Berlin, 2013.

[111] Thomas Eisenbarth, Timo Kasper, Amir Moradi, Christof Paar, Mahmoud Salmasizadeh, and Mohammad T. Manzuri Shalmani. On the power of power analysis in the real world: A complete break of the KeeLoq code hopping scheme. In David Wagner, editor, *Proceedings of CRYPTO 2008*, volume 5157 of *Lecture Notes in Computer Science*, pages 203–220. Springer, 2008.

[112] Thomas Eisenbarth, Sandeep Kumar, Christof Paar, Axel Poschmann, and Leif Uhsadel. A survey of lightweight-cryptography implementations. *IEEE Design and Test of Computers*, 24(6):522–533, November 2007.

[113] Saber El Aydi. *An Introduction to Difference Equations*. Springer Verlag, New York, 2005.

[114] M. Abdelaziz Elaabid and Sylvain Guilley. Practical Improvements of Profiled Side-Channel Attacks on a Hardware Crypto-Accelerator. In Daniel J. Bernstein and Tanja Lange, editors, *AFRICACRYPT*, volume 6055 of *Lecture Notes in Computer Science*, pages 243–260. Springer, 2010.

[115] S.E. Eldridge and C.D. Walter. Hardware implementation of Montgomery's modular multiplication algorithm. *Computers, IEEE Transactions on*, 42(6):693–699, June 1993.

[116] Euclid and Thomas Heath. *The Thirteen Books of the Elements Volume 2*. Dover Publications, Mineola, 2012.

[117] G. Everest. *An Introduction to Number Theory*. Springer Verlag, New York, 2006.

[118] Graham Everest, Alf van der Poorten, Igor Shparlinski, and Thomas Ward. *Recurrence Sequences*. American Mathematical Society, Providence, 2003.

[119] D. Fainstein, S. Rosenblatt, A. Cestero, N. Robson, T. Kirihata, and S.S. Iyer. Dynamic intrinsic chip ID using 32nm high-K/metal gate SOI embedded DRAM. In *VLSI Circuits (VLSIC), 2012 Symposium on*, pages 146–147. IEEE, 2012.

[120] H. Feistel. Block Cipher Cryptographic System. U.S. Patent No. 3,798,359 (Filed June 30, 1971).

[121] Martin Feldhofer, Sandra Dominikus, and Johannes Wolkerstorfer. Strong authentication for rfid systems using the aes algorithm. In Marc Joye and Jean-Jacques Quisquater, editors, *Cryptographic Hardware and Embedded Systems: CHES 2004*, volume 3156 of *Lecture Notes in Computer Science*, pages 357–370. Springer Berlin Heidelberg, 2004.

[122] William Feller. The hypergeometric series. In *An Introduction to Probability Theory and Its Applications, Third Edition*, volume 1, pages 41–45. Wiley, New York, 1968.

[123] N. Ferguson and B. Schneier. *Practical cryptography*, volume 23. Wiley New York, 2003.

[124] Niels Ferguson, Stefan Lucks, Bruce Schneier, Doug Whiting, Mihir Bellare, Tadayoshi Kohno, Jon Callas, and Jesse Walker. The Skein Hash function family: Submission to NIST (Round 3). http://ehash.iaik.tugraz.at/wiki/Skein, 2010.

[125] Benjamin Fine and Gerhard Rosenberger. *Number Theory: An Introduction Via the Distribution of Primes*. Birkhauser, Boston, 2006.

[126] V. Fischer. A closer look at security in random number generators design. In *Constructive Side-Channel Analysis and Secure Design – COSADE 2012*, pages 167–182. Springer, 2012.

[127] V. Fischer, F. Bernard, N. Bochard, and M. Varchola. Enhancing security of ring oscillator-based RNG implemented in FPGA. In *Field-Programable Logic and Applications – FPL 2008*, pages 245–250. IEEE, 2008.

[128] Wieland Fischer and Berndt M. Gammel. Masking at gate level in the presence of glitches. In Josyula R. Rao and Berk Sunar, editors, *Cryptographic Hardware and Embedded Systems: CHES 2005*, volume 3659 of *Lecture Notes in Computer Science*, pages 187–200. Springer, 2005.

[129] Domenic Forte and Ankur Srivastava. On improving the uniqueness of silicon-based physically unclonable functions via optical proximity correction. In *IEEE/ACM Design Automation Conference*, pages 96–105. IEEE, 2012.

[130] John Fraleigh. *A First Course in Abstract Algebra*. Addison-Wesley, Boston, 2002.

[131] H. Fujiwara, M. Yabuuchi, H. Nakano, H. Kawai, K. Nii, and K. Arimoto. A Chip-ID generating circuit for dependable LSI using random address errors on embedded SRAM and on-chip memory BIST. In *VLSI Circuits (VLSIC), 2011 Symposium on*, pages 76–77, 2011.

[132] Lisl Gaal. *Classical Galois Theory*. American Mathematical Society, Providence, 1998.

[133] Kris Gaj, Ekawat Homsirikamol, and Marcin Rogawski.   Fair and comprehensive methodology for comparing hardware performance of fourteen round two SHA-3 candidates using FPGAs. In *Cryptographic Hardware and Embedded Systems: CHES 2010*, pages 264–278. Springer, 2010.

[134] Karine Gandolfi, Christophe Mourtel, and Francis Olivier. Electromagnetic analysis: Concrete results. In Ç. Koç, D. Naccache, and C. Paar, editors, *Proceedings of the 3rd International Workshop on Cryptographic Hardware and Embedded Systems: CHES 2001*, volume 2162 of *Lecture Notes in Computer Science*, pages 251–261. Springer-Verlag, 2001.

[135] Blaise Gassend. Physical random functions. Master's thesis, Massachusetts Institute of Technology, Dept. of Electrical Engineering and Computer Science, 2003.

[136] P. Gauravaram, L. R. Knudsen, K. Matusiewicz, F. Mendel, C. Rechberger, M. Schlaffer, and S. S. Thomsen. Groestl — A SHA-3 candidate: Submission to NIST (Round 3). http://ehash.iaik.tugraz.at/wiki/Groestl, 2011.

[137] Benedikt Gierlichs, Lejla Batina, Pim Tuyls, and Bart Preneel. Mutual information analysis. In Elisabeth Oswald and Pankaj Rohatgi, editors, *Cryptographic Hardware and Embedded Systems: CHES 2008*, volume 5154 of *Lecture Notes in Computer Science*, pages 426–442. Springer, 2008.

[138] Oded Goldreich. *Foundations of Cryptography II*. Cambridge University Press, Cambridge, 2004.

[139] Oded Goldreich. *Foundations of Cryptography I*. Cambridge University Press, Cambridge, 2006.

[140] Oded Goldreich. *Computational Complexity: A Conceptual Perspective*. Cambridge University Press, Cambridge, 2008.

[141] Oded Goldreich. *Studies in Complexity and Cryptography*. Springer Verlag, New York, 2011.

[142] J. Golz, J. Safran, Bishan He, D. Leu, Ming Yin, T. Weaver, A. Vehabovic, Yan Sun, A. Cestero, B. Himmel, G. Maier, C. Kothandaraman, D. Fainstein, J. Barth, N. Robson, T. Kirihata, Ken Rim, and S. Iyer. 3D stackable 32nm High-K/Metal Gate SOI embedded DRAM prototype. In *VLSI Circuits (VLSIC), 2011 Symposium on*, pages 228–229. IEEE, 2011.

[143] Fernando Quadros Gouvea. *p-Adic Numbers: An Introduction*. Springer Verlag, New York, 2003.

[144] Tim Grembowski, Roar Lien, Kris Gaj, Nghi Nguyen, Peter Bellows, Jaroslav Flidr, Tom Lehman, and Brian Schott. Comparative analysis of the hardware implementations of hash functions SHA-1 and SHA-512. In *Information Security*, pages 75–89. Springer, 2002.

[145] Jorge Guajardo, Sandeep S. Kumar, Geert-Jan Schrijen, and Pim Tuyls. FPGA intrinsic PUFs and their use for IP protection. In *International Workshop on Cryptographic Hardware and Embedded Systems: CHES 2007*, pages 63–80. Springer-Verlag, Berlin, Heidelberg, 2007.

[146] S. Gueron. Intel advanced encryption standard (AES) new instructions set. Intel Corporation, September 2012.

[147] N. Guillermin. A high speed coprocessor for elliptic curve scalar multiplications over Fp. In *Advances in Cryptology—Cryptographic Hardware and Embedded Systems: CHES 2010*, Lecture Notes in Computer Science, pages 48–64. Springer, 2010.

[148] T. Guneysu. True random number generation in block memories of reconfigurable devices. In *Proceedings of the International Conference on Field-Programmable Technology*, pages 200–207. IEEE, 2010.

[149] Tim Güneysu and Amir Moradi. Generic Side-Channel Countermeasures for Reconfigurable Devices. In Bart Preneel and Tsuyoshi Takagi, editors, *Cryptographic Hardware and Embedded Systems: CHES 2011*, volume 6917 of *Lecture Notes in Computer Science*, pages 33–48. Springer, 2011.

[150] X. Guo, M. Srivistav, S. Huang, D. Ganta, M. Henry, L. Nazhandali, and P. Schaumont. Silicon implementation of SHA-3 finalists: BLAKE, Grostl, JH, Keccak and Skein. In *Proceedings of ECRYPT II Hash Workshop*. CRYPTO, 2011.

[151] Frank K. Gürkaynak, Kris Gaj, Beat Muheim, Ekawat Homsirikamol, Christoph Keller, Marcin Rogawski, Hubert Kaeslin, and Jens-Peter Kaps. Lessons learned from designing a 65nm ASIC for evaluating third round SHA-3 candidates. In *Third SHA-3 candidate conference*, 2012.

[152] Gal Hachez and Jean-Jacques Quisquater. Montgomery exponentiation with no final subtractions: Improved results. In CetinK. Koç and Christof Paar, editors, *Cryptographic Hardware and Embedded Systems: CHES 2000*, volume 1965 of *Lecture Notes in Computer Science*, pages 293–301. Springer Berlin Heidelberg, 2000.

[153] P. Haddad, Y. Teglia, F. Bernard, and V. Fischer. On the assumption of mutual independence of jitter realizations in P-TRNG stochastic models. In *Design, Automation & Test in Europe Conference & Exhibition (DATE), 2014*. IEEE, 2014.

[154] A. Hajimiri, S. Limotyrakis, and T.H. Lee. Jitter and phase noise in ring oscillators. *Solid-State Circuits, IEEE Journal of*, 34(6):790–804, 1999.

[155] Ghaith Hammouri, Aykutlu Dana, and Berk Sunar. CDs have fingerprints too. In *International Workshop on Cryptographic Hardware and Embedded Systems: CHES 2009*, pages 348–362. Springer-Verlag, 2009.

[156] Harris Hancock. *Development of the Minkowski Geometry of Numbers: Volume 1*. Dover Phoenix, Mineola, 2005.

[157] Harris Hancock. *Development of the Minkowski Geometry of Numbers: Volume 2*. Dover Phonenix, Mineola, 2005.

[158] Helena Handschuh. Hardware-anchored security based on SRAM PUFs, part 1. *Security Privacy, IEEE*, 10(3):80–83, May 2012.

[159] Helena Handschuh. Hardware-anchored security based on SRAM PUFs, part 2. *Security Privacy, IEEE*, 10(4):80–81, July 2012.

[160] Darrel Hankerson, Alfred Menezes, and Scott Vanstone. *Guide to Elliptic Curve Cryptography*. Springer Verlag, New York, 2004.

[161] G. H. Hardy and E. M. Wright. *An Introduction to the Theory of Numbers*. Oxford University Press, New York, 1980.

[162] D. Harris, R. Krishnamurthy, M. Anders, S. Mathew, and S. Hsu. An improved unified scalable radix-2 Montgomery multiplier. In *Computer Arithmetic, 2005. ARITH-17 2005. 17th IEEE Symposium on*, pages 172–178. IEEE, 2005.

[163] Joe Harris. *Algebraic Geometry: A First Course*. Springer Verlag, New York, 1995.

[164] Owen Harrison and John Waldron. AES encryption implementation and analysis on commodity graphics processing units. In Pascal Paillier and Ingrid Verbauwhede, editors, *Cryptographic Hardware and Embedded Systems: CHES 2007*, volume 4727 of *Lecture Notes in Computer Science*, pages 209–226. Springer, 2007.

[165] Robin Hartshorne. *Algebraic Geometry*. Springer Verlag, New York, 1997.

[166] Wei He, Eduardo de la Torre, and Teresa Riesgo. An interleaved EPE-immune PA-DPL structure for resisting concentrated EM side channel attacks on FPGA implementation. In Werner Schindler and Sorin A. Huss, editors, *COSADE*, volume 7275 of *Lecture Notes in Computer Science*, pages 39–53. Springer, 2012.

[167] Christoph Herbst, Elisabeth Oswald, and Stefan Mangard. An AES smart card implementation resistant to power analysis attacks. In Jianying Zhou, Moti Yung, and Feng Bao, editors, *ACNS*, volume 3989 of *Lecture Notes in Computer Science*, pages 239–252. Springer, 2006.

[168] Haruzo Hida. *Elliptic Curves and Arithmetic Invariants*. Springer Verlag, New York, 2013.

[169] H. Hidaka. Evolution of embedded flash memory technology for MCU. In *IC Design Technology (ICICDT), 2011 IEEE International Conference on*, pages 1–4. IEEE, 2011.

[170] James William Peter Hirschfeld, G. Korchmaros, and F. Torres. *Algebraic Curves over a Finite Field*. Princeton University Press, Princeton, 2008.

[171] Jeffrey Hoffstein, Jill Pipher, and Joseph Silverman. *Introduction to Mathematical Cryptography*. Springer Verlag, New York, 2008.

[172] Daniel E. Holcomb, Wayne Burleson, and Kevin Fu. Initial SRAM state as a fingerprint and source of true random numbers for RFID tags. In *Proceedings of the Conference on RFID Security*, 2007.

[173] Daniel E. Holcomb, Wayne P. Burleson, and Kevin Fu. Power-up SRAM state as an identifying fingerprint and source of true random numbers. *IEEE Transactions on Computers*, 58(9):1198–1210, September 2009.

[174] Audun Holme. *A Royal Road to Algebraic Geometry*. Springer Verlag, New York, 2011.

[175] W. G. Horner. A new method of solving numerical equations of all orders, by continuous approximation. *Philosophical Transactions of the Royal Society of London*, 109(1):308–335, 1819.

[176] Gabriel Hospodar, Roel Maes, and Ingrid Verbauwhede. Machine learning attacks on 65nm Arbiter PUFs: Accurate modeling poses strict bounds on usability. In *IEEE International Workshop on Information Forensics and Security*, 2012.

[177] T. C. Hu and M. T. Shing. *Combinatorial Algorithms*. Dover Publications, Mineola, 2002.

[178] Thomas Hungerford. *Algebra*. Springer Verlag, New York, 1980.

[179] Dale Husemöller. *Elliptic Curves*. Springer Verlag, New York, 2003.

[180] IEEE Computer Society. IEEE std 1363-2000: IEEE standard specifications for public-key cryptography. IEEE, 2000.

[181] IEEE Computer Society. IEEE std 1363a-2004 (Amendment to IEEE std 1363-2000): IEEE standard specifications for public-key cryptography — Amendment 1: Additional techniques. IEEE, 2004.

[182] Ronald S. Indeck and Marcel W. Muller. Method and apparatus for fingerprinting magnetic media, US Patent No. 5365586, 1994.

[183] Nisha Jacob, Sirote Saetang, Chien-Ning Chen, Sebastian Kutzner, San Ling, and Axel Poschmann. Feasibility and practicability of standardized cryptography on 4-bit micro controllers. In Lars R. Knudsen and Huapeng Wu, editors, *Proceedings of SAC*, volume 7707 of *Lecture Notes in Computer Science*, pages 184–201. Springer, 2013.

[184] Nathan Jacobson. *Basic Algebra I*. Dover Publications, Mineola, 2009.

[185] Nathan Jacobson. *Basic Algebra II*. Dover Publications, Mineola, 2009.

[186] Gordon James and Martin Liebeck. *Representations and Characters of Groups*. Cambridge University Press, Cambridge, 2001.

[187] Nan Jiang and D. Harris. Quotient pipelined very high radix scalable Montgomery multipliers. In *Signals, Systems and Computers, 2006. ACSSC '06. Fortieth Asilomar Conference on*, pages 1673–1677. IEEE, 2006.

[188] Nan Jiang and D. Harris. Parallelized radix-2 scalable Montgomery multiplier. In *Very Large Scale Integration, 2007. VLSI - SoC 2007. IFIP International Conference on*, pages 146–150. IEEE, 2007.

[189] Gareth Jones and Josephine Jones. *Elementary Number Theory*. Springer Verlag, New York, 1998.

[190] Jing Ju, J. Plusquellic, R. Chakraborty, and R. Rad. Bit string analysis of physical unclonable functions based on resistance variations in metals and transistors. In *Hardware-Oriented Security and Trust (HOST), 2012 IEEE International Symposium on*, pages 13–20. IEEE, 2012.

[191] Ari Juels and Stephen A. Weis. Authenticating pervasive devices with human protocols. In *Advances in Cryptology - CRYPTO 2005: International Cryptology Conference*, volume 3621 of *Lecture Notes in Computer Science*, pages 293–308. Springer, 2005.

[192] Stasys Jukna. *Extremal Combinatorics*. Springer Verlag, New York, 2011.

[193] Stasys Jukna. *Boolean Function Complexity: Advances and Frontiers*. Springer Verlag, New York, 2012.

[194] David Kahn. *The Codebreakers*. Scribner, New York, 1996.

[195] J. Kamp, A. Rao, S. Vadhan, and D. Zuckerman. Deterministic extractors for small-space sources. *J. Comput. Syst. Sci.*, 77(1):191–220, January 2011.

[196] A. Karatsuba and Y. Ofman. Multiplication of multidigit numbers on automata. *Soviet Physics-Doklady*, 7(7):595–596, January 1963.

[197] Anatoli Karatsuba and Yuri Ofman. Multiplication of many-digital numbers by automatic computers. *Proceedings of the USSR Academy of Sciences*, 145:293–294, 1962.

[198] Oleg Karpenkov. *Geometry of Continued Fractions*. Springer Verlag, New York, 2013.

[199] Emilia Käsper and Peter Schwabe. Faster and timing-attack resistant AES-GCM. In Christophe Clavier and Kris Gaj, editors, *Cryptographic Hardware and Embedded Systems: CHES 2009*, volume 5747 of *Lecture Notes in Computer Science*, pages 1–17. Springer, 2009.

[200] Markus Kasper, Timo Kasper, Amir Moradi, and Christof Paar. Breaking KeeLoq in a flash: On extracting keys at lightning speed. In Bart Preneel, editor, *Proceedings of AFRICACRYPT 2009*, volume 5580 of *Lecture Notes in Computer Science*, pages 403–420. Springer-Verlag, 2009.

[201] Jonathan Katz and Yehuda Lindell. *Introduction to Modern Cryptography*. Chapman & Hall / CRC, Boca Raton, 2008.

[202] Stefan Katzenbeisser, Ünal Kocabaş, Vladimir Rožić, Ahmad-Reza Sadeghi, Ingrid Verbauwhede, and Christian Wachsmann. PUFs: Myth, fact or busted? A security evaluation of physically unclonable functions (PUFs) cast in silicon. In *International Conference on Cryptographic Hardware and Embedded Systems: CHES 2012*, pages 283–301. Springer-Verlag, 2012.

[203] Stefan Katzenbeisser, Ünal Kocabaş, Vincent van der Leest, Ahmad-Reza Sadeghi, Geert-Jan Schrijen, and Christian Wachsmann. Recyclable PUFs: logically reconfigurable PUFs. *Journal of Cryptographic Engineering*, 1(3):177–186, 2011.

[204] S. Kawamura, M. Koike, F. Sano, and A. Shimbo. Cox-Rower architecture for fast parallel montgomery multiplication. In *Advances in Cryptology — EUROCRYPT*, Lecture Notes in Computer Science, pages 523–538. Springer, 2000.

[205] S. S. Keller. NIST-recommended random number generator based on ANSI X9.31 Appendix A.2.4: Using the 3-key triple DES and AES algorithms, 2005. Online. Available at: http://csrc.nist.gov/groups/STM/cavp/documents/rng/931rngext.pdf.

[206] K. Kelley and D. Harris. Parallelized very high radix scalable Montgomery multipliers. In *Signals, Systems and Computers, 2005. Conference Record of the Thirty-Ninth Asilomar Conference on*, pages 1196–1200. IEEE, 2005.

[207] Kyle Kelley and David Harris. Very high radix scalable Montgomery multipliers. *System-on-Chip for Real-Time Applications, International Workshop on*, 0:400–404, 2005.

[208] J. Kelsey, B. Schneier, and N. Ferguson. Yarrow-160: Notes on the design and analysis of the Yarrow cryptographic pseudorandom number generator. In *Sixth Annual Workshop on Selected Areas in Cryptography*, pages 13–33. Springer, 1999.

[209] Stephanie Kerckhof, Francois Durvaux, Nicolas Veyrat-Charvillon, Francesco Regazzoni, Guerric Meurice de Dormale, and Francois-Xavier Standaert. Compact FPGA implementations of the five SHA-3 finalists. In *Smart Card Research and Advanced Applications (CARDIS)*, pages 217–233. Springer, 2011.

[210] W. Killmann and W. Schindler. AIS 31: Functionality classes and evaluation methodology for true (physical) random number generators, version 3.1. Online. Available at: https://www.bsi.bund.de, 2001.

[211] W. Killmann and W. Schindler. A design for a physical RNG with robust entropy estimators. In Elisabeth Oswald and Pankaj Rohatgi, editors, *Cryptographic Hardware and Embedded Systems: CHES 2008*, volume 5154 of *Lecture Notes in Computer Science*, pages 146–163. Springer, 2008.

[212] W. Killmann and W. Schindler. A proposal for: Functionality classes for random number generators. Online. Available at: https://www.bsi.bund.de, 2011.

[213] Daeik Kim, Choongyeun Cho, Jonghae Kim, Jean-Olivier Plouchart, Robert Trzcinski, and David Ahlgren. CMOS mixed-signal circuit process variation sensitivity characterization for yield improvement. In *IEEE Custom Integrated Circuits Conference*, pages 365–368. IEEE, 2006.

[214] Inyoung Kim, Abhranil Maiti, Leyla Nazhandali, Patrick Schaumont, Vignesh Vivekraja, and Huaiye Zhang. From statistics to circuits: Foundations for future physical unclonable functions. In Ahmad-Reza Sadeghi and David Naccache, editors, *Towards Hardware-Intrinsic Security*, Information Security and Cryptography, pages 55–78. Springer Berlin Heidelberg, 2010.

[215] R. King. Businessweek. http://www.businessweek.com/technology/special_reports/20100302ceo_guide_to_counterfeit_tech.htm.

[216] T. Kirihata and S. Rosenblatt. Dynamic intrinsic chip ID for hardware security. In *VLSI: circuits for emerging applications*. CRC Press, 2014.

[217] Toshiaki Kirihata. High performance embedded dynamic random access memory in nano-scale technologies. In Krzysztof Iniewski, editor, *CMOS Processors and Memories*, Analog Circuits and Signal Processing, pages 295–336. Springer Netherlands, 2010.

[218] P. Kitsos and N. Sklavos. On the hardware implementation efficiency of sha-3 candidates. In *Proceedings of 17th IEEE International Conference on Electronics, Circuits, and Systems, (IEEE ICECS'10)*, pages 1240–1243. IEEE, 2010.

[219] Achim Klenke. *Probability Theory*. Springer Verlag, New York, 2013.

[220] Anthony Knapp. *Advanced Algebra*. Birkhauser, Boston, 2007.

[221] Anthony Knapp. *Basic Algebra*. Birkhauser, Boston, 2007.

[222] N. Koblitz. Elliptic curve cryptosystems. *Mathematics of Computation*, 48(177):203–209, 1985.

[223] N. Koblitz, A. Menezes, and S. Vanstone. The state of elliptic curve cryptography. *Designs, Codes and Cryptography*, 19(2):173–193, 2000.

[224] Neal Koblitz. *Algebraic Aspects of Cryptography*. Springer Verlag, Berlin, 1998.

[225] Neal Koblitz. *A Course in Number Theory and Cryptography*. Springer Verlag, New York, 2001.

[226] Neal Koblitz. Elliptic curve cryptosystems. *Mathematics of Computation*, 48(177):203–209, January 1987.

[227] Neal Koblitz. The uneasy relationship between mathematics and cryptography. *Notices of the American Mathematical Society*, 54(8):972–979, September 2007.

[228] Cetin Kaya Koc, editor. *Cryptographic Engineering*. Springer, 2009.

[229] Paul C. Kocher. Timing attacks on implementations of Diffie-Hellman, RSA, DSS, and other systems. In Neal Koblitz, editor, *Advances in Cryptology - CRYPTO '96*, volume 1109 of *Lecture Notes in Computer Science*, pages 104–113. Springer-Verlag, 1996.

[230] Paul C. Kocher, Joshua Jaffe, and Benjamin Jun. Differential power analysis. In Michael J. Wiener, editor, *Advances in Cryptology — CRYPTO '99*, volume 1666 of *Lecture Notes in Computer Science*, pages 388–397. Springer, 1999.

[231] Francois Koeune and Jean jacques Quisquater. A timing attack against Rijndael. Technical Report CG-1999/1, Université catholique de Louvain, 1999.

[232] Robert Könighofer. A fast and cache-timing resistant implementation of the AES. In Tal Malkin, editor, *Proceedings of CT-RSA*, volume 4964 of *Lecture Notes in Computer Science*, pages 187–202. Springer, 2008.

[233] J.B. Kuang, A. Mathews, J. Barth, F. Gebara, T. Nguyen, J. Schaub, K. Nowka, G. Carpenter, D. Plass, E. Nelson, I. Vo, W. Reohr, and T. Kirihata. An on-chip dual supply charge pump system for 45nm PD SOI eDRAM. In *Solid-State Circuits Conference, 2008. ESSCIRC 2008. 34th European*, pages 66–69, 2008.

[234] Raghavan Kumar, Harikrishnan K. Chandrikakutty, and Sandip Kundu. On improving reliability of delay based physically unclonable functions under temperature variations. In *IEEE Symposium on Hardware-Oriented Security and Trust*, pages 142–147. IEEE, 2011.

[235] Raghavan Kumar, Siva Dhanuskodi, and Sandip Kundu. On manufacturing aware physical design to improve the uniqueness of silicon-based physically unclonable functions. In *International Conference on VLSI Design*, pages 381–386. IEEE, 2014.

[236] Raghavan Kumar, Vinay C. Patil, and Sandip Kundu. On design of temperature invariant physically unclonable functions based on ring oscillators. In *IEEE Annual Symposium on VLSI*, pages 165–170. IEEE, 2012.

[237] H. Kuo, I. Verbauwhede, and P. Schaumont. A 2.29Gbit/sec, 56mW, non-pipelined Rijndael AES encryption IC in a 1.8V, 0.18μm CMOS technology. In *Proceedings of the IEEE 2002 Custom Integrated Circuits Conference*, pages 147–150. IEEE, 2002.

[238] Serge Lang. *Elliptic Curves: Diophantine Analysis*. Springer Verlag, New York, 1978.

[239] Serge Lang. *Algebra*. Springer Verlag, New York, 2005.

[240] Serge Lang. *Undergraduate Algebra*. Springer Verlag, New York, 2005.

[241] D.-U. Lee, A. A. Gaffar, R. C. C. Cheung, O. Mencer, W. Luk, and G. A. Constantinides. Accuracy-guaranteed bit-width optimization. *IEEE Transactions on Computer-Aided Design of Integrated Circuits and Systems*, 25(10):1990–2000, 2006.

[242] Yong Ki Lee, Herwin Chan, and Ingrid Verbauwhede. Iteration bound analysis and throughput optimum architecture of SHA-256 (384,512) for hardware implementations. In *Information Security Applications*, pages 102–114. Springer, 2007.

[243] Mario Lefebvre. *Basic Probability Theory With Applications*. Springer Verlag, New York, 2009.

[244] Hendrik Lenstra. Factoring integers with elliptic curves. *Annals of Mathematics*, 126(3):649–673, November 1987.

[245] P. H. W. Leong and I. K. H. Leung. A microcoded elliptic curve processor using FPGA technology. *IEEE Transactions on Very Large Scale Integration (VLSI) Systems*, 10(5):550–559, 2002.

[246] Rudolf Lidl and Gunther Pilz. *Applied Abstract Algebra*. Springer Verlag, New York, 1997.

[247] Roar Lien, Tim Grembowski, and Kris Gaj. A 1 Gbit/s partially unrolled architecture of hash functions SHA-1 and SHA-512. In *Topics in Cryptology–CT-RSA 2004*, pages 324–338. Springer, 2004.

[248] D. Lim, J.W. Lee, B. Gassend, G.E. Suh, M. van Dijk, and S. Devadas. Extracting secret keys from integrated circuits. *Very Large Scale Integration (VLSI) Systems, IEEE Transactions on*, 13(10):1200–1205, Oct 2005.

[249] Daihyun Lim. Extracting secret keys from integrated circuits. Master's thesis, Massachusetts Institute of Technology, Dept. of Electrical Engineering and Computer Science, 2004.

[250] Daihyun Lim, Jae W. Lee, Blaise Gassend, Edward Suh, Marten van Dijk, and Srinivas Devadas. Extracting secret keys from integrated circuits. *IEEE Transactions on Very Large Scale Integration (VLSI) Systems*, 13(10):1200 1205, Oct. 2005.

[251] Lang Lin and Wayne Burleson. Analysis and mitigation of process variation impacts on power-attack tolerance. In *ACM/IEEE Design Automation Conference*, DAC '09, pages 238–243. ACM, 2009.

[252] Lang Lin, Dan Holcomb, Dilip Kumar Krishnappa, Prasad Shabadi, and Wayne Burleson. Low-power sub-threshold design of secure physical unclonable functions. In *ACM/IEEE International Symposium on Low-Power Electronics and Design*, pages 43 –48. IEEE, 2010.

[253] Lang Lin, Sudheendra Srivathsa, Dilip Kumar Krishnappa, Prasad Shabadi, and Wayne Burleson. Design and validation of arbiter-based pufs for sub-45-nm low-power security applications. *IEEE Transactions on Information Forensics and Security*, 7(4):1394 –1403, Aug. 2012.

[254] E. Lindholm, J. Nickolls, S. Oberman, and J. Montrym. NVIDIA Tesla: A unified graphics and computing architecture. *IEEE Micro*, 28(2):39–55, 2008.

[255] San Ling, Huaxiong Wang, and Chaoping Xing. *Algebraic Curves in Cryptography*. CRC Press, Boca Raton, 2011.

[256] N. Liu, S. Hanson, D. Sylvester, and D. Blaauw. OxID: On-chip one-time random ID generation using oxide breakdown. In *VLSI Circuits (VLSIC), 2010 IEEE Symposium on*, pages 231–232, 2010.

[257] Qing Liu. *Algebraic Geometry and Algebraic Curves*. Oxford University Press, Oxford, 2006.

[258] K. Lofstrom, W.R. Daasch, and D. Taylor. IC identification circuit using device mismatch. In *Solid-State Circuits Conference, 2000. Digest of Technical Papers. ISSCC. 2000 IEEE International*, pages 372–373, 2000.

[259] P. Longa and C. Gebotys. Analysis of efficient techniques for fast elliptic curve cryptography on x86-64 based processors. IACR Cryptology ePrint Archive 335, 2010.

[260] Yingxi Lu, Keanhong Boey, Philip Hodgers, and Máire O'Neill. Lightweight DPA resistant solution on FPGA to counteract power models. In Jinian Bian, Qiang Zhou, Peter Athanas, Yajun Ha, and Kang Zhao, editors, *FPT*, pages 178–183. IEEE, 2010.

[261] Yingxi Lu, Máire O'Neill, and John V. McCanny. Evaluation of random delay insertion against DPA on FPGAs. *TRETS*, 4(1):11, 2010.

[262] K.H. Lundberg. Noise sources in bulk CMOS. Available at: http://dandelion-patch.mit.edu/people/klund/papers/UNP_noise.pdf, 2002.

[263] R. Maes, V. Rozic, I. Verbauwhede, P. Koeberl, E. van der Sluis, and V. van der Leest. Experimental evaluation of physically unclonable functions in 65 nm CMOS. In *ESSCIRC (ESSCIRC), 2012 Proceedings of the*, pages 486–489, 2012.

[264] Roel Maes, Pim Tuyls, and Ingrid Verbauwhede. Low-overhead implementation of a soft decision helper data algorithm for sram pufs. In *International Workshop on Cryptographic Hardware and Embedded Systems*, volume 5747 of *Lecture Notes in Computer Science*, pages 332–347. Springer, 2009.

[265] Roel Maes and Ingrid Verbauwhede. Physically unclonable functions: A study on the state of the art and future research directions. In Ahmad-Reza Sadeghi and David Naccache, editors, *Towards Hardware-Intrinsic Security*, Information Security and Cryptography, pages 3–37. Springer Berlin Heidelberg, 2010.

[266] M. Majzoobi, F. Koushanfar, and S. Devadas. FPGA-based true random number generation using circuit metastability with adaptive feedback control. In *Cryptographic Hardware and Embedded Systems*, volume 6917 of *Lecture Notes in Computer Science*, pages 17–32. Springer, 2011.

[267] Mehrdad Majzoobi, Farinaz Koushanfar, and Miodrag Potkonjak. Testing techniques for hardware security. In *IEEE International Test Conference*, pages 1 –10. IEEE, 2008.

[268] Stefan Mangard. A simple power-analysis (SPA) attack on implementations of the AES key expansion. In Pil Joong Lee and Chae Hoon Lim, editors, *Proceedings of the 5th International Conference on Information Security and Cryptology — ICISC'02*, volume 2587 of *Lecture Notes in Computer Science*, pages 343–358. Springer, 2003.

[269] Stefan Mangard, Maria Elisabeth Oswald, and Thomas Popp. *Power Analysis Attacks: Revealing the Secrets of Smart Cards*. Springer, 2007.

[270] Yu. I. Manin and Alexei A. Panchishkin. *Introduction to Modern Number Theory: Fundamental Problems, Ideas, and Theories*. Springer Verlag, Berlin, 2005.

[271] G. Marsaglia. DIEHARD: Battery of Tests of Randomness. Online. Available at: http://stat.fsu.edu/pub/diehard/, 1996.

[272] George Martin. *Transformation Geometry: An Introduction to Symmetry*. Springer Verlag, New York, 1996.

[273] George Martin. *Counting: The Art of Enumerative Combinatorics*. Springer Verlag, New York, 2001.

[274] Jacques Martinet. *Perfect Lattices in Euclidean Space*. Springer Verlag, New York, 2010.

[275] S. Mathew, S. Satpathy, V. Suresh, M. Anders, H. Kaul, A. Agarwal, S. Hsu, G. Chen, and R. Krishnamurthy. 340mV-1.1V, 289Gbps/W, 2090-gate NanoAES hardware accelerator with area-optimized encrypt/decrypt $GF(2^4)^2$ polynomials in 22nm tri-gate CMOS. In *2014 IEEE Symposium on VLSI Circuits Digest of Technical Papers*, pages 166–167. IEEE, 2014.

[276] Sanu Mathew, Farhana Sheikh, Michael Kounavis, Shay Gueron, Amit Agarwal, Steven Hsu, Himanshu Kaul, Mark Anders, and Ram Krishnamurthy. 53Gbps native $GF(2^4)^2$ composite-field AES-encrypt/decrypt accelerator for content-protection in 45nm high-performance microprocessors. *IEEE Journal of Solid-State Circuits*, 46(4):767–776, 2011.

[277] Jiri Matousek. *Lectures on Discrete Geometry*. Springer Verlag, New York, 2002.

[278] Mitsuru Matsui. Linear cryptoanalysis method for DES cipher. In Tor Helleseth, editor, *Proceedings of EUROCRYPT*, volume 765 of *Lecture Notes in Computer Science*, pages 386–397. Springer, 1993.

[279] Mitsuru Matsui and Junko Nakajima. On the power of bitslice implementation on Intel Core2 processor. In Pascal Paillier and Ingrid Verbauwhede, editors, *Cryptographic Hardware and Embedded Systems:*

CHES 2007, volume 4727 of *Lecture Notes in Computer Science*, pages 121–134. Springer, 2007.

[280] S. Matsuo, M. Knezevic, P. Schaumont, I. Verbauwhede, A. Satoh, K. Sakiyama, and K. Ota. How can we conduct fair and consistent hardware evaluation for SHA-3 candidate? In *Proceedings of Second SHA-3 Candidate Conference*. NIST SHA-3 Workshop, 2010.

[281] Robert P. McEvoy, Francis M. Crowe, Colin C. Murphy, and William P. Marnane. Optimisation of the SHA-2 family of hash functions on FP-GAs. In *Emerging VLSI Technologies and Architectures, 2006. IEEE Computer Society Annual Symposium on*, pages 317–322. IEEE, 2006.

[282] A.J. Menezes, P.C. Van Oorschot, and S.A. Vanstone. *Handbook of Applied Cryptography*. CRC Press, 1997. Online. Available at: http://www.cacr.math.uwaterloo.ca/hac/.

[283] Alfred Menezes. *Elliptic Curve Public Key Cryptosystems*. Kluwer Academic Publishers, Boston, 1993.

[284] Alfred Menezes et al. *Applications of Finite Fields*. Springer Verlag, New York, 2010.

[285] Alfred J. Menezes, Paul C. Van Oorschot, and Scott A. Vanstone. *Handbook of Applied Cryptography*. CRC Press, Inc., 1996.

[286] Nele Mentens, Lejla Batina, Bart Preneel, and Ingrid Verbauwhede. A systematic evaluation of compact hardware implementations for the Rijndael S-Box. In Alfred Menezes, editor, *Proceedings of CT-RSA*, volume 3376 of *Lecture Notes in Computer Science*, pages 323–333. Springer, 2005.

[287] Thomas S. Messerges, Ezzy A. Dabbish, and Robert H. Sloan. Investigations of power analysis attacks on smartcards. In *USENIX Workshop on Smartcard Technology*, pages 151–162. USENIX Association, 1999.

[288] Daniele Micciancio and Shafi Goldwasser. *Complexity of Lattice Problems: A Cryptographic Perspective*. Kluwer Academic Publishers, Boston, 2002.

[289] V. S. Miller. Use of elliptic curves in cryptography. In H. Williams, editor, *Lecture Notes in Computer Science: Advances in Cryptology - CRYPTO 1985 Proceedings*, pages 417–426. Springer Berlin, 1985.

[290] Victor Miller. Use of elliptic curves in cryptography. In *Advances in Cryptology—CRYPTO '85 Proceedings*, volume 218 of *Lecture Notes in Computer Science*, pages 417–426. Springer, 1986.

[291] P. V. A. Mohan. *Residue Number Systems: Algorithms and Architectures*. Kluwer Academic Publishers, Norwell, MA, 2002.

[292] H. Gregor Molter, Marc Stöttinger, Abdulhadi Shoufan, and Falko Strenzke. A simple power analysis attack on a McEliece cryptoprocessor. *Journal of Cryptographic Engineering*, 1:29–36, January 2011.

[293] P. L. Montgomery. Modular multiplication without trial division. *Mathematics of Computation*, 44(170):519–521, 1985.

[294] Peter Montgomery. Modular multiplication without trial division. *Mathematics of Computation*, 44(170):519–521, April 1985.

[295] Amir Moradi, Alessandro Barenghi, Timo Kasper, and Christof Paar. On the vulnerability of FPGA bitstream encryption against power analysis attacks – Extracting keys from Xilinx Virtex-II FPGAs. In George Danezis and Vitaly Shmatikov, editors, *Proceedings of ACM CCS 2011*, pages 111–124. ACM Press, 2011.

[296] Amir Moradi, Oliver Mischke, and Thomas Eisenbarth. Correlation-enhanced power analysis collision attack. In Stefan Mangard and François-Xavier Standaert, editors, *Cryptographic Hardware and Embedded Systems: CHES 2010*, volume 6225 of *Lecture Notes in Computer Science*, pages 125–139. Springer, 2010.

[297] Amir Moradi, Axel Poschmann, San Ling, Christof Paar, and Huaxiong Wang. Pushing the limits: A very compact and a threshold implementation of AES. In Kenneth G. Paterson, editor, *EUROCRYPT*, volume 6632 of *Lecture Notes in Computer Science*, pages 69–88. Springer, 2011.

[298] Stefan Moser. *A Student's Guide to Coding and Information Theory*. Cambridge University Press, Cambridge, 2012.

[299] Andrew Moss, Elisabeth Oswald, Dan Page, and Michael Tunstall. Automatic insertion of DPA countermeasures. *IACR Cryptology ePrint Archive*, 2011:412, 2011.

[300] Andrew Moss, Elisabeth Oswald, Dan Page, and Michael Tunstall. Compiler assisted masking. In *Cryptographic Hardware and Embedded Systems: CHES 2012*, pages 58–75. Springer, 2012.

[301] Gary Mullen and Daniel Panario. *Handbook of Finite Fields*. Chapman & Hall/CRC, Boca Raton, 2013.

[302] Paul Nahin. *An Imaginary Tale: The Story of $\sqrt{-1}$*. Princeton University Press, Princeton, 1998.

[303] S. Narasimha et al. 22nm high-performance SOI technology featuring dual-embedded stressors, epi-plate high-K deep-trench embedded DRAM and self-aligned via 15LM BEOL. In *Electron Devices Meeting (IEDM), 2012 IEEE International*, pages 3.3.1–3.3.4. IEEE, 2012.

[304] Wladyslaw Narkiewicz. *The Development of Prime Number Theory: From Euclid to Hardy and Littlewood*. Springer Verlag, New York, 2010.

[305] National Bureau of Standards. Data Encryption Standard. FIPS-Pub.46. National Bureau of Standards, U.S. Department of Commerce, Washington D.C., January 1977.

[306] National Institute of Standards and Technology. Announcing Development of a Federal Information Processing Standard for Advanced Encryption Standard. http://csrc.nist.gov/archive/aes/pre-round1/aes_9701.txt, 1997.

[307] National Institute of Standards and Technology. Specification for the Advanced Encryption Standard (AES) - FIPS PUB 197. FIPS-Pub.46. National Bureau of Standards, U.S. Department of Commerce, Washington D.C., 2001.

[308] National Institute of Standards and Technology (NIST). Cryptographic hash algorithm competition. http://www.csrc.nist.gov/groups/ST/hash/sha-3/.

[309] National Institute of Standards and Technology (NIST). Announcing the standard for secure hash standard, FIPS 180-1. Technical report, National Institute of Standards and Technology, April 1995.

[310] National Institute of Standards and Technology (NIST). The keyed-hash message authentication code (HMAC), FIPS 198. Technical report, National Institute of Standards and Technology, March 2002.

[311] National Institute of Standards and Technology (NIST). Fips-180-3: Secure hash standard. http://www.itl.nist.gov/fipspubs/, October 2008.

[312] National Institute of Standards and Technology (NIST). Digital signature standard (dss) - fips pub 186-3. NIST website, 2009.

[313] Phong Nguyen and Brigitte Vallée. *The LLL Algorithm: Survey and Applications*. Springer Verlag, New York, 2009.

[314] Andre Nies. *Computability and Randomness*. Oxford University Press, Oxford, 2009.

[315] Svetla Nikova, Christian Rechberger, and Vincent Rijmen. Threshold implementations against side-channel attacks and glitches. In Peng Ning, Sihan Qing, and Ninghui Li, editors, *ICICS*, volume 4307 of *Lecture Notes in Computer Science*, pages 529–545. Springer, 2006.

[316] Viacheslav Nikulin. *Geometries and Groups*. Springer Verlag, New York, 2002.

[317] H. Nozaki, M. Motoyama, A. Shimbo, and S. Kawamura. Implementation of RSA algorithm based on RNS montgomery multiplication. *Advances in Cryptology—Cryptographic Hardware and Embedded Systems: CHES 2001*, pages 364–376, 2001.

[318] Carl Olds, A. Rockett, and P. Szusze. *Continued Fractions*. Mathematical Association of America, Washington, DC, 1992.

[319] Holger Orup. Simplifying quotient determination in high-radix modular multiplication. In *Computer Arithmetic, Proceedings of the 12th Symposium on*, pages 193–199. IEEE, 1995.

[320] David Oswald and Christof Paar. Breaking Mifare DESFire MF3ICD40: Power analysis and templates in the real world. In Bart Preneel and Tsuyoshi Takagi, editors, *Cryptographic Hardware and Embedded Systems: CHES 2011*, volume 6917 of *Lecture Notes in Computer Science*, pages 207–222. Springer-Verlag, 2011.

[321] Elisabeth Oswald, Luke Mather, and Carolyn Whitnall. Choosing distinguishers for differential power analysis attacks. In *Non-Invasive Attack Testing Workshop*, pages 1–14, 2011.

[322] Ravikanth Srinivasa Pappu. *Physical One-way Functions*. PhD thesis, Dept. of Electrical Engineering and Computer Science, Massachusetts Institute of Technology, 2001.

[323] Z. Paral and S. Devadas. Reliable PUF value generation by pattern matching. http://www.google.com/patents/US20120183135, US Patent App. 13/009,205, July 2012.

[324] Eric Peeters. Side-channel cryptanalysis: A brief survey. In *Advanced DPA Theory and Practice*, pages 11–19. Springer, 2013.

[325] Eric Peeters, François-Xavier Standaert, and Jean-Jacques Quisquater. Power and electromagnetic analysis: Improved model, consequences and comparisons. *Integr. VLSI J.*, 40(1):52–60, January 2007.

[326] N. Perlroth, J. Larson, and S. Shane. N.S.A. able to foil basic safeguards of privacy on Web. Available at: http://ctvoterscount.org/CTVCdata/13/09/NYTimes20130905.pdf, 2013.

[327] Daniel Perrin. *Algebraic Geometry: An Introduction*. Springer Verlag, New York, 2007.

[328] N. Pinckney and D. Harris. Parallelized radix-4 scalable Montgomery multipliers. *Journal of Integrated Circuits and Systems*, 3(1):39–45, 2008.

[329] Nathaniel Pinckney, Philip Amberg, and David Money Harris. Parallelized booth-encoded radix-4 Montgomery multipliers. In *IFIP/IEEE International Conference on Very Large Scale Integration (VLSI-SoC)*. IEEE, 2008.

[330] Thomas Popp, Mario Kirschbaum, Thomas Zefferer, and Stefan Mangard. Evaluation of the masked logic style MDPL on a prototype chip. In Pascal Paillier and Ingrid Verbauwhede, editors, *Cryptographic Hardware and Embedded Systems: CHES 2007*, volume 4727 of *Lecture Notes in Computer Science*, pages 81–94. Springer, 2007.

[331] Thomas Popp and Stefan Mangard. Masked dual-rail pre-charge logic: DPA-resistance without routing constraints. In Josyula R. Rao and Berk Sunar, editors, *Cryptographic Hardware and Embedded Systems: CHES 2005*, volume 3659 of *Lecture Notes in Computer Science*, pages 172–186. Springer, 2005.

[332] Thomas Popp, Stefan Mangard, and Elisabeth Oswald. Power analysis attacks and countermeasures. *IEEE Design and Test of Computers*, 24(6):535–543, 2007.

[333] N. Potlapally. Hardware security in practice: Challenges and opportunities. In *Hardware-Oriented Security and Trust (HOST), 2011 IEEE International Symposium on*, pages 93–98. IEEE, 2011.

[334] G. Provelengios, P. Kitsos, N. Sklavos, and C. Koulamas. FPGA-based design approaches of Keccak Hash function. In *Proceedings of 15th Euromicro Conference on Digital Systems (DSD'12)*. EUROMICRO, 2012.

[335] PUB FIPS. FIPS 140-1: Security requirements for cryptographic modules. Online. Available at: http://csrc.nist.gov/publications/fips/fips140-1/fips1401.pdf, 1994.

[336] PUB FIPS. FIPS 140-2: Security requirements for cryptographic modules. Online. Available at: http://csrc.nist.gov/publications/fips/fips140-2/fips1402.pdf, 2001.

[337] PUB FIPS. FIPS 186-2: Digital Signature Standard (DSS). Online. Available at: http://csrc.nist.gov/publications/fips/archive/fips186-2/fips186-2-change1.pdf, 2001.

[338] Jean-Jacques Quisquater and David Samyde. Electromagnetic analysis (EMA): Measures and counter-measures for smart cards. In Isabelle Attali and Thomas P. Jensen, editors, *Smart Card Programming and Security - E-smart 2001*, volume 2140 of *Lecture Notes in Computer Science*, pages 200–210. Springer-Verlag, 2001.

[339] Christian Rechberger and Elisabeth Oswald. Practical template attacks. In Chae Hoon Lim and Moti Yung, editors, *Information Security Applications, 5th International Workshop, WISA 2004*, volume 3325 of *Lecture Notes in Computer Science*, pages 440–456. Springer-Verlag, 2004.

[340] Francesco Regazzoni, Alessandro Cevrero, François-Xavier Standaert, Stéphane Badel, Theo Kluter, Philip Brisk, Yusuf Leblebici, and Paolo

Ienne. A design flow and evaluation framework for DPA-resistant instruction set extensions. In Christophe Clavier and Kris Gaj, editors, *Cryptographic Hardware and Embedded Systems: CHES 2009*, volume 5747 of *Lecture Notes in Computer Science*, pages 205–219. Springer, 2009.

[341] O. Regev. On lattices, learning with errors, random linear codes, and cryptography. In *Proceedings of 37th ACM Symposium on Theory of Computing (STOC)*, pages 84–93. ACM, 2005.

[342] Miles Reid. *Undergraduate Algebraic Geometry*. Cambridge University Press, Cambridge, 1989.

[343] Paulo Ribenboim. *The Little Book of Big Primes*. Springer Verlag, New York, 1991.

[344] Paulo Ribenboim. *The New Book of Prime Number Records*. Springer Verlag, New York, 1995.

[345] Hans Riesel. *Prime Numbers and Computer Methods for Factorization*. Springer Verlag, New York, 2011.

[346] Vincent Rijmen. Efficient Implementation of the Rijndael SBox. http://www.esat.ku-leuven.ac.be/~rijmen/rijndael/, 2000.

[347] Ronald L. Rivest, Adi Shamir, and Leonard M. Adleman. A method for obtaining digital signatures and public- key cryptosystems. *Communications of the ACM*, 21(2):120–126, 1978.

[348] Alain Robert. *A Course in p-Adic Analysis*. Springer Verlag, New York, 2010.

[349] N. Robson, J. Safran, C. Kothandaraman, A. Cestero, Xiang Chen, R. Rajeevakumar, A. Leslie, D. Moy, T. Kirihata, and S. Iyer. Electrically programmable fuse (eFUSE): From memory redundancy to autonomic chips. In *Custom Integrated Circuits Conference, 2007. CICC '07. IEEE*, pages 799–804. IEEE, 2007.

[350] F. Rodriguez-Henriquez, N. A. Saqib, A. Diaz Pérez, and C. K. Koc. *Cryptographic Algorithms on Reconfigurable Hardware*. Springer, New York, 2007.

[351] Kenneth Rosen. *Elementary Number Theory and Its Applications*. Addison-Wesley, Boston, 2000.

[352] S. Rosenblatt, S. Chellappa, A. Cestero, N. Robson, T. Kirihata, and S.S. Iyer. A self-authenticating chip architecture using an intrinsic fingerprint of embedded DRAM. *Solid-State Circuits, IEEE Journal of*, 48(11):2934–2943, Nov 2013.

[353] S. Rosenblatt, D. Fainstein, A. Cestero, J. Safran, N. Robson, T. Kirihata, and S.S. Iyer. Field tolerant dynamic intrinsic chip ID using 32 nm high-K/metal gate SOI embedded DRAM. *Solid-State Circuits, IEEE Journal of*, 48(4):940–947, April 2013.

[354] Michael Rosing. *Implementing Elliptic Curve Cryptography*. Manning Publications, Greenwich, 1998.

[355] Jorg Rothe. *Complexity Theory and Cryptology: An Introduction to Crypto-complexity*. Springer Verlag, Berlin, 2005.

[356] Joseph Rotman. *An Introduction to the Theory of Groups*. Springer Verlag, New York, 1999.

[357] Louis Halle Rowen. *Algebra: Groups, Rings, and Fields*. A K Peters, Wellesley, 1994.

[358] Atri Rudra, Pradeep K. Dubey, Charanjit S. Jutla, Vijay Kumar, Josyula R. Rao, and Pankaj Rohatgi. Efficient Rijndael encryption implementation with composite field arithmetic. In Çetin Kaya Koç, David Naccache, and Christof Paar, editors, *Cryptographic Hardware and Embedded Systems: CHES 2001*, volume 2162 of *Lecture Notes in Computer Science*, pages 171–184. Springer, 2001.

[359] Ulrich Rührmair, Frank Sehnke, Jan Sölter, Gideon Dror, Srinivas Devadas, and Jürgen Schmidhuber. Modeling attacks on physical unclonable functions. In *ACM Conference on Computer and Communications Security*, CCS '10, pages 237–249. ACM, 2010.

[360] Antonio Lloris Ruiz, Encarnación Castillo Morales, Luis Parrilla Roure, and Antonio García Ríos. *Algebraic Circuits*. Springer Verlag, Berlin, 2014.

[361] A. Rukhin, J. Soto, J. Nechvatal, M. Smid, E. Barker, S. Leigh, M. Levenson, M. Vangel, D. Banks, A. Heckert, J. Dray, and S. Vo. A Statistical Test Suite for Random and Pseudorandom Number Generators for Cryptographic Applications – NIST SP 800-22, rev. 1a, 2010.

[362] S. Rusu, Simon Tam, H. Muljono, J. Stinson, D. Ayers, Jonathan Chang, R. Varada, M. Ratta, and S. Kottapalli. A 45nm 8-core enterprise Xeon® processor. In *Solid-State Circuits Conference - Digest of Technical Papers, 2009. ISSCC 2009. IEEE International*, pages 56–57. IEEE, 2009.

[363] J. Safran, A. Leslie, G. Fredeman, C. Kothandaraman, A. Cestero, Xiang Chen, R. Rajeevakumar, Deok-kee Kim, Yan Zun Li, D. Moy, N. Robson, T. Kirihata, and S. Iyer. A compact eFUSE programmable array memory for SOI CMOS. In *VLSI Circuits, 2007 IEEE Symposium on*, pages 72–73. IEEE, 2007.

[364] Bruce Sagan. *The Symmetric Group: Representations, Combinatorial Algorithms, and Symmetric Functions*. Springer Verlag, New York, 2010.

[365] F. Saidak. A new proof of euclid's theorem. *American Mathematical Monthly*, 113(10):937–938, 2006.

[366] David Samyde, Sergei Skorobogatov, Ross Anderson, and Jean-Jacques Quisquater. On a new way to read data from memory. In *Proceedings of the First International IEEE Security in Storage Workshop - SISW 2002*, pages 65–69. IEEE Computer Society, 2002.

[367] József Sándor, Dragoslav S. Mitrinovic, and Borislav Crstici. *Handbook of Number Theory I*. Springer Verlag, New York, 2005.

[368] József Sándor, Dragoslav S. Mitrinovic, and Borislav Crstici. *Handbook of Number Theory II*. Springer Verlag, New York, 2005.

[369] A. Satoh and T. Inoue. ASIC hardware focused comparison for hash functions MD5, RIPEMD-160, and SHS. In *Information Technology: Coding and Computing, 2005. ITCC 2005. International Conference on*, volume 1, pages 532–537. IEEE, 2005.

[370] Akashi Satoh, Sumio Morioka, Kohji Takano, and Seiji Munetoh. A compact Rijndael hardware architecture with S-Box optimization. In Colin Boyd, editor, *Proceedings of ASIACRYPT*, volume 2248 of *Lecture Notes in Computer Science*, pages 239–254. Springer, 2001.

[371] Hal Schenck. *Computational Algebraic Geometry*. Cambridge University Press, Cambridge, 2003.

[372] Werner Schindler. On the optimization of side-channel attacks by advanced stochastic methods. In Serge Vaudenay, editor, *PKC*, volume 3386 of *Lecture Notes in Computer Science*, pages 85–103. Springer, 2005.

[373] Werner Schindler, Kerstin Lemke, and Christof Paar. A stochastic model for differential side channel cryptanalysis. In Josyula R. Rao and Berk Sunar, editors, *Cryptographic Hardware and Embedded Systems: CHES 2005*, volume 3659 of *Lecture Notes in Computer Science*, pages 30–46. Springer-Verlag, 2005.

[374] D. M. Schinianakis, A. P. Fournaris, H. E. Michail, A. P. Kakarountas, and T. Stouraitis. An RNS implementation of an Fp elliptic curve point multiplier. *IEEE Transactions on Circuits and Systems I: Regular Papers*, 56(6):1202–1213, 2009.

[375] A. Schonage and V. Strassen. Schnelle multiplikation grober zahlen. *Computing*, 7:281–292, 1971.

[376] Kai Schramm, Thomas J. Wollinger, and Christof Paar. A new class of collision attacks and its application to DES. In Thomas Johansson, editor, *Fast Software Encryption, 10th International Workshop, FSE - 2003*, volume 2887 of *Lecture Notes in Computer Science*, pages 206–222. Springer-Verlag, 2003.

[377] G.-J. Schrijen and V. van der Leest. Comparative analysis of SRAM memories used as PUF primitives. In *Design, Automation Test in Europe Conference Exhibition (DATE), 2012*, pages 1319–1324. IEEE, 2012.

[378] Manfred Schroeder. *Number Theory in Science and Communications: With Applications In Cryptography, Physics, Digital Information, Computing, and Self-Similarity*. Springer Verlag, New York, 2008.

[379] Peter Seibt. *Algorithmic Information Theory: Mathematics of Digital Information Processing*. Springer Verlag, New York, 2006.

[380] G. Selimis, M. Konijnenburg, M. Ashouei, J. Huisken, H. de Groot, V. van der Leest, G.-J. Schrijen, M. van Hulst, and P. Tuyls. Evaluation of 90nm 6T-SRAM as physical unclonable function for secure key generation in wireless sensor nodes. In *Circuits and Systems (ISCAS), 2011 IEEE International Symposium on*, pages 567–570. IEEE, 2011.

[381] Igor Shafarevich. *Basic Notions of Algebra*. Springer Verlag, New York, 2005.

[382] Igor Shafarevich. *Basic Algebraic Geometry 1*. Springer Verlag, New York, 2013.

[383] Igor Shafarevich. *Basic Algebraic Geometry 2*. Springer Verlag, New York, 2013.

[384] M. Shand and J. Vuillemin. Fast implementations of RSA cryptography. In *Computer Arithmetic, 1993, Proceedings, 11th Symposium on*, pages 252–259. IEEE, 1993.

[385] Claude Shannon. Communication theory of secrecy systems. *Bell System Technical Journal*, 28:656–715, 1949.

[386] Claude Shannon and Warren Weaver. *The Mathematical Theory of Communication*. University of Illinois Press, Urbana, 1998.

[387] Claude E. Shannon. A mathematical theory of communication. *Bell System Technical Journal*, 27:379–423, 1948.

[388] Malik Umar Sharif, Rabia Shahid, Marcin Rogawski, and Kris Gaj. Use of embedded FPGA resources in implementations of five round three SHA-3 candidates. In *Proceedings of Ecrypt II Hash Workshop*. ECRYPT II Hash, 2011.

[389] A. P. Shenoy and R. Kumaresan. Fast base extension using a redundant modulus in RNS. *IEEE Transactions on Computers*, 38(2):292–297, 1989.

[390] Zhijie Shi, Chujiao Ma, Jordan Cote, and Bing Wang. Hardware implementation of hash functions. In *Introduction to Hardware Security and Trust*, pages 27–50. Springer, 2012.

[391] Abdulhadi Shoufan, Falko Strenzke, H. Gregor Molter, and Marc Stöttinger. A timing attack against Patterson algorithm in the McEliece PKC. In Donghoon Lee and Seokhie Hong, editors, *ICISC*, volume 5984 of *Lecture Notes in Computer Science*, pages 161–175. Springer, 2010.

[392] Victor Shoup. *A Computational Introduction to Number Theory and Algebra*. Cambridge University Press, Cambridge, 2005.

[393] Joseph Silverman. *Advanced Topics in the Arithmetic of Elliptic Curves*. Springer Verlag, New York, 1994.

[394] Joseph Silverman. *The Arithmetic of Elliptic Curves*. Springer Verlag, New York, 2009.

[395] Joseph Silverman and John Tate. *Rational Points on Elliptic Curves*. Springer Verlag, New York, 2010.

[396] P. Simons, E. van der Sluis, and V. van der Leest. Buskeeper PUFs, a promising alternative to D Flip-Flop PUFs. In *Hardware-Oriented Security and Trust (HOST), 2012 IEEE International Symposium on*, pages 7–12. IEEE, 2012.

[397] Simon Singh. *The Code Book*. Anchor, New York, 2000.

[398] N. Sklavos and P. Kitsos. BLAKE HASH function family on FPGA: From the fastest to the smallest. In *Proceedings of IEEE Computer Society Annual Symposium on VLSI (IEEE ISVLSI'10)*, pages 139–142. IEEE, 2010.

[399] Nicolas Sklavos. Multi-module hashing system for sha-3 and fpga integration. In *Proceedings of 21st International Conference on Field Programmable Logic (FPL'11) and Applications*. FPL, 2011.

[400] Nicolas Sklavos and O. Koufopavlou. On the hardware implementations of the SHA-2 (256, 384, 512) hash functions. In *Circuits and Systems, 2003. ISCAS'03. Proceedings of the 2003 International Symposium on*, volume 5, pages V–153–V–156. IEEE, 2003.

[401] Nicolas Sklavos and Odysseas Koufopavlou. Implementation of the SHA-2 hash family standard using FPGAs. *The Journal of Supercomputing*, 31(3):227–248, 2005.

[402] Sergei P. Skorobogatov and Ross J. Anderson. Optical fault induction attacks. In B. Kaliski, Ç. Koç, and C. Paar, editors, *Proceedings of the 4th International Workshop on Cryptographic Hardware and Embedded Systems: CHES 2002*, volume 2523 of *Lecture Notes in Computer Science*, pages 2–12. Springer-Verlag, 2002.

[403] Karen Smith, Lauri Kahanpää, Pekka Kekäläinen, and William Traves. *An Invitation to Algebraic Geometry*. Springer Verlag, Berlin, 2010.

[404] R.T. Smith, J.D. Chlipala, J.F.M. Bindels, R.G. Nelson, F.H. Fischer, and T.F. Mantz. Laser programmable redundancy and yield improvement in a 64K DRAM. *Solid-State Circuits, IEEE Journal of*, 16(5):506–514, Oct 1981.

[405] M. A. Soderstrand, W. K. Jenkins, G. A. Jullien, and F. J. Taylor. *Residue Number System Arithmetic: Modern Applications in Digital Signal Processing*. IEEE Press, Piscataway, NJ, 1986.

[406] W. Stallings. *Cryptography and Network Security, Fourth Edition*. Pearson Prentice Hall, Upper Saddle River, NJ, 2006.

[407] François-Xavier Standaert, Tal Malkin, and Moti Yung. A unified framework for the analysis of side-channel key recovery attacks. In Antoine Joux, editor, *EUROCRYPT*, volume 5479 of *Lecture Notes in Computer Science*, pages 443–461. Springer, 2009.

[408] Richard Stanley. *Enumerative Combinatorics: Advances and Frontiers*. Cambridge University Press, Cambridge, 2012.

[409] William Stein. *Elementary Number Theory: Primes, Congruences, and Secrets—A Computational Approach*. Springer Verlag, New York, 2008.

[410] Benjamin Steinberg. *Representation Theory of Finite Groups*. Springer Verlag, New York, 2011.

[411] Ian Stewart. *Galois Theory*. Chapman & Hall / CRC, Boca Raton, 1990.

[412] Marc Stöttinger, Felix Madlener, and Sorin A. Huss. Procedures for securing ECC implementations against differential power analysis using reconfigurable architectures. In Marco Platzner, Jürgen Teich, and Norbert Wehn, editors, *Dynamically Reconfigurable Systems - Architectures, Design Methods and Applications*, pages 305–321. Springer, 2009.

[413] Chih-Yuang Su, Shih-Arn Hwang, Po-Song Chen, and Cheng-Wen Wu. An improved Montgomery's algorithm for high-speed RSA public-key cryptosystem. *Very Large Scale Integration (VLSI) Systems, IEEE Transactions on*, 7(2):280–284, June 1999.

[414] Y. Su, J. Holleman, and B. Otis. A 1.6pJ/bit 96% stable Chip-ID generating circuit using process variations. In *Solid-State Circuits Conference, 2007. ISSCC 2007. Digest of Technical Papers. IEEE International*, pages 406–611. IEEE, 2007.

[415] K. R. Sudha, A. Chandra Sekhar, and Prasad Reddy. Cryptography protection of digital signals using some recurrence relations. *International Journal of Computer Science and Network Security*, 7(5):203–207, May 2007.

[416] G. Edward Suh and Srinivas Devadas. Physical unclonable functions for device authentication and secret key generation. In *IEEE/ACM Design Automation Conference DAC '07*, pages 9 –14. IEEE, 2007.

[417] Daisuke Suzuki, Minoru Saeki, and Tetsuya Ichikawa. Random switching logic: A countermeasure against DPA based on transition probability. *IACR Cryptology ePrint Archive*, 2004:346, 2004.

[418] G. Swaminathan. Random number generators (RNG) VHDL package. Online. Available at: http://www.ittc.ku.edu/EE-CS/EECS\ _546/magic/files/vlsi/vhdl/random.pkg, 1992.

[419] R. Szerwinski and T. Gneysu. Exploiting the power of GPUs for asymmetric cryptography. In E. Oswald and P. Rohatgi, editors, *Advances in Cryptology—Cryptographic Hardware and Embedded Systems: CHES 2008*, Lecture Notes in Computer Science, pages 79–99. Springer, 2008.

[420] M. Takeuchi, K. Inoue, M. Sakao, T. Sakoh, T. Kitamura, S. Arai, T. Iizuka, T. Yamamoto, H. Shirai, Y. Aoki, M. Hamada, R. Kubota, and S. Kishi. A 0.15μm logic based embedded DRAM technology featuring 0.425μm^2 stacked cell using MIM (metal-insulator-metal) capacitor. In *VLSI Technology, 2001. Digest of Technical Papers. 2001 Symposium on*, pages 29–30, 2001.

[421] John Talbot and Dominic Welsh. *Complexity and Cryptography: An Introduction*. Cambridge University Press, Cambridge, 2006.

[422] G. Taylor and G. Cox. Behind Intel's new random-number generator. *IEEE Spectrum*, August 2011. Online. Available at: http://spectrum.ieee.org/computing/hardware/behind-intels-new-randomnumber-generator.

[423] A.F. Tenca and C.K. Koc. A scalable architecture for modular multiplication based on Montgomery's algorithm. *Computers, IEEE Transactions on*, 52(9):1215–1221, Sept 2003.

[424] Gérald Tenenbaum and Michel Mendès France. *The Prime Numbers and Their Distribution*. American Mathematical Society, Providence, 2000.

[425] Qizhi Tian, Annelie Heuser, and Sorin A. Huss. A novel analysis method for assessing the side-channel resistance of cryptosystems. In *IIHMSP*, pages 126–129. IEEE, 2012.

[426] S. Tillich, M. Feldhofer, M. Kirschbaum, T. Plos, J.-M. Schmidt, and A. Szekely. High-speed hardware implementations of BLAKE, Blue Midnight Wish, CubeHash, ECHO, Fugue, Groestl, Hamsi, JH, Keccak, Lu_a, Shabal, Shavite-3, SIMD, and Skein. Cryptology ePrint Archive, Report 2009/510, http://eprint.iacr.org/, 2009.

[427] A. L. Toom. The complexity of a scheme of functional element realizing the multiplication of integers. *Soviet Mathematics*, 3:714–716, 1963.

[428] L. Trevisan and S. Vadhan. Extracting randomness from samplable distributions. In *Proceedings of the 41-st Annual Symposium on Foundations of Computer Science (FOCS'00)*, pages 32–42. IEEE, 2000.

[429] Elena Trichina, Tymur Korkishko, and Kyung-Hee Lee. Small size, low power, side channel-immune AES coprocessor: Design and synthesis results. In Hans Dobbertin, Vincent Rijmen, and Aleksandra Sowa, editors, *AES Conference*, volume 3373 of *Lecture Notes in Computer Science*, pages 113–127. Springer, 2004.

[430] G. Uhlmann, T. Aipperspach, T. Kirihata, K. Chandrasekharan, Yan Zun Li, C. Paone, B. Reed, N. Robson, J. Safran, D. Schmitt, and S. Iyer. A commercial field-programmable dense eFUSE array memory with 99.999% sense yield for 45nm SOI CMOS. In *Solid-State Circuits Conference, 2008. ISSCC 2008. Digest of Technical Papers. IEEE International*, pages 406–407. IEEE, 2008.

[431] B. Valtchanov, A. Aubert, F. Bernard, and V. Fischer. Modeling and observing the jitter in ring oscillators implemented in FPGAs. In *Design and Diagnostics of Electronic Circuits and Systems, 2008. DDECS 2008. 11th IEEE Workshop on*, pages 1–6. IEEE, 2008.

[432] Vincent van der Leest, Bart Preneel, and Erik van der Sluis. Soft decision error correction for compact memory-based PUFs using a single enrollment. In Emmanuel Prouff and Patrick Schaumont, editors, *Cryptographic Hardware and Embedded Systems: CHES 2012*, volume 7428 of *Lecture Notes in Computer Science*, pages 268–282. Springer Berlin Heidelberg, 2012.

[433] Vincent van der Leest, Erik van der Sluis, Geert-Jan Schrijen, Pim Tuyls, and Helena Handschuh. Efficient implementation of true random number generator based on SRAM PUFs. In David Naccache, editor, *Cryptography and Security: From Theory to Applications*, volume 6805 of *Lecture Notes in Computer Science*, pages 300–318. Springer Berlin Heidelberg, 2012.

[434] B. L. van der Waerden. *Algebra: Volume I*. Springer Verlag, New York, 1991.

[435] B. L. van der Waerden. *Algebra: Volume II*. Springer Verlag, New York, 2003.

[436] J. H. van Lint and R. M. Wilsom. *A Course in Combinatorics*. Cambridge University Press, Cambridge, 2001.

[437] Joel VanLaven, Mark Brehob, and Kevin J. Compton. Side channel analysis, fault injection and applications — A computationally feasible SPA attack on AES via optimized search. In Ryôichi Sasaki, Sihan Qing, Eiji Okamoto, and Hiroshi Yoshiura, editors, *Security and Privacy in the Age of Ubiquitous Computing - SEC 2005*, pages 577–588. Springer-Verlag, 2005.

[438] M. Varchola and M. Drutarovsky. New High Entropy Element for FPGA Based True Random Number Genetators. In *Cryptographic Hardware and Embedded Systems: CHES 2010*, volume 6225 of *Lecture Notes in Computer Science*, pages 351–365. Springer Verlag, 2010.

[439] Rajesh Velegalati and Jens-Peter Kaps. Improving security of SDDL designs through interleaved placement on Xilinx FPGAs. In *FPL*, pages 506–511. IEEE, 2011.

[440] I. Verbauwhede, J. Balasch, S. S. Roy, and A. Van Herrewge. Circuit challenges from cryptography. In *Proceedings of 2015 IEEE International Solid-State Circuits Conference*, pages 1–2. IEEE, 2015.

[441] Ingrid Verbauwhede and Henry Kuo. Architectural optimization for 1.82Gb/s VLSI implementation of the AES Rijndael algorithm. In C.K. Koc, D. Naccache, and C. Paar, editors, *Cryptographic Hardware and Embedded Systems: CHES 2001*, volume 2162 of *Lecture Notes in Computer Science*, pages 51–64. Springer, 2001.

[442] Vignesh Vivekraja and Leyla Nazhandali. Feedback based supply voltage control for temperature variation tolerant PUFs. In *IEEE International Conference on VLSI Design*, VLSID '11, pages 214–219, Washington, DC, 2011. IEEE Computer Society.

[443] J. von Neumann. Various techniques used in connection with random digits, Notes by G. E. Forsythe. *National Bureau of Standards Applied Math Series*, 12:36–38, 1951.

[444] Sartorius von Waltershausen. Gauss zum Gedächtniss, 1856.

[445] Jesse Walker, Farhana Sheikh, Sanu K. Mathew, and Ram Krishnamurthy. A Skein-512 hardware implementation. http://csrc.nist.gov/groups/ST/hash/sha-3/Round2/Aug2010/ documents/papers/WALKER_skein-intel-hwd.pdf, August 2010.

[446] C.D. Walter. Montgomery exponentiation needs no final subtractions. *Electronics Letters*, 35(21):1831–1832, Oct 1999.

[447] Sheng-Hong Wang, Wen-Ching Lin, Jheng-Hao Ye, and Ming-Der Shieh. Fast scalable radix-4 Montgomery modular multiplier. In *Circuits and Systems (ISCAS), 2012 IEEE International Symposium on*, pages 3049–3052. IEEE, 2012.

[448] Wenping Wang, Vijay Reddy, Bo Yang, Varsha Balakrishnan, Srikanth Krishnan, and Yu Cao. Statistical prediction of circuit aging under process variations. In *IEEE Custom Integrated Circuits Conference*, pages 13–16. IEEE, 2008.

[449] X. Wang, Y. L. Yin, and H. Yu. Collision search attacks on SHA-1. http://www.c4i.org/erehwon/shanote.pdf, February 2005.

[450] J. Warnock, Y.H. Chan, H. Harrer, D. Rude, R. Puri, S. Carey, G. Salem, G. Mayer, Yiu-Hing Chan, M. Mayo, A. Jatkowski, G. Strevig, L. Sigal, A. Datta, A. Gattiker, A. Bansal, D. Malone, T. Strach, Huajun Wen, Pak-Kin Mak, Chung-Lung Shum, D. Plass, and C. Webb. 5.5GHz system z microprocessor and multi-chip module. In *Solid-State Circuits Conference Digest of Technical Papers (ISSCC), 2013 IEEE International*, pages 46–47. IEEE, 2013.

[451] Lawrence Washington. *Elliptic Curves, Number Theory, and Cryptography*. Chapman & Hall/CRC, Boca Raton, 2008.

[452] S.W. Wedge. Predicting random jitter-exploring the current simulation techniques for predicting the noise in oscillator, clock, and timing circuits. *Circuits and Devices Magazine, IEEE*, 22(6):31–38, 2006.

[453] André Weimerskirch and Christof Paar. Generalizations of the karatsuba algorithm for polynomial multiplication. Unpublished— available from International Association for Cryptologic Research (eprint.iacr.org), March 2002.

[454] D.F. Wendel, R. Kalla, J. Warnock, R. Cargnoni, S.G. Chu, J.G. Clabes, D. Dreps, D. Hrusecky, J. Friedrich, S. Islam, J. Kahle, J. Leenstra, G. Mittal, J. Paredes, J. Pille, P.J. Restle, B. Sinharoy, G. Smith, W.J. Starke, S. Taylor, J. Van Norstrand, S. Weitzel, P.G. Williams, and V. Zyuban. POWER7^{TM}, a highly parallel, scalable multi-core high end server processor. *Solid-State Circuits, IEEE Journal of*, 46(1):145–161, Jan 2011.

[455] Johannes Wolkerstorfer, Elisabeth Oswald, and Mario Lamberger. An ASIC implementation of the AES SBoxes. In Bart Preneel, editor, *Proceedings of CT-RSA*, volume 2271 of *Lecture Notes in Computer Science*, pages 67–78. Springer, 2002.

[456] H. Wu. The Hash Function Jha: Submission to NIST (Round 3). http://ehash.iaik.tugraz.at/wiki/JH, 2010.

[457] Hans Wussing. *The Genesis of the Abstract Group Concept: A Contribution to the History of the Origin of Abstract Group Theory*. Dover Publications, Mineola, 2007.

[458] Lu Xiao and Howard M. Heys. A simple power analysis attack against the key schedule of the camellia block cipher. *Inf. Process. Lett.*, 95(3):409–412, 2005.

[459] T. Yamamoto, K. Uwasawa, and T. Mogami. Bias temperature instability in scaled p^+ polysilicon gate p-MOSFET's. *Electron Devices, IEEE Transactions on*, 46(5):921–926, May 1999.

[460] Ching-Chao Yang, Tian-Sheuan Chang, and Chein-Wei Jen. A new RSA cryptosystem hardware design based on Montgomery's algorithm. *Circuits and Systems II: Analog and Digital Signal Processing, IEEE Transactions on*, 45(7):908–913, Jul 1998.

[461] Meng-Day Yu and Srinivas Devadas. Secure and robust error correction for physical unclonable functions. *IEEE Design Test of Computers*, 27(1):48–65, 2010.

[462] Meng-Day Yu, R. Sowell, A. Singh, D. M'Raihi, and S. Devadas. Performance metrics and empirical results of a PUF cryptographic key generation ASIC. In *Hardware-Oriented Security and Trust (HOST), 2012 IEEE International Symposium on*, pages 108–115. IEEE, 2012.

[463] Don Zagier. The first 50 million prime numbers. *The Mathematical Intelligencer*, 1(1 Supplement):7–19, December 1977.

[464] Michael Zohner, Marc Stöttinger, Sorin A. Huss, and Oliver Stein. An adaptable, modular, and autonomous side-channel vulnerability evaluator. In *HOST*, pages 43–48. IEEE, 2012.

Index